Battlespace Technologies

Network-Enabled
Information Dominance

This book is part of the Artech House *Intelligence and Information Operations Series*. For a list of recent related titles, turn to the back of this book.

Battlespace Technologies

Network-Enabled Information Dominance

Richard S. Deakin

ARTECH
HOUSE

BOSTON | LONDON
artechhouse.com

Library of Congress Cataloging-in-Publication Data

A catalog record for this book is available from the U.S. Library of Congress.

British Library Cataloguing in Publication Data

A catalogue record for this book is available from the British Library.

Cover design by Vicki Kane

Text design by Darrell Judd

ISBN: 978-1-59693-337-8

Cover photo of two F-22 Raptors by Senior Master Sergeant Thomas Meneguin © U.S. Air Force.

© 2010 ARTECH HOUSE
685 Canton Street
Norwood, MA 02062

10 9 8 7 6 5 4 3 2 1

To Commander David Deakin LVO, OBE, RN

Contents

Introduction   **1**

Principles and Evolution of Network-Enabled Warfare   **31**

NEC Concepts   **195**

NEC Techniques and Technologies 265

Future Trends in Network-Enabled Capabilities 437

Western Coalitions 457

Link-16 Network Message Sets and Network Participation
Groups   461

Modulation Techniques   465

Frequency Classifications   469

The Kill Chain   473

The Defend Chain   477

Acknowledgments

In writing this book over several years, I have had much support and encouragement.

In particular I would like to express my thanks to Ian Poole from Adrio Communications who was of invaluable assistance both with pushing the project forward in its early stages and with providing material for the section on communications, and for Barry Trimmer for his help with reading the draft and pointing me in the right direction on a number of technical aspects.

I would also like to make a special mention of those companies and individuals that have been particularly helpful with the provision of illustrations for the book, particularly Mike Kurth at Boeing U.K., Ron Cook at L3 U.K., Chris Trippick at Lockheed Martin U.K., Armin Papperger at Rheinmetall, Matt Pothecary and Kevin Swales at Thales, Karen Thomas and Virginie Brizzard at Thales Raytheon Systems, Karen Rice at Raytheon U.K., and Darren Coe at QinetiQ.

Jane Pittaway and Jennifer Davies at Wragge & Co. and Sam Keayes at Thales U.K. were also particularly helpful in their advice regarding the copyright aspects of producing this book.

The production of this book would not have been possible without the professionalism and hard work of many people at Artech House, particularly Darrell Judd for the production and editorial management, Erin Donahue for the editing, Vicki Kane for the creativity on the artwork, and Jack Stone for the management of the project.

A final special mention goes to Marian, Laurence, and Rosie for their patience over the past few years.

An F-22 Raptor from Elmendorf Air Force Base, Alaska, takes gas from a KC-135 Stratotanker July 27, 2010. By flying with a tanker, the fighter could routinely fill its fuel tanks to ensure it had enough fuel to reach a safe base in the event it had to land. The tanker is assigned to the 100th Air Refueling Wing at Royal Air Force Mildenhall, England. (U.S. Air Force photo/Staff Sgt. Austin M. May.)

Introduction

The advent of the information age has had a significant and direct impact on the way we live. Information technology has grown at an exponential rate and significantly shapes all aspects of the world around us. In both civilian and military environments, computers and information technology drive a diverse range of capabilities and accelerate the impact on our environment and society in general. Perhaps one of the most significant changes generated by computing technology is the ability to manipulate data, turning information into a commodity in its own right and introducing new dimensions to virtually all aspects of our society and daily lives.

Some notable impacts include:

- Facilitating the easy and rapid accessing and sharing of vast amounts of information around the world;
- Making near-instantaneous global communication between people and computers an everyday reality;
- Shifting the balance of power in societies to favor those who can best access and control information;
- Creating new communications networks (and vulnerabilities) across vital information-based aspects of our lives;
- Making the "global village" a reality for those who have access to an Internet network and effectively eliminating distance as a barrier to communication;
- Creating new service-based opportunities for wealth creation through the organization and control of information;
- Blurring national and cultural boundaries and facilitating the creation of new allegiances and virtual communities that are linked by means of the Web;
- Sharing images and messages on a global stage using Web-based technologies;
- Introducing the concept of information dominance and driving the need to assimilate, process, and distribute vast amounts of data around the globe;
- Speeding up the tempo of our lives and that of the world around us.

The significant changes in the dynamics of societies and their infrastructures, heralded by the arrival of new technologies, have created new tensions and threats. Equally, they have placed new interpretations on the way that nations perceive and address global security issues. Technology has allowed the creation of many new threats that are frequently unsophisticated and unpredictable and differ significantly in scope and scale to those for which most armed forces have trained and equipped. However, these advances have also enabled the creation of new methods of addressing this increasingly diverse range of threats. The reliance on information technology by modern societies has also made it possible to intervene at an earlier stage in global events through the exploitation of the technology to gather information at an earlier stage in the cycle. Again, such technology creates the possibility to influence an opponent's environment and capabilities by means of information dominance rather than military force. Indeed, warfare in the information age places new challenges on political and military leaderships who now have to operate in increasingly open, information-based societies and whose every move is exposed to global scrutiny.

The change in opportunites follows from those able to recognize the complexity and at least start to understand the effect their actions will have in such a complex environment.

—Barry Trimmer, Technical Director, Thales Aerospace UK

Information technology has introduced the ability to rapidly and directly influence populations and their leaders far away from the area of conflict, enabling the swift communication and assimilation of information and increasingly making use of influence rather than force as the preferred means of settling disputes. Even in the public domain, recent conflicts in Kosovo, Iraq, Afghanistan, and Chad have demonstrated how the "CNN effect," the free flow of real-time information and shortened news cycles, has a huge impact on public opinion, placing some items at the top of the public agenda that might otherwise warrant a lower priority [1].

The changing pace of warfare

Throughout the history of warfare, the strategies employed have evolved in accordance with the military technology available. As societies have progressed (with that progress frequently being driven by advances in technology), so have armed forces had to adapt their strategies to exploit the new technologies available to them.

When examining the influence of technology on the way that armies have evolved, it is important to consider the military environment in which such technologies are deployed in order to be able to understand both the requirements for technology solutions and their impact on command and control structures.

Military historians have categorized the evolution of such strategies by looking at the way technology has influenced both the weapons and the command and control structures employed along with the way in which advantages conferred have been integrated into military thinking. As is evident from Figure 1.1, the pace at which significant change has taken place in military strategy has increased as technology has made earlier generations of warfare obsolete. Accordingly, armed forces around the world have had to face rapid changes in the technology of warfare, exploit those changes to their own advantage, and, at the same time, work out how to counter the use of new technology by opposing forces.

The tactics and tools of warfare have generally evolved in line with the progress in industrial technology that has facilitated the development of new

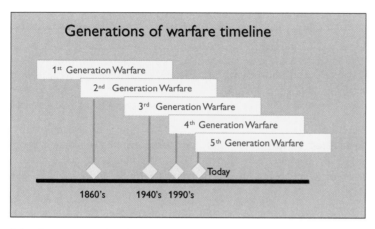

Figure 1.1   Generations of warfare timeline.

weapons and the tactics for their employment. In addition, improvements in communications have significantly changed the approach to command and control as the ability to communicate with dispersed forces across the battlespace with speed and accuracy has become both feasible and easier.

Progress within the first to third generations of warfare has been largely characterized by developments in mechanized warfighting, and although this has been significant, the building blocks that have been available to military commanders have remained constrained by the ability to sense the environment and to make decisions in real time. As a result tactics employed by commanders in the era of first and second generation warfare were also significantly limited by the ability to both accurately sense what was taking place and to easily communicate with units deployed in the field. This began to change with the evolution of technologies associated with third generation warfare, which allowed improved communications and the ability—albeit in a limited sense—to report back (through the command and control network) on unfolding events. Such flexibility permitted smaller, more mobile units to be deployed and enabled them to find and exploit weaknesses in the enemy's positions before they were able to react. Given the limited communications available, such unstructured actions were extremely difficult to counter, even assuming awareness of a move by the opponent at a stage early enough to respond effectively.

From a technology perspective, the significant difference between fourth generation warfare and earlier generations is not the advent of new weapons, but the transition to the concept of information dominance and the recognition of its role as a weapon in its own right. Although third generation warfare included the benefits of early developments in sensors and technologies that exploited the electromagnetic spectrum, it was not until the advent of fourth generation warfare that the emphasis switched

"Every few hundred years throughout Western History, a sharp transformation has occurred. In a matter of a few decades, society altogether rearranges itself, its world views, its social and political structure, its arts, its key institutions. Fifty years later a New World exists. And the people born into that world cannot even imagine the world in which their grandparents lived and into which their own parents were born."

—Peter Drucker [2], The New Society of Organizations

Early Generations of Warfare [17]

First Generation Warfare: Short-range, smoothbore weapons; line and column formations; rigid command and control approach typically with single-point top-down control. Training and élan could often enable one side to close with and defeat the enemy before absorbing overwhelming casualties. Examples: Austerlitz, 1805; Waterloo, 1815.

Second Generation Warfare: Mechanized warfare, rifled automatic weapons, heavy artillery, people replaced by machines as the force multiplier, high attrition on both sides, rigid command and control approach. Attempts to use élan to overcome firepower are now suicidal. Examples: Crimea, 1853–1856; Sedan, 1870, World War I, 1914–1918.

Third Generation Warfare: Mechanized warfare, increasingly flexible tactics, smaller units; dynamic rather than static battles, combined use of air and naval power to deliver effects, advent of electronic sensors as a force multiplier. Examples: World War II, 1939–1945; Korean War 1950–1953.[1]

from sensors and networks as support tools to a recognition of both their centrality to warfare and the delivery of military strategy itself. The concepts surrounding network-enabled capability (NEC; this is often somewhat erroneously termed network-centric warfare or NCW) are at the heart of fourth generation warfare and have become the cornerstone of military advantage in the information age.

Warfare in the information age

The transition to fourth generation warfare (together with the development of information age concepts and strategies) has engendered more seismic shifts in approaches to warfare than have ever been seen before. The situational awareness that can be delivered through information age technology has also had a significant impact on the organization and deployment of military forces. The potential military battlespace has also expanded in size with the increasing range, lethality, and coordination of sensors and weapons systems driving forces and platforms to disperse to improve survivability. As a result of increased situational awareness and the increasing size of the potential battlespace, engagement capabilities will drive both offensive and defensive engagements out to increasingly long ranges [4].

Although there is no consensus on the definition of fifth generation warfare, it will undoubtedly offer new dimensions and perspectives on the options available to military and political commanders. Fifth generation warfare will certainly be heavily influenced by information dominance and the recognition that the defeat or coercion of any enemy can be achieved through the economic impact of suitable nonlethal effects rather than a singular reliance on military force.

Fourth Generation Warfare: Warfare in the Information Age

Fourth generation warfare focuses on psychology and belief frameworks at a personal level, where it works to establish the moral superiority of the cause for which it is fighting. It seeks to destabilize established political frameworks infiltrating state organizations and social structures to undermine and change the established status quo of a society.

Unlike traditional mechanized warfare, fourth generation warfare seeks to exploit the unpredictability of its actions and its fragmented, covert, and fleeting engagements and, through its nonlinear organization and lines of battle, cause confusion by exploiting the difficulty in determining the real enemy.

This concept was first proposed by William Lind [3].

Information age technology has not only driven a change in the tools of war, but also has significant implications for the command and control strategies employed within the battlespace. As new technology has provided seamless remote links between sensors and participants, the ability to produce a more accurate picture of unfolding events in the battlespace has revolutionized the way that forces are deployed and controlled. Highly accurate, near real-time views of the disposition and predictive intentions of the enemy, together with the location of neutral and friendly forces, are now feasible for most types of theater. Such an ability allows for the creation of a three-dimensional picture of the evolving battlespace and has facilitated a rapid shift away from large rigid formations (with the rigid, centralized, and hierarchical approach to command and control that is associated with the industrial age) and a task-driven approach to an approach where small military units are organized to achieve the commander's intent, provide near real-time feedback on progress, and continuously adjust their objectives to meet the overall needs of the commander. Such a transformation (which has been enabled by information age technology) has also created the need for new technologies to collect, process, and disseminate the vast amounts of information that flow across the decision-making processes of devolved, networked command and control systems.

Information age warfare has distributed the process of collection, dissemination, processing, and information storage to the global stage and, as a consequence, allows a much higher proportion of the command and control organization to be fielded away from the front line. As a result (and although sensors are generally positioned in or near the area where the information is required to be collected), the supporting infrastructure that is required to manage, process, and disseminate information age data is often located far away from areas of engagement with the enemy.

Fifth Generation Warfare: Information-Centric Warfare

Fifth generation warfare will pitch a nation against political, religious, or social interest groups and against other nations. It will focus on the destruction of opponents through the domination and disruption of information systems and networks that control and manage a nation's infrastructure.

Neither centralized nor decentralized in its approach, fifth generation warfare will rely on being able to infiltrate and disrupt complex network systems that lie at the heart of modern society. Massive disruption to infrastructure systems such as transportation, banking, communications, logistics, healthcare, and manufacturing will be caused by penetration and disruption of networks and information systems without causing any physical damage.

Fifth generation warfare will seek to destabilize established industrial and social infrastructures, causing unrest and the disintegration of social frameworks.

The role of technology in influencing military infrastructure is noticeable in many other areas as technology increasingly replaces manpower. By way of example, many of the front-line roles and functions that are traditionally associated with armed forces personnel are being replaced by semiautonomous, unmanned systems that allow for precision engagement by remote supporting infrastructures (Figure 1.2). Smaller, more flexible, and

Figure 1.2 Unmanned air vehicles are now taking on the role of many manned systems previously employed for dull, difficult, or dangerous missions. Here an MQ-9B Reaper Unmanned Aerial Vehicle (UAV) taxies into Creech Air Force Base, Nevada, home to the USAF 432nd Wing. The 432nd Wing consists of six operations squadrons and a maintenance squadron for the Air Force fleet of 60 MQ-1 Predator and six MQ-9 Reaper UAVs. Note that current U.S. terminology generally refers to unmanned air systems (UAS) rather than unmanned air vehicles (UAV). However, for the purpose of discussing networked systems, it is important to distinguish between a system and a vehicle, which may be part of the system.

more capable information-enabled air, land, and sea strike forces are now able to be deployed against hostile infrastructure rather than the commander having to rely on massed formation attacks. Even the capabilities and role of the foot soldier have been transformed by the situational awareness achieved in a networked sensor environment. Accordingly, the threats we face and the way we gather intelligence and fight have changed forever due to rapid advances in information technology and networking technologies.

Several key concepts associated with information age warfare are important to master if one is to fully appreciate the flexibility of choices available to the combatants in the information-rich fourth and fifth generation battlespace. Specifically, these concepts (which are explained in more detail later) comprise:

- **Network-enabled capability:** The added dimension to military capability that is enabled through near real-time communications and data access across global multinational networks.
- **Spectrum dominance:** The exploitation of the electromagnetic spectrum to the advantage of friendly forces and the denial of its use by hostile forces.
- **Information superiority:** A relative measure that refers to the collection, processing, and dissemination of timely and accurate information to create a shared understanding of threats and their disposition and intentions within a three-dimensional picture of the battlespace.
- **Decision superiority:** A relative measure that refers to the use of superior information to produce more accurate and faster decision making within the command and control cycle.
- **Effects-based operations:** The use of synchronized military effects (from nonlethal to lethal) that are appropriate to the intended military outcome. Effects need to be timely, proportionate, and precise and delivered in a safe, secure, and legal manner against confirmed targets with minimal collateral damage.

Information age technology is creating a rapid change in military strategy and operations as situational awareness, agility, speed, and precision become tools in their own right. Such a situation can only transform the way in which armed forces fight and will continue to do so into the foreseeable future.

Increasingly diverse challenges and threats in the information age

The increasing pace of progress in industrial technology (and its use in military applications) has also heralded significant change in the nature of the threat, in the way in which wars are fought, and in the organization of military structures, with the whole taking place within increasingly short time spans. While such changes have always characterized the evolution between one generation of warfare and the next, such rapid change in such a short period of time has produced self-evident difficulties in many theaters where large mechanized forces face elusive enemies who are hard to identify and whose form and appearance break all the rules of traditional mechanized warfare (Figure 1.3).

Recent advances in sensor technology, data fusion, and networking have brought about a radical transformation in the nature of the threats and the

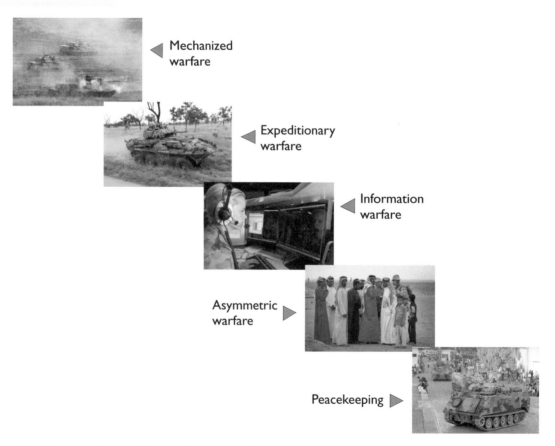

Figure 1.3   The many-faceted challenges of modern warfare.

environment in which military operations take place. The sophistication and accessibility of modern electronic systems have not only improved the ability to combat a diverse range of threats, but also have generated a new range of threats that are often referred to as *asymmetric* and that operate outside the norms of traditional mechanized warfare. Warfare in our information-rich age requires a fundamental change in many aspects of military operations, particularly as targets become harder to identify and are typically smaller, more fleeting, and ephemeral in nature (Figure 1.4).

With information collection, accessibility, and distribution becoming increasingly important at all stages of conflict (whether high intensity warfighting or a peacekeeping operation), the diversity of threats and the means to counter them are rapidly evolving. Already, traditional mechanized forces seem ill-equipped to undertake many of the global conflicts we see today. Tanks (once the mainstay of the offensive fighting force) are ever more vulnerable to modern weapons systems that are equipped with ever more sophisticated sensors and targeting systems. Again, the need for manned aircraft is increasingly being challenged as the situational awareness

The Urban Operations Challenge

The scale of an urban battle is a human scale, measured in meters rather than kilometers. It is a scale that demands superb spatial precision, where weapons must deliver precise destructive force. It is a scale where space and time also shrink.

Information becomes obsolete the moment an adversary or a blue force moves from one city block to the next, one building to the next. Time for decision making and acting on those decisions is measured in seconds.

Soldiers from the 87th Infantry Regiment lead a patrol down during a village assessment in the Jalrez Valley of Afghanistan's Wardak province, March 12, 2009.

Figure 1.4 The urban operations challenge [5].

achievable with UAV-mounted integrated and networked electronic sensors far surpasses that of a human pilot. Unmanned land, air, and underwater vehicles are becoming progressively more reliable and their command and control functions can now be remote, obviating the need to accommodate expensive and complex humans in hostile environments.

In terms of defense planning, the challenge for many countries frequently centers on the diversity of the threat, with the options including asymmetric warfare, potential massed armored attack, homeland defense, and expeditionary warfare. While today's preoccupation is asymmetric warfare and the War on Terror, the reader should not lose sight of the potential rise of countries capable of taking on the dominant position of the United States and the implications that such a scenario has for allies who would be involved in such conflicts. Between the two lies a mean that would see force structures that would be capable of operating far away from home in an expeditionary warfare capacity and that would be supported by the complex infrastructure needed to support such deployed operations.

The common feature of all these scenarios is the increasing use of information systems to achieve information dominance over the opposing side. As will be described later, information dominance is a relative measure that relies heavily on electronic combat systems, sensors, and networks to provide situational awareness, decision-making processes, and, if necessary, the ability to select and deliver an appropriate effect to the target. Elsewhere, one of the greatest challenges of the information age is the need to counter information warfare, whose effects are frequently so difficult to detect as to render it uncertain as to whether or not one is actually under attack. Again, the dynamics of modern conflict are so heavily influenced by information age technology that we are truly on the cusp of change within military technology. Authors such as Martin van Creveld [6] have gone so far as to suggest that the security environment is changing so radically that today's armed forces will soon become obsolete and be replaced by qualitatively different organizations.

The strategic context

The United States Department of Defense's Joint Vision (JV) 2020 strategy document outlines the key aspects of military capability that will all the U.S. armed forces to adapt their information superiority capabilities to achieve spectrum dominance in the forecast global security environment (Figure 1.5). JV 2020 represents a consensus across America's armed services as to what are the likely threats that will be faced by the United States and its allies in the period to 2020 [7]. Key points include:

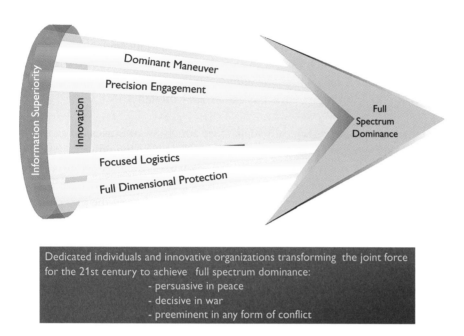

Figure 1.5 Joint vision 2020.

- The recognition that the United States will increase its involvement in an expanding global economy, leading to greater interaction with a variety of regional entities;
- Potential adversaries being able to access much of the technology and commercial commodities available to U.S. forces, including information and information-based capabilities;
- Potential adversaries adapting "niche" capabilities in the face of U.S. technological advantages that will enable them to target American capabilities where the disparity in strength is not as great or can be addressed by asymmetric approaches.

Of course, unlike warfare in the mechanized age, information age adversaries will not necessarily employ easily recognizable armed forces or tactics and hence will not be as easy to detect, identify, or track. Unlike threats from the age of mechanized warfare, fourth generation threats are no longer identifiable solely by geographical boundaries or military insignia and new adversaries will increasingly look to adapt their strategies to exploit relative weaknesses in their opponent's capabilities through the use of *asymmetric* techniques.

The term asymmetric warfare implies an asymmetry between opponents on any one of a number of parameters including structure, military tactics, equipment, and their associated command and control structures. In such an environment, the challenge is to find ways to use of network-enabled technologies against such threats and as enablers to facilitate rapid reconfiguration of capabilities and approaches to engage asymmetric forces and to develop new approaches to counter new threats.

Asymmetric opponents typically focus their attacks in such as way as to make the strengths of an opponent largely irrelevant to the outcome of the engagement. For many opponents around the world, the military strength of NATO and its members' domestic security organizations will continue to dictate the use of an asymmetric approach to negate the traditional combat power required for large-scale mechanized conflict.

Asymmetric warfare is just one example of how the environment in which the military must operate is changing (Figure 1.6). The surroundings in which military operations are carried out have moved from open landscapes with clearly identifiable opponents to operations in dense urban environments where friend and foe blend into a scenery of people, buildings, and everyday patterns of life. Rather than finding large tank formations, we now need to find and follow targets in the form of individuals if we are to determine their intent and level of threat. The ability to identify a nontraditional adversary from the background noise and determine its

Figure 1.6   Modern warfare will continue to become ever more complex as the range of threats and adversaries continues to challenge standard military doctrines. Here two U.S. Marine Corps CH-53E Super Stallion helicopters assigned to Marine Heavy Helicopter Squadron-772 (HMM-772) receive fuel from a KC-130 Hercules.

capabilities and intentions is among the greatest challenges that we face in the information age.

Increasingly, adversaries operating by blending in to the civilian infrastructure and are relocatable, mobile, and unpredictable, and, by their very nature, need to be accurately targeted very rapidly once they have been identified. In such scenarios, traditional military strengths such as sheer weight of force and ability to deliver massive destructive power become outmoded, being replaced by the need for accurate and rapid delivery of precise effects at the sharp end of network-enabled actions.

Additional threats have evolved with the advent of global information technology networks, resulting in the ability to attack an opponent's critical civil and military information systems without, in some cases, the target even realizing that they are under attack. Such attacks (often referred to as *information warfare*) can be structured or unstructured, with the former being frequently state-sponsored and involving a systematic exploration of the weaknesses and exploitable opportunities within an opponent's networks. Unstructured attacks usually originate from small groups with limited funding, but can often prove as much of a challenge (if not more so) due to their erratic nature, methods of attack, and timing.

Information and influence: the new weapons of war

The information age has created a completely new strategic infrastructure and an accompanying set of information-related targets. For traditional physical targets, network-enabled warfare has meant that the immense improvements in intelligence, surveillance, and reconnaissance (ISR) capabilities will certainly enable the prioritization and targeting of easily visible physical assets—stationary or mobile—in the early stages of any conflict.

This being said, the dominance of today's battlespace is heavily dependent upon information superiority rather than physical assets, with targeting priorities being increasingly focused around the nodes and networks essential for information dominance (Figure 1.7). The loss of information networks and command and control links render physical assets largely irrelevant to the outcome of a conflict.

The achievement of information dominance involves the compilation of an accurate and timely intelligence picture for friendly forces and the denial or adverse manipulation of intelligence-related information to the enemy. Information superiority and dominance of the information domain are

"To fight and conquer in all your battles is not supreme excellence; supreme excellence consists in breaking the enemy's resistance without fighting."

—Sun Tzu, *The Art of War*

Figure 1.7 Information infrastructure targets such as this Russian satellite terminal discovered near Kabul, Afghanistan, during Operation Enduring Freedom are a high priority in the opening stages of any conflict.

essential in achieving overall control of the battlespace and subsequently successful military outcomes. More frequently, kinetic-based, hard-kill warfare is being replaced by information-based, soft-kill techniques that permit the application of an increased and more economical range of specific and deliberate effects that are appropriate to the desired outcome.

If we associate network-enabled capability (NEC) with speed of decision making and the precise deployment of men, material, assets, and effects, the approach enables warfare to be conducted on terms that are most favorable to the force that is better network-enabled. This increase in tempo provides the ability to be able to react faster with a more appropriate and accurate effect than the opposition. Again, the exploitation of an NEC approach constrains opposing forces to fight on terms dictated by the friendly force. It renders them less able to react and respond to changing environments than a network-enabled opponent, who has a better view of the unfolding scene and who can deploy forces using a complex network of nodes rather than being constrained to traditional rigid military structures that rely on a centralized command and control approach.

Recent conflicts have shown us that asymmetric organizations employing well-organized irregular forces present a significant identification and targeting challenge. Such forces often choose to avoid conflict on terms that do not optimize their strengths and can prove extremely difficult to engage with traditional military tactics designed for large-scale open conflict. In this environment, NEC enables an asymmetric approach to be countered by the use of multiple sensors, sensor types, and intelligence systems and the ability to react almost instantly to the threat once it has been detected. Such a capability allows engagements to take place within very short windows of opportunity.

Information dominance is usually delivered via *information warfare*, defined as the ability to overtly or covertly influence an adversary's information collection, information-based processes, dissemination systems, and computer-based networks while protecting one's own information and processes, control systems, and computer-based networks. A coordinated approach to these activities is essential to achieve dominance of the most critical asset that any commander can wish to acquire: information.

Whether the threat is characterized by a modern, well-organized opposing force or a disparate and covert asymmetric opponent, the outcome of the battle will be heavily weighted in favor of the force that masters the collection of information, its distribution, associated decision making, and the delivery of appropriate effects in the most timely manner, while at the same time protecting its own information collection, processing, and dissemination capabilities. Whatever the threat, the ability to access accurate and timely intelligence information and to use this to deliver a faster tempo of decision making and action is a key capability in determining the outcome of any conflict. Increased tempo is important because it enables decisions to be made and actions taken before the opponent is aware of the unfolding intentions of its adversary. Where decisions are made at a faster tempo, they make the force operating at this tempo appear unpredictable and able to counter opposing moves without the enemy being able to understand the pattern of reaction. This, in turn, forces the opposing force to make poor-quality and/or hasty decisions based on partial information and, as a result, generate confusion and uncertainty.

Information-based warfare offers many offensive and defensive capabilities and nonlethal options that are outside the traditional spectrum of warfare. Information warfare and its ability to influence specific audiences and shape the sociopolitical environment (in particular during pre- and postcombat operations) provide additional options to shorten conflict and tension or to reduce their intensity. To achieve maximum impact, information operations must be employed in a structured manner across all phases of an engagement and must be integrated in such a way as to support effective command and control decision making. In particular, the challenge is to ensure that network-based intelligence is shared across the entire network and is channeled directly into targeting and maneuvering decisions rather than being used in a centralized manner to direct command and control.

Influence Operations

Influence operations are employment of capabilities to affect behaviors, protect operations, communicate commander's intent, and project accurate information to achieve desired effects across the cognitive domain.

Psychological operations (PSYOPS) seek to induce, influence, or reinforce the perceptions, attitudes, reasoning, and behavior of foreign leaders, groups, and organizations in a manner favorable to friendly national and military objectives.

Military deception (MILDEC) misleads or manages the perception of adversaries, causing them to act in accordance with friendly objectives.

Operations security (OPSEC) is an activity that helps prevent our adversaries from gaining and exploiting critical information.

Counterintelligence is defined as information gathered and activities conducted to protect against espionage, other intelligence activities, sabotage, or assassinations conducted by or on behalf of foreign governments or elements thereof, foreign organizations or foreign persons, or international terrorist activities.

Public affairs operations are a key component of informational flexible deterrent options and build commanders' predictive awareness of the international public information environment and the means to use information to take offensive and preemptive defensive actions.

—Adapted from [8]

Collecting information on our opponents and using that to our advantage in information warfare is an essential tool of modern warfare. However, the only way to defeat an enemy is to really understand how they think, act, and interpret the information around them. It is essential therefore to understand the political, economic, socio-political, and military context in which the information is collected and used. There is no substitute for human intelligence (HUMINT) collection and careful analysis of its sources and applications in which that information may be used. Electronic sensors can aid in the intelligence collection process, but the information gathered needs to be interpreted along with other intelligence sources if it is to have maximum value.

One further note of caution is that we need to ensure that we do not make the mistake of assuming that although an adversary's information warfare capabilities may be relatively unsophisticated, this does not mean that they are not viable or effective, especially in environments where the opponent is on home territory, operating in an asymmetric environment.

The role of electronic combat systems in supporting information operations

The collection, processing, and sharing of information are fundamental to achieving a successful outcome in any military operation. Whether the threat is characterized by an overt, modern, well-organized opposing force or disparate and covert asymmetric opponents, the battle will be won by the force that best masters the collection of information, its distribution, and its timely use to achieve decision superiority. At the same time significant challenges need to be overcome to protect the information collection, processing, storage, and dissemination capabilities of friendly forces and to enable that information to be shared in a secure manner among coalition forces.

General Hal Hornburg, USAF, noted that information operations should be separated into three areas [9]: manipulation of public perception, computer network attack, and electronic warfare. Of these, Hornburg noted that only electronic warfare should be the exclusive domain of the military. It is worth noting, however, that electronic combat systems in their broadest sense, or at least their associated technologies, have a role to play across all three areas. Whether that is setting up TV and radio broadcast systems in an immediate postconflict period to keep the population informed of events or the processing and sharing of human intelligence (HUMINT) to support information operations, electronic technologies play a vital supporting role. Across all phases of an engagement, electronic combat systems are required to support specific tasks across theater-wide networks and to connect the information and command and control aspects of operations back to the national headquarters from which those forces are deployed.

Electronic combat and information operations play a dominant role across all phases of an engagement in times of both tension and conflict (Table 1.1). In most cases, the boundaries between each phase are unclear and the sequence cannot be predefined, with operational phases having the potential to move forward in the sequences shown, or even backwards.

Behind the sharp end of a military engagement comes a long logistics and support chain, where the effectiveness of the enabling elements also relies heavily on the speed of response and decision-making. This in itself is heavily dependent on the fundamental need to share information across military networks in a common data format.

Each of these phases, as indistinct as they can sometimes be, will have their own requirements that determine the performance requirements of electronic combat and information warfare systems and the need to protect

Phase	Predominant Role of Electronic Combat Systems
Prevention stabilize deter coerce contain	Information operations Intelligence, surveillance, target acquisition; reconnaissance; communications; threat warning; electronic intelligence: detection, geolocation, classification, identification; knowledge superiority
Intervention penetrate disrupt destroy defeat	Information operations; intelligence, surveillance, target acquisition, reconnaissance; electronic warfare; sensors to support kill-chain processes: find, track, fix, target, engage, assess; communications; decision superiority
Stabilization contain protect control communicate	Information operations, threat warning, information dominance, knowledge superiority
Construction establish build withdraw	Information operations

Source: [10].

Table 1.1   The Predominant Role of Electronic Combat Systems in Security Engagement Phases

them from intentional or unintentional interference. Threats to friendly Command, Control, Communications, Computing, and Intelligence (C4I) systems vary in complexity and potential impact from phase to phase, as does the need to counter information warfare-based attacks on one's own systems.

During periods of tension (where the aim is to avoid conflict) the emphasis will be very much focused on information operations, with a particular emphasis on collecting information that will be useful to friendly forces should intervention be required, and there is a need to influence sociopolitical frameworks and local populations in order to align their support for the desired outcome of military intervention. During the intervention phase (and where combat operations become necessary), information operations will continue, with the emphasis switching to electronic warfare to enable dominance of the electromagnetic spectrum to be achieved both as an enabler to maximize the effectiveness of military effects and to protect the intervening forces. Postconflict (the stabilization period), the emphasis typically switches back to information operations to support a wide range of tasks associated with eventual withdrawal.

However, irrespective of the phase of the operation or type of threat, the key element to master across all of these operations is the rapid reconfiguration of the assets, sensors, and networks that determine the effective capability of the deployed forces as each phase unfolds. Such rapid reconfiguration, often called *agility*, is essential in meeting the rapidly changing threat environment found in most modern conflicts, many of which are low intensity compared with the conflicts for which most armed forces have been shaped.

Across all phases of conflict, the ability to rapidly collect, assimilate, and distribute intelligence across the battlespace has lead to a dramatic improvement in the effectiveness of joint forces and their ability to deploy to counter emerging threats and, if necessary, to deliver precision effects in a timely manner. In order to dominate the battlespace through knowledge rather than strength, it is necessary to master a number of key concepts, the effectiveness of which is heavily dependent upon the timeliness and quality of the available information. Specifically, these elements comprise:

- **Intelligence:** The timely provision of information relating to the disposition, capabilities, strength, and intentions of opposing forces.
- **Persistence:** The ability to provide the necessary intelligence information in any unfolding scenario on a 24/7 basis. Nothing happens without it being noticed.
- **Agility:** Delivering rapid effects, appropriate to the threat, in a range of environments.
- **Adaptability:** Seizing emerging opportunities to dominate the battlespace.
- **Precision:** The ability to deliver the desired effect with the minimum of assets while at the same time, minimizing collateral damage.
- **Speed:** Defines the ability to operate within the enemy's OODA loop and to bring the desired effect to bear through rapid decision making and communication across the kill chain.
- **Accessibility:** Getting the right information to the right place at the right time across the shared data environment.
- **Interconnectivity:** The ability to share information in a standard format over different networks.

All of these interdependent concepts are essential if information dominance is to be achieved and an effective NEC is to be provided. Each element, in turn, has significant implications for the methods by which data is collected, processed, and disseminated across the network.

The integrity of these parameters and the interaction between them determine the effectiveness that sharing information across the decision-making processes has in achieving decision superiority and, ultimately, information superiority. Effective processing of high-quality data from electronic combat systems (when combined with the emphasis on rapid decision-making across the entire military system) has the ability to greatly increase not only the effectiveness of the decision-making process, but also the general tempo (from logistics to the delivery of effects) of operations themselves. Increased tempo is in itself another significant dimension of information-age warfare.

Information superiority and information dominance not only concerns friendly decision-making processes, but also encompasses the ability to overtly or covertly influence an adversary's information collection, information-based processing, dissemination, and computer-based networks while protecting one's own information and processes, control systems, and computer-based networks. The denial of information gathering sources and communications channels to the opposing forces is another key objective of information dominance. Without timely and accurate information, the enemy is at a significant disadvantage in the decision-making process where timeliness and decision-making quality will be significantly degraded by even a modest impairment in the information collection and dissemination processes.

The traditional view is that improvements in weaponry, transportation, and communications have each, in turn, heralded a transformational step in the way that wars have been fought. We are now entering a new era, where information dominance—characterized by the timeliness and accuracy of information—will lead to dramatic improvements in the speed and quality of decision making. The ability to process information faster, integrate multiple information sources, and, most critically, identify the information nuggets within the integrated intelligence output, make associations, identify patterns, and draw conclusions about the enemy's actions and intentions and will undoubtedly herald a new generation of information warfare, which may be better described as knowledge-based warfare rather than information-based warfare.

From data to decision: the role of sensors in networked decision making

System versus system warfare is heavily reliant on information dominance. In turn, information dominance requires the capabilities of Intelligence, Surveillance, Target Acquisition and Reconnaissance (ISTAR) and Command, Control, Communications, and Computers (C4) technology to acquire and exploit the information needed to dominate and defeat adversary forces while simultaneously profiting from the information networks of friendly forces to achieve information superiority. As such, it includes the ability to acquire and disseminate near real-time awareness of the disposition of friendly, coalition, enemy, and neutral forces and civilians throughout the battlespace.

Three-Dimensional Intelligence

"...the technology that is available to the US military today and now in development can revolutionize the way we conduct military operations. That technology can give us the ability to see a 'battlefield' as large as Iraq or Korea—an area 200 miles on a side—with unprecedented fidelity, comprehension and timeliness; by night or day, in any kind of weather, all the time. In a future conflict, that means an Army corps commander in his field headquarters will have instant access to a live, three-dimensional image of the entire battlefield..."

—Former Vice Chairman of the U.S. Joint Chiefs of Staff Admiral Bill Owens [11]

Most Command, Control, Communications (C3) and ISTAR information is collected by a relatively limited range of sensor types, all of which exploit different sections of the sonic or electromagnetic spectrum. From initial target detection or surveillance mapping, sensor information provides the foundations for essential intelligence input into the decision-making process. Combined with information processing (and dissemination to automatic or man-in-the-loop decision-making processes), sensors that are linked across a network can then be directed to collect data on specific targets or areas of interest that are of high priority to the network rather than to the specific platform alone.

Figure 1.8 shows the data gathering role of sensors in the battlespace. The process is a continuous loop that results in constant data collection and adjustment to the intelligence collection process that is needed to suit the unfolding requirements for situational awareness: detect, collect, process, and disseminate.

The cycle starts with the collection of data from a specific sensor that is tuned in to look for particular characteristics in the battlespace. To enhance detection chances, the input from this sensor will be typically fused with a diverse range of other sensor inputs that cover multiple bands within the electromagnetic spectrum.

Functions required for determining situational awarness

- Detect
- Collect
- Process
- Disseminate

Figure 1.8 Functions required to deliver ISTAR capability in a NEC environment.

A typical example of this sort of sensor fusion aboard a multisensor platform such as the Lockheed Martin F-35 Lightning II (Figure 1.9) would be the initial detection of a target using the aircraft's Electro-Optical Targetting System (EOTS) and its identification by means of the advanced radar processing techniques inherent in its AN/APG-81 AESA radar. Additional information may also be provided through off-platform networked sensors such as supplementary target data from the Airborne Warning and Control System (AWACS) or a ground-based radar. Data fusion may occur on- or off-board the collecting platform, but either way the challenge is to ensure that it is placed on the network in a format that can be easily accessed by the participating units.

The ability to integrate data from multiple networks in a timely manner greatly increases the knowledge and accuracy of the battlespace picture. This, in turn, means that more precise effects can be delivered to a specific

(a) F-35 platform

(b) Pilot's sensor fusion display

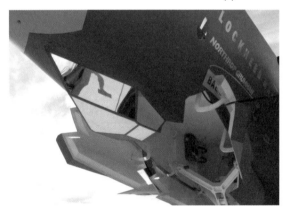

(c) Electro-optical targeting system.

Figure 1.9   (a–c) On the Lockheed Martin F-35 Lighting II the pilot is presented with a comprehensive situational awareness picture delivered through multisensor data fusion.

location, thereby avoiding the risk of attacking a target that has moved or changed, minimizing collateral damage, and ensuring that precision weapons can be utilized to maximum effect on a precision target location.

Once the data has been collected, the challenge is to turn it in to meaningful information to assist in the identification and classification of the target and to feed this information to the command and control decision-making process. In turn, the processing node needs to disseminate the decision to the other elements of the network that will need to take the information into account in their actions. Information from battlespace sensors is used to provide the primary source of real-time situational awareness and coordination of forces through the command and control (C2) decision-making process and to support subsequent offensive or defensive actions. Dissemination of data and decisions across various networks takes place across the network by means of hard-wire or radio frequency connectivity. As will be described later, the challenge is to ensure that the many coalition data links found in today's battlespace can all utilize the information provided by the sensor networks.

Depending on the decisions that have been made, the sensor networks will then be directed to the areas of interest in the battlespace that may require further surveillance or may, for example, be required as part of a guidance system to assist in delivering an effect on to the target. The integration of this information network is increasingly dependent upon a secure and uninterruptible high-bandwidth C4 network that links all coalition and friendly forces and provides common awareness of the current situation throughout the battlespace area.

Battlespace sensors are evolving towards an integrated mix of cooperative sensor types that together provide varying levels of spatial coverage and data resolution. Rather than having sensors, processing systems, and effects systems all located on one platform, networks are evolving such that these capabilities will be distributed as independent nodes within a network, with each node being accessible by others on the network. The strength of such an arrangement lies in its ability to link sensors across multiple networks to provide a fused, high-resolution picture that reflects targeting priorities and the resolution accuracy needed to meet the specific situational awareness requirement

Sensor types can be classified in many ways, with the most common being the active/passive distinction, where an active sensor illuminates a target through emissions and listen for reflections at a particular part of the electromagnetic spectrum, while passive equipment emits no radiation and simply listens to signals in specific bands of the electromagnetic spectrum.

Within this environment, radar and electronic warfare sensors are key elements in delivering what is called a *military effect* on the desired target. The effect delivered to the target can be destructive or nonlethal as is appropriate to achieve the desired military outcome. Sensor networks also provide inputs for information warfare systems to assist in the provision of a near real-time data and information-sharing network and a precision strike capability. The choice of sensors, their capabilities, and their emission signatures will have covert and overt implications for the intelligence-gathering picture, its accuracy, latency, and likelihood of sensor detection.

Historically, the technical challenge surrounding electronic combat systems has been to produce ever more sensitive reliable and accurate sensors. With rapid improvements in technology in recent decades, sensors have become increasingly accurate, cheap, and prolific. This in itself has shifted the technical challenge to that of assimilating and processing sensor data from multiple sources to present knowledge of the unfolding situation in a suitable format for near real-time decision making. Information dominance across the battlespace is therefore dependent upon dominance of the electromagnetic spectrum, with radar and electronic warfare being only elements of the mix.

Achieving information superiority in warfighting requires advanced sensor technology and efficient networks, the ability to translate increasingly large amounts of information into useful knowledge for decision makers and the development of new operational concepts to exploit real-time situational awareness. These challenges are amplified in coalition operations, where the challenge of data compatibility, real-time information exchange, the delivery of information to the relevant decision-makers, the handling of classified information across shared networks, and the compatibility of information grids need to be addressed well before the onset of combat operations.

Networked warfare: myth and reality

Network technologies have enabled sensors and weapons systems (often called *effects systems* in this context) to form a broad link between all assets in the battlespace, allowing tactical decision making to be decentralized and plans rapidly adapted to meet the changing requirements and unfolding threat scenarios in any part of the network. To refer to network technologies as network-centric is therefore misleading. Network technologies have created quite the opposite effect of delivering decision making right across the network rather than centralizing it as the term would suggest [12].

In a descriptive sense, networks are more about power across the network rather than "power to the edge" [12] or power to the center. Such descriptions certainly help to contrast the inflexibility and centralized decision-making approaches associated with rigid command and control structures found in earlier generations of warfare and are most certainly counter to the concept of networked warfare, which has the capability to connect all nodes across a network rather than disproportionately at the edge or within the centre of the network itself. The network cannot be at the center of any capability; it is certainly an important element and enabler, but by definition cannot be the central function within a distributed network environment.

Networks span geographical and cyber domains like a web that contains centers or nodes (of varying influence and functionality) that operate seamlessly together in a coordinated manner to achieve the same intent. We shall therefore refer to the concept as NEW or NEC, terms that are more representative of the model in a real environment.

NEC is more about networking than networks. NEC is more than simply improving communications and installing point-to-point data links between elements in the battlespace. It concentrates more on the challenges of networking than networks and more on the transformation of data rather than its collection. The real challenge with NEC lies in the conversion of masses of data into useful information and, in turn, converting that information into timely knowledge that can aid the decision-making process. It is clear that the areas of data processing and information access in a command and control infrastructure are still in their infancy, and as the amounts of network-enabled information available to commanders increases dramatically, the challenge will become focused on turning the data into relevant knowledge that is presentable to the decision makers across the network in a timely manner. NEC recognizes that sharing information becomes power in its own right and that military doctrine must adapt itself to exploit these new dimensions if it is to adapt its capabilities to meet emerging information-based threats (Figure 1.10).

NEW is not a panacea for addressing all military threats, but it does significantly improve the speed of response to an unfolding situation, enabling increased reaction times. Against this, NEW is heavily dependent on the quality of information fed in to the network when it comes to delivering an appropriate response. Even the most advanced intelligence networks cannot know (and are certainly are not able to access in a timely manner) all key facts about the threat's activities and intentions. In a traditional mechanized war, NEW will therefore provide more visible benefits than in an asymmetric war where the enemy is harder to detect and is characterized by fleeting targets taking on many nontraditional forms and behaviors. Outside of direct contact with opposing forces, NEW can also

Figure 1.10   A modern network-enabled environment makes use of a multiple intelligence feed. Here the combined air operations center at Headquarters Air Command at the Royal Australian Air Force Base Glenbrook, Blue Mountains, Australia, collates multinational intelligence feeds and directs airborne assets during Exercise Pitch Black in August 2004.

NEC Aspiration

NEC [13]: "encompasses the elements required to deliver controlled and precise military effect rapidly and reliably. At its heart are three elements: sensors (to gather information); a network (to fuse, communicate and exploit information); and strike assets to deliver military effect. The key is the ability to collect, fuse and disseminate accurate, timely and relevant information with much greater rapidity (sometimes only a matter of minutes or even in 'real time') to help provide a common understanding among commanders at all levels."

greatly assist in operations such as peacekeeping and deterrence operations by more rapidly being able to distribute intelligence information to obtain a shared picture of the situation and intentions of the enemy.

Adapting to networked warfare

Increasingly, conflicts can be seen as a battle of system versus system, rather than node versus node, as was the case in mechanized warfare. In today's battlespace, if people and assets are outside the network and the associated decision making occurs outside the system, the effects on the overall outcome are limited. Alternatively, if decisions are taken within a networked system, their effectiveness can be maximized to the benefit of the entire system.

Although the theoretical benefits of NEC are recognized by many, there is a general recognition that there is some way to go before the principles set out in this book—both with regard to networks and sensor technology—become reality. Indeed, the embodiment of these concepts has been described as [14]: "...a monumental task (which) will span a quarter century or more."

The challenges in implementing NEC are not simply technology-orientated; they are also related to the infrastructure and support mechanisms surrounding the technology that will help transform theory in to reality. All lines of development and doctrine need to evolve to maximize the benefits of NCW. As NEC evolves and new applications are found that can be enhanced through a networked approach, we need to be open to conceiving entirely new ways of doing things and to be able to respond to the resulting doctrinal, organizational, and operational challenges.

As armed forces start to integrate the benefits of NEC into their doctrines (and benefit from the resulting military advantages), future conflicts will evolve towards contests in adaptability [15], where the winning force will be the one that adapts itself more effectively to the evolving situation with greater speed, faster tempo of operations, and better decision making (whether for combat operations or for winning the peace), and from a doctrinal perspective, there will be an increasing recognition of the importance of the information dimension in the calculation of combat potential.

In summary then, it is inevitable that there will continue to be conflict of varying degrees in the years ahead. Unlike previous generations of warfare (which were generally predictable in nature), future conflicts will be global in character, unpredictable in their nature, and complex in their patterns. Despite the advances in sensor technology and networks, there is no easy military solution to address the described range of future conflict diversity.

NEC Thinking

- Think globally, not locally.
- Think need to share, not need to know.
- Think collaboration.
- Think many-to-many instead of point-to-point connectivity.
- Think what not to optimize, not just what to optimize.
- Think IP data formats and languages.
- Think synchronized knowledge, not synchronized requests.
- Think near real time.
- Think speed of automation.
- Think subsidiarity of decision making.
- Think loosely coupled flexibility, not tightly coupled efficiency.
- Think 3D seamless battlespace.

Adapted from: [16].

Success will be achieved by those who are able to harness the power of information warfare by deploying sensor networks to rapidly understand the strategic scenario, to process the information to accelerate effective decision-making, to increase the tempo of warfighting, and then to adapt as the situation changes. In all of these challenges, electronic sensors and networks will be at the heart of an effective information warfare capability: understanding how they operate and interact is the key to harnessing new dimensions of military power in the information age.

References

[1]   Nye, Jr., Joseph S., [Dean of the Kennedy School of Government and former U.S. Assistant Secretary of Defense for International Security Affairs (1994–1995)], "Redefining NATO's Mission in the Information Age," *NATO Review*, Web edition, Vol. 47, No. 4, Winter 1999, pp. 12–15.

[2]   Drucker, Peter, "The New Society of Organizations," *Harvard Business Review*, September/October 1992.

[3]   Lind, William, "The Changing Face of War: Into the Fourth Generation," *Marine Corps Gazette,* 1989.

[4]   "FORCEnet—A Functional Concept for the 21st Century," U.S. DoD, 2005.

[5]   Dr. Robert Leheny, Deputy Director, DARPA, DARPA's Urban Operations Programs, http://www.darpa.mil/DARPATech2005/presentations/diro/leheny.pdf.

[6]   van Creveld, Martin, *The Transformation of War*, New York: The Free Press, 1991.

[7]   U.S. DoD "Joint Vision 2020," May 2000, updated November 2006.

[8]   USAF Information Operations Air Force Doctrine, Document 2-5, January 11, 2005.

[9]   Fulghum, David, "USAF Redefining Boundaries of Computer Attack," *Aviation Week and Space Technology*, No. 9, March 3, 2003, p. 33.

[10] U.K. MoD High Level Operating Concept (HLOC) and Future Maritime Operational Concept (FMOC), 2005.

[11] Owens, Bill, *Lifting the Fog of War*, New York: Farrar, Straus and Giroux, 2000.

[12] Alberts, David S., and Richard E. Hayes, *Power to the Edge: Command and Control in the Information Age*, Washington, D.C.: CCRP Publication Series, 2003.

[13] U.K. Secretary of State for Defence, *Strategic Defence Review New Chapter*, July 2002.

[14] *NCW Report to the U.S. Congress*, U.S. DoD, July 2001.

[15] U.S. Net-Centric Operational Environment Joint Integrating Concept, Version 1.0, Joint Staff, Washington, D.C., October 31, 2005.

[16] USAF ESC Strategic Technical Plan (STP), v2.0, 2005.

[17]   Van Creveld, Martin, "Conflict in the Years Ahead," edition 7.1, August 2006. Presentation originally prepared for the Royal Norwegian Navy's staff course in Bergen.

Endnote

1.   For further information see *Neither Shall the Sword*, by Chet Richards, on 4EW; *Transformation of War*, by Martin Van Creveld, on Boyd's general theory; and *Science, Strategy and War: The Strategic Theory of John Boyd*, by Frans Osinga.

The guided-missile destroyer USS Benfold (DDG 65) maneuvers ahead of the guided-missile cruiser USS Chosin (CG 65) as it fires a surface-to-air missile off the coast of Hawaii during Rim of the Pacific (RIMPAC) 2010 exercises. RIMPAC is a biennial, multinational exercise designed to strengthen regional partnerships and improve multinational interoperability. (U.S. Navy photo by Mass Communication Specialist 2nd Class Mark Logico.)

Principles and Evolution of Network-Enabled Warfare

2

For most of the past century, the era of mechanized warfare has been dominated by the deployment of physical military assets such as tanks, ships, and aircraft, the strength of which has been measured by the type and number of deployed platforms.

In the era of mechanized warfare, platform-centric approaches have led to the coordination of military assets in a synchronized manner in accordance with the appropriate C2 philosophy of the day. Platform-centric thinking provided systems to accomplish defined tasks independently of other systems within the battlespace. Conceptually, this led to systems that housed all the necessary sensors, decision-makers, and weapons on dedicated platforms, but resulted in individual platforms that could not easily work together [1]. In this era, platform coordination was effected by radio-based voice communications and, to a lesser extent in the later stages of the mechanized era, by limited data communications. Partly because of the difficulties of integration between platforms, each platform was very much operated as an independent asset in that its capabilities were assigned to the accomplishment of a preplanned specific mission. In the most complex and demanding environments, with limited communications abilities, platforms tended to operate in a semiautonomous manner, with only intermittent communications received and sent to the coordinating command unit (Figure 2.1). Planning was therefore undertaken in accordance with the capabilities embedded in each individual platform, and this approach was often referred to as platform-centric warfare.

The concept of network-enabled capability (NEC) and network-centric warfare (NCW), on the other hand, focuses on flexible, cooperative, dynamic, and real-time collaboration between participants. NEC is an expression of a whole new way of enabling command and control in an environment where every platform can be linked together and subsequently every sensor or effects system can be regarded as simply a node on a network, with the platform, at least conceptually, being considered as a means of transportation for the sensors and effects systems, rather than as an entity in its own right. NCW is made possible through the technology and concepts associated with NEC.

NCW is the warfighting aspect of NEC and envisages the coordination of many assets, each with their own intelligence gathering or effects capabilities, in a near real-time framework. The key enablers for NCW are

U.K. MoD Definition of NEC

"The ability to gather knowledge; to share it in a common and comprehensible form with our partners; to assess and refine it to turn into knowledge; to pass it to the people who need it in an edited, focussed form; and to do it in a timescale necessary to enable relevant decisions to be made in the most economic and efficient manner."

—DCDS (EC), November 8, 2001

Figure 2.1 (a) Platform-centric approach: HMS *Vanguard*, 1941–1946. In the era of mechanized warfare, platform-centric approaches provided stand-alone platforms and force structures housing all the sensors, decision-makers, and weapons systems in a single location. This provided a high degree of autonomy, but limited communications systems

Figure 2.1 (b) Mission-centric approach: F3 Tornado and E-3D Sentry AWACS. Mission-centric approaches used dedicated integrated systems with limited networking capability to accomplish a specific mission; the Tornado F3, E-3D AWACS, and U.K. Air Defence Ground Environment (UKADGE), for example, provided a dedicated air defense system for the United Kingdom. (Image courtesy of U.K. MoD.)

agility, reconfigurability, currency, and accuracy of information and the subsequent quality of decision making and speed of response.

As will be explained later in this chapter, the advent of networked capabilities has introduced the biggest single change to military strategy and

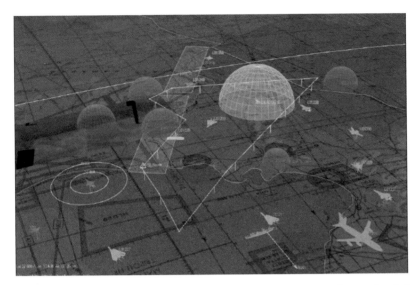

Figure 2.1 (c)   Network-centric approach. A network-centric approach uses data links to share sensor information, to cue effects systems, and to build a shared real-time situational awareness picture. The asset becomes the network rather than the individual platforms or systems. Many functionally different networks may interact with the sensor data priorities. (*Adapted from:* [1].)

U.S. DoD Definition of NCW

"Net-centric warfare (NCW) is a concept of military operations in which information superiority through networked interoperable platforms leads to military victory, and is achieved through the timely integration of finder, decider, and shooter information and decision making. This results in greater shared awareness, increased speed of command, higher tempo of operations, greater precision and therefore lethality, increased survivability, reduced fratricide, and unaided self-synchronization that leads to an overall increase in combat power."

—*U.S. ESC Strategic Technical Plan* (STP), v2.0, 2005

tactics in the past century of modern warfare. Networked approaches, for example, have fundamentally changed the way in which command and control (C2) is effected in the modern battlespace, and have introduced new dimensions and concepts to military power such as information superiority and decision superiority while at the same time enabling military forces to adapt their C2 approaches to take advantage of these new approaches through the use of the same networked communications technologies.

NCW has itself been defined as: "an information superiority-enabled concept of operations that generates increased combat power by networking sensors, decision makers, and shooters to achieve shared awareness, increased speed of command, higher tempo of operations, greater lethality, increased survivability, and a degree of self-synchronization" [2].

Although the term network-centric warfare tends to be in more common use than network-enabled warfare, it is important to recognize that networks are an enabler rather than a central feature of any network-based doctrine. By their very nature, networks permeate across capabilities and systems rather than forming any sort of centric point within the overall concept.

NEC describes an integrated force approach to modern warfare enabled by the cohesion of communications and computer networks, sensors, intelligence-gathering assets, and databases integrated with the necessary command and control (C2) processes. The decisions enabled by these new

U.K. SDR Definition of NEC

NEC "encompasses the elements required to deliver controlled and precise military effect rapidly and reliably. At its heart are three elements: sensors (to gather information); a network (to fuse, communicate and exploit information); and strike assets to deliver military effect. The key is the ability to collect, fuse and disseminate accurate, timely and relevant information with much greater rapidity (sometimes only a matter of minutes or even in 'real time') to help provide a common understanding among commanders at all levels."

—Secretary of State for Defence, *Strategic Defence Review*, New Chapter, July 2002

structures are linked to effects systems (or effectors) (see Chapter 3 for a definition of effects-based operations) in a dynamic and seamless decision-making process by technologies and approaches described as network-enabled capability (NEC).

In appearance, network-enabled capability is most in evidence in the physical capabilities of a military force, However, the construction of a network-enabled capability is a long journey and touches on many aspects of military doctrine, not only the hardware aspects, but also the leadership, procedures, training, and logistics aspects of an efficient fighting force (Figure 2.2).

Setting the scene for network-enabled capabilities

While the technological advantages of networks can easily be analyzed and mapped, it is important to recognize that all networks operate in a complex environment in which many factors, some tangible and others intangible or opaque, influence the perception of the battlespace. While the information gained from sharing data across networks is hugely beneficial, it needs to be interpreted within the context in which it is set. Information from one scenario, for example, may have a completely different implication in another scenario, or may be relevant only for a very limited window of time.

NEW takes place across several conceptual domains, all of which interact together to varying degrees according to the scenario concerned (Figure 2.3). Common within and across all layers is the exchange of information in various guises. These interactions take place at different frequencies depending on the media concerned and the speed of processing within and across the layers. The interactive patterns created by the

Network-Centric Warfare and Network-Enabled Capability [3]

- NCW is considered to be resource-driven, while NEC is resource-limited.
- NCW considers the network to be the primary driver, while NEC views the network as an enabler only.
- NCW is considered a doctrine, while NEC is considered part of a gradual improvement in force effectiveness.
- NCW is a planned and structured development of technology roll-out, while NEC is expected to evolve through networking battlefield entities.
- NCW is limited, by definition, to warfare, while NEC is expected to be applied more widely to operations other than war (OOTW).

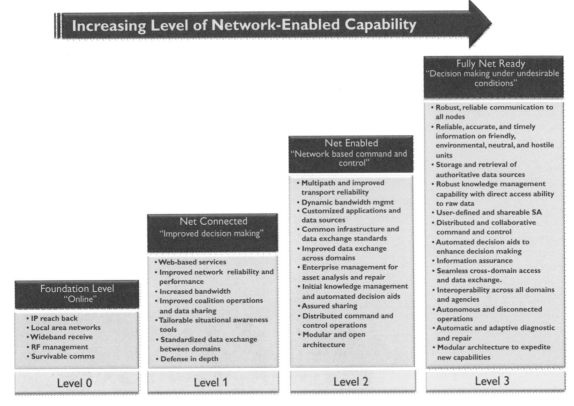

Figure 2.2 While there is no standard definition of capability levels, the model proposed by the U.S. Navy's FORCEnet program gives a useful reference for the physical aspects of the capability journey. (*Adapted from:* [4].)

information, processing, and decision exchanges vary according to the problem or mission being addressed and will determine the speed and quality of the desired outcome (Figure 2.4).

Although various models exist to set these factors in context, one of the clearest models is that which describes multiple state layers (Figure 2.5). This model enables the multiple layers that contribute to the understanding of the complex battlespace to be clearly envisaged and the complexities of interpretation and decision making—even with the advantages of a networked capability—can be visibly understood. The model is particularly useful in that it recognizes that accurate interpretation very much depends on the interpretation of the data rather than just the fusion of the data itself.

Decision making and decision superiority

Key to understanding the advantages inherent in a networked approach to modern warfare is the concept of the OODA loop. The OODA loop

"The security challenges of the 21st century are characterized by change and uncertainty. Operations vary widely and partners cannot be anticipated. However, we are confronting that uncertainty by becoming more agile. Greater levels of agility rest upon leveraging the power of information—the centerpiece of today's Defense transformation to net-centric operations (NCO)."

—Department of Defense, *Global Information Grid, Vision for a Net-Centric, Service-Oriented DoD Enterprise*, Version 1.0, June 2007

Physical Domain
where strike, protect, and maneuver take place across different environments
Information Domain
where information is created, manipulated, and shared
Cognitive Domain
where perceptions, awareness, beliefs, and values reside and where, as a result of sense making, decisions are made
Social Domain
set of interactions between and among force entities

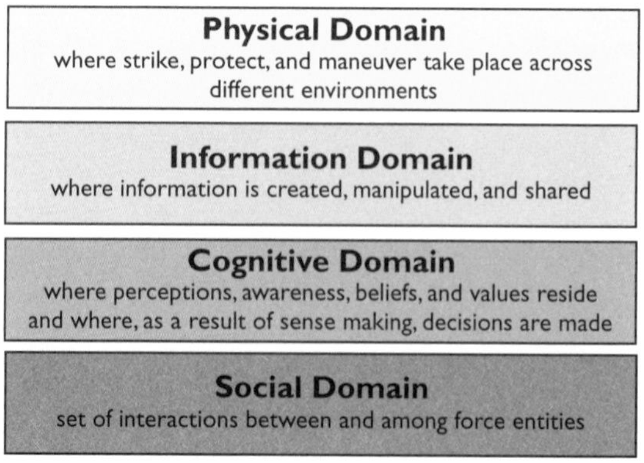

Figure 2.3 In NEW, the holistic interaction of multiple domains determines the effectiveness of the overall capability [5].

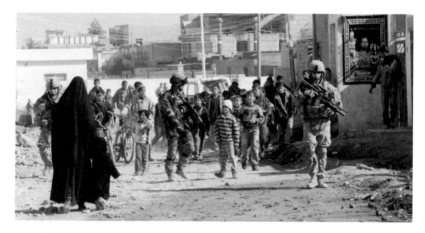

Figure 2.4 Information collected and distributed across network systems needs to be carefully interpreted against the political, economic, cultural, and social environment against which it is collected.

represents the military decision-making cycle first introduced by USAF Colonel John Boyd and remains at the heart of the military process of gathering intelligence and decision making to achieve the commander's intent.

The OODA loop enables the military commander to quickly understand the environment, generate options, decide on the most effective course of action, and implement the action faster than the opposition. It breaks down the decision-making process into a closed loop of four interacting steps: *observe, orient, decide,* and *act.*

The OODA loop describes how information is assimilated from sources that will affect decisions, how the options are assessed, and how a decision is

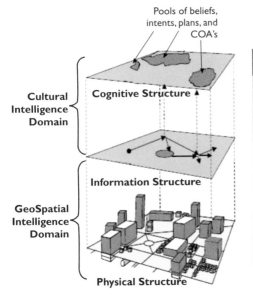

DOMAIN	ANALYSIS	ROLE	SOURCE
Cognitive	Perception, ideas, patterns, beliefs, thoughts, interpretation	Perceive, assess, decide, act	Culture, training, will
Informational	Logic, recognition, matching, linking	Process, symbolize, transform, fuse	SIGINT, NETINT, HUMINT
Physical	Analysis, measurement, symbology	Sense	IMINT, MASINT, OSINT

Figure 2.5 Multiple state layers. (*After:* [6, 7, 60].)

made about the most appropriate course of action. The OODA loop also provides a useful framework for understanding the importance of speed and agility in the decision-making process and how that contributes to the concept of *decision superiority*.

In more detail, the OODA loop principles are as follows (Figure 2.6):

- **Observe:** This stage describes the process of collecting information about the surrounding environment including the disposition, capabilities, and intentions of enemy, friendly, and noncombatant forces. It requires a diverse range of intelligence sources including intelligence, surveillance, target acquisition, and reconnaissance (ISTAR) sensors operating across the spectrum to provide the data for the orientate stage.

- **Orient:** This stage is primarily concerned with the compilation of a complete real-time situational awareness picture through the correlation, processing, evaluation, and analysis of multiple sources of data, both real time and other sources such as database and human intelligence sources. The quality of the situational awareness picture produced in a networked environment is largely dependent up on the C4I datalinks and protocols connecting assets across the battlespace. At this stage sensor and effects assets may be positioned to maximize the intelligence and choices available for the decide stage.

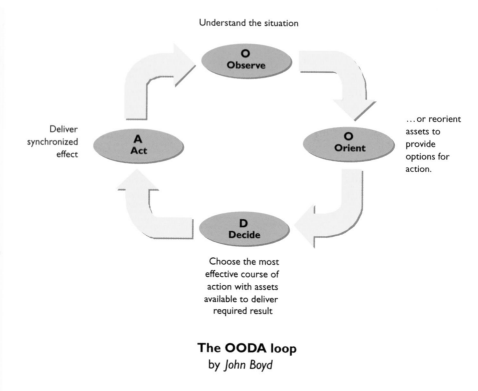

The OODA loop
by *John Boyd*

Figure 2.6   Key principles of the OODA loop.

- **Decide:** The decision step involves the presentation of assimilated information from various sources to form knowledge, assessing the options available against the desired objectives and taking account of the assets available for those options to provide what is called *decision superiority*. A superior flow of information in to the decision-making process along with the incorporation of assets and intelligence across the network also results in a better judgment of the options (and likely impact) that will deliver the desired military outcome. Decision superiority also aims to achieve higher-quality and more frequent decisions than those made by the opposing force. In a networked environment the choices open to the decision-makers will also need to take into account the competing priorities across the network.

- **Act:** Once a decision is made it needs to be communicated across the network to the military commanders and teams who will execute the planned task or tasks such as an air tasking order or an engagement order. The outcome of this action then feeds into the observe step once more to ensure that the objective of the engagement has been achieved.

The OODA loop is, in fact, a series of interlinking processes containing many inputs and decision steps that will have a bearing on the outcome of the engagement. It is not a series of discrete steps, but rather a continuous loop, with many simultaneous decisions running in parallel around the OODA loop at the same time. Information may be injected or extracted from the cycle at any point; indeed, one of the challenges of information-based warfare is how to manage and process large volumes of rapidly updated information when trying to achieve timely and pertinent military decisions.

Within a NEC context, the challenge is to transform information into readily accessible formats and make it available in a timely manner to all nodes on the network, including decision-makers who may need to use it in the OODA loop. The timely availability of such information will often be a decisive factor in the quality and speed of the decision-making process (Figure 2.7).

The key to understanding the contribution that NEC can bring to achieving the desired outcome is the recognition that decisions made within the OODA loop in turn contribute to other processes such as the kill chain (see Appendix E for an explanation of the kill-chain concept), which enables the

Typical Contents of an Adequate Attack Order

- Geographical extent of the authority
- Time duration for the authority
- Types of targets authorized
- The standard for target identification
- Types of weapons that can be used
- Further conditions that must be met,
  – such as coordination between the
    shooter and friendly forces in the
    target vicinity
- The level of risk to the delivery platform
  that these targets warrant

Figure 2.7 An F-15E Strike Eagle deployed from the 492nd Fighter Squadron, Royal Air Force Lakenheath, England, releases a GBU-28 Bunker Buster laser-guided bomb. The GBU-28 was designed and first used during the Persian Gulf War in 1991 to destroy hardened targets such as bunkers and underground command centers [59].

desired effect to be delivered to the target. The speed at which these processes are undertaken is increasingly important in order to ensure that information is collected, options assessed, and decisions made at a pace faster than that of the opponent. This enables more effective decision making to be employed against an enemy and the ability to defeat potentially larger forces where the emphasis lies in the agility and speed of decision making allows precision effects to be delivered before the opposing force is able to react or assimilate the implications of the action being taken against it. Even in an asymmetric environment, where targets are increasingly fleeting in nature, the speed of response is essential in reacting appropriately to neutralize the threat.

In addition, the ability to make faster, more accurate decisions, and to act on those decisions faster than one's opponent does, implies that an opponent will be making decisions based on out-of-date information, against an incorrect understanding of the unfolding situation, and in response to actions that have occurred well before the opposing force was aware of them. Actions taken against the opposing force will therefore appear unpredictable and seemingly unrelated to the emerging pattern of events, causing the opposing force to react inappropriately as it struggles to comprehend the uncertain and seemingly erratic nature of events unfolding before it. The ability to seize and maintain the initiative through these NEC principles therefore forces the opponent to react to the unfolding situation rather than being able to set the agenda or determine the course of the battle and its outcome (Figure 2.8).

The speed at which information is collected and processed within the OODA loop will also enable a better situational awareness of the opposing force to be generated and, when combined with faster decision making, will also serve to limit or confuse the ability of the opposing force to collect accurate and timely information about the actions and intentions of the forces ranged against it.

The challenge for any military commander is to ensure that the decision-making process occurs at a much more rapid pace than that of the opponent so that many more decisions concerning action and reaction can be taken within a given time period compared to those of an adversary, ideally so that many more OODA decision cycles can be achieved within the time that it takes the enemy to achieve just one cycle. If this can be consistently achieved, it gives the commander a significant advantage in the speed of response, the generation of tactical options, and subsequently his ability to out-think and outmaneuver his opponent.

This advantage is often referred to as *decision superiority*. In essence, decision superiority is a relative advantage in the quality and speed of decision-making

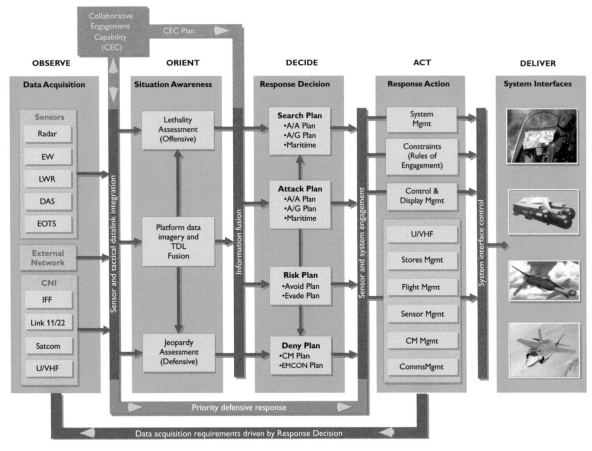

Figure 2.8 The principles of the OODA loop can be applied to ensure optimal systems design for network-enable systems such as the F-35 Lighting II Sensor & threat processing architecture shown here.

processes of one side over another. As it is a relative advantage, further benefits can be achieved through a doctrinal awareness of the decision-making process employed by the opposing force, providing the ability to better analyze, influence, and exploit the weak points in the adversary's OODA loop or equivalent processes (Figure 2.9).

The speed and accuracy with which an efficient OODA loop can be processed can open up many tactical advantages enabling many more engagements to be planned and undertaken not just against traditional targets, but also against fleeting and transient targets where the speed of the decision-making and engagement process will determine whether the target can be attacked before it disappears.

Efficient and timely decision making enables higher quality and therefore more effective decisions to be made as it buys time for the decision process. If those decisions are also implemented faster than an opponent, a tempo is

Speed, accuracy, and options in the decision-making process are key to defeating an adversary.

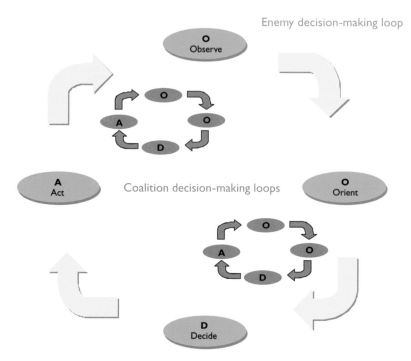

Figure 2.9   Faster decision making permits many more decisions to be made within the time it takes for the opposition to make a single decision.

established that limits the ability of the opposing force to shape their options or react effectively to changes in the battlespace or the actions of their opponent (Table 2.1).

Operating at a slower tempo reduces the opponent to making decisions using outdated information about a situation that will have changed by the time he decides what action is necessary. As a result, the decisions that an adversary makes are therefore less effective, and his options are reduced, leading to a

Integrity of information	Richness, collection, fusion, analysis, completeness, prioritization.
Accuracy	Timeliness of information, security of information, interpretation of information, certainty of information relevance, effectiveness of decision making.
Agility	The ability to be able to continuously form and reform almost infinitely variable patterns of intelligence gathering and effects systems to adapt to rapidly changing battlespace requirements.
Tempo	Fast decision making, increased tempo of operations, faster mission execution.
Persistence	24/7 accurate intelligence provision for the decision-making cycle.

Adapted from *Command and Control Joint Integrating Concept*, Final Version 1.0, September 1, 2005

Table 2.1   Factors Affecting Decision Superiority [1]

pattern where the adversary's C2 processes are simply overwhelmed by the pace with which the situation unfolds and the lack of appropriate outcomes from his decisions. In fact, it is clear to see that the engagement will rapidly force the opponent into a situation where he can only react rather than effectively plan ahead, and even where the adversary force is much larger, an efficient OODA loop process will provide a significant advantage to a smaller opposing force.

The OODA loop process (Figure 2.10) must therefore work faster, with more accuracy and with faster decision implementation than the adversary's decision cycle if the information warfare and decision-making processes are to be won.

Maximizing the speed and accuracy of the OODA loop process is very much dependent upon the quality and speed of the information that feeds into each step and the transfer of processed data between subsequent steps.

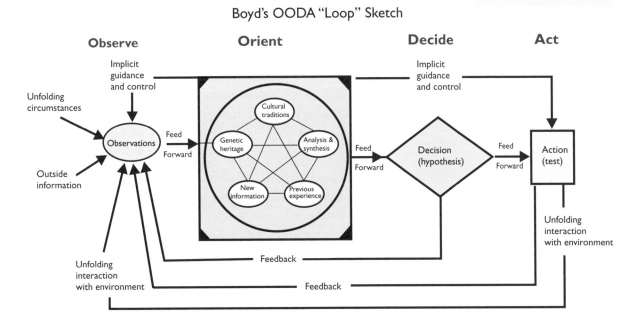

Boyd's OODA "Loop" Sketch

Note how orientation shapes observation, decision, action, and in turn is shaped by the feedback and other phenomena coming into our sensing or observing window.

Also note how the entire "loop"(not just orientation) is an ongoing many sided implicit cross-referencing process of projection, empathy, correlation, and rejection.

From "The Essence of Winning and Losing," John R. Boyd, January 1996.

Figure 2.10 The OODA loop as drawn by John Boyd. Note the many linkages within and between each decision step that are enhanced through networked communications. (Figure courtesy of Defense and National Interest, http://www.d-n-i.net, 2006.)

Network-enabled approaches emphasize speed and tempo, the speed of information gathering and transmission, the speed of data processing, and the speed of integration of information from multiple sources to create knowledge to support rapid decision making and implementation.

With the inherent emphasis on speed, a network-enabled approach can contribute to the effectiveness of the OODA loop in the following ways.

Observe:

- Tempo: more rapid decision making and deployment of effects;
- More accurate intelligence and information gathering;
- Broader base of intelligence sources from across the network—both real time and reference;
- Simultaneous collection of intelligence from multiple sources;
- Ability to adapt observations to suit the priorities of the network at the time the observations are being made.

Orient:

- Ability to fuse data from multiple sources to enhance the accuracy of the observations;
- Rapid transformation of data to knowledge;
- More dynamic and accurate situational awareness.

Decide:

- Decision-making based on timely "quality" information;
- Decisions made in a more timely manner where the assessed situation is as near to the actual situation as possible;
- Possibility that more time is available to consider the course of action to take due to the time saved in the observe and orient phases.

Act:

- Presentation of fused information in a clear picture from which priorities can be discerned;
- Communicating those priorities and associated tasks across the network;
- Faster delivery of effects and assessment of whether those effects have achieved the desired result.

Across all steps of the OODA loop, the ability to communicate information in near real-time and decide and act swiftly provides the following benefits to the outcome of the engagement:

- **Tempo:** Faster pace of decision making leading to improved operational tempo, more effective outcomes in a shorter period of time. With a relatively higher tempo of operations, the opposing force may not be able to comprehend the pattern actions that are being undertaken against them or respond effectively to counter them.
- **Agility:** The dynamic regrouping and retasking of assets across the network to achieve the desired outcome, independent of any preplanned scenario.
- **Accuracy:** The probability that an appropriate effect is delivered to the specific target in a timely manner with minimum of collateral damage.
- **Survivability:** Increased situational awareness leading to much improved survivability.

There are, however, plenty of risks in this process that need to be carefully considered in each step of the OODA loop. Heavy reliance is placed on the accuracy and integrity of the information being inserted into the OODA loop and consequently on the outputs reached as a result. Many erroneous decisions have been made in recent conflicts where data fed into the OODA loop was found to have been inaccurate or untimely, and the decision made as a result was therefore ineffective, or at best suboptimal in determining the successful outcome of the commander's intent.

Additional complexities can also be introduced by the potentially significant increase in information available to the commander through the decision-making process. In a network-enabled environment, this information is continuously collected and of course does not wait for discrete steps before it is made available. The decision-making process therefore becomes much more dynamic in nature and careful consideration needs to be given about how to feed such rapidly changing information into the OODA loop.

Furthermore, although the OODA loop is meant to enable more rapid decision making, there is always a risk that the desire to wait for information from all intelligence sources prior to a decision being made could, in fact, slow the process to an unacceptable level. The question that every commander needs to ask is: "Could an 'adequate' decision be made with current information?"

The utility of network-enabled capability

NEC is an enabling concept that connects sensors, decision makers, and effects systems across the battlespace in a timely and seamless manner to permit the battle to be fought through a networked multisensor approach rather than by autonomous platforms each taking independent decisions to achieve the desired outcome (Figure 2.11). The effectiveness and speed of operational decision-making processes very much revolve around this concept of networked connectivity.

NEC relies on the rapid fusion of timely, relevant, accurate, and standardized information from networked sensors in order to prioritize threats and allocate the most appropriate forces and weapons systems on the network to engage priority targets. NEC emphasizes collaboration across platforms and across multiple networks in environments where many forces and coalition nations may be operating together to achieve a common objective through coordinated effects. It promotes the need for a common real-time, or near real-time, intelligence picture, which in turn necessitates collaborative data fusion, high bandwidth data transmission, synchronization, and automation of data handling and decision-making processes within the information domain. NEC looks to exploit information while denying its effective use of information to the adversary.

Figure 2.11 NEC enables one platform to take control of the effects systems of another platform. Here the guided missile cruiser USS *Vicksburg* (CG 69) and the guided missile destroyers USS *Roosevelt* (DDG 80), USS *Carney* (DDG 64), and USS *The Sullivans* (DDG 68) launch a coordinated volley of missiles during an exercise. The exercise is designed to practice the network-enabled interception of hostile missiles with remote-controlled drones simulating the hostile missiles.

The building blocks of NEC are therefore those technologies that help to acquire data, process and exploit it in a timely manner, and enable it to be shared and protected, while denying the same to the adversary.

The development of NCW and NEC concepts stems from the cold war days when means were sought to use NATO's technology advantages as a counterweight to the numbers' superiority of the Warsaw Pact countries. The thinking behind these concepts was to develop a rapidly reconfigurable C2 infrastructure that was flexible enough to be formed at the required time to meet the needs of a rapidly changing threat and help NATO make up for the lack of superiority in numbers through superiority in information dominance and associated improvement in tasking. Much effort was placed on stand-alone sensor technologies, reconnaissance, and precision engagement capabilities, which provided a technological advantage, but, largely due to technology limitations, had a limited capability to then exploit the acquired battlespace information in a truly networked manner.

Operations during this era were characterized by limited communications, military strength realized through physical assets, hierarchical decision making, and command structures supporting a centralized C2 approach and information that was difficult to obtain and even harder to share in a timely manner [9].

Towards the end of the cold war, it became clear that the threat facing NATO countries was no longer confined to mass attacks by relatively unsophisticated forces of the Warsaw Pact countries. At the same time, information technology and computing power were making rapid strides in capability and processing speed, and it was recognized that these new technologies could be harnessed to provide a shared near real-time picture of the battlespace, enhance precision engagement capabilities, and improve the command and control (C2) infrastructure.

Early network systems such as Link 4 and Link 11 started to provide a networked picture through the use of specific data categories, but suffered from limitations of the type of data that could be conveyed and the speed with which data could be transmitted. These limitations imposed significant restrictions on the utility of the data through data latency and limited operational applications. To overcome these constraints, early networked approaches focused on the development of dedicated systems where dedicated platform and equipment types were engineered to work together to accomplish specific but limited battlespace missions (Figure 2.12).

It is only in recent years that the technology to support a near real-time multiplatform, network-enabled, shared data environment has emerged. Given the dynamic and unpredictable nature of future threats, the key role

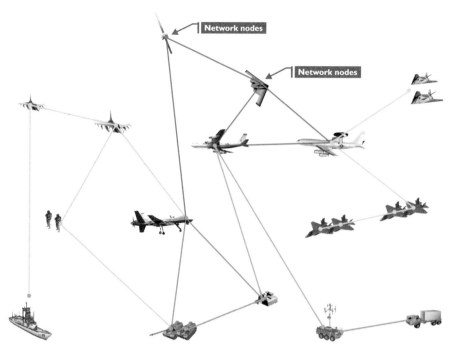

Figure 2.12 In an NEC environment, many different networks interact together and each node may act as a connection point for multiple networks.

for NEC must be to enable military planners to move away from formal hierarchical structures towards enabling the formation of ad hoc task-based groups through the rapid integration of building blocks using reconfigurable sensor and effects systems.

NEC is an approach to capability enhancement that aims to improve operational effectiveness by permitting a more efficient sharing and exploitation of information within the military, paramilitary, and government departments and across coalition partners. In an ideal world, NEC should ensure that the concept of operations employed across all parties will guarantee that the information-sharing, decision-making, and engagement authorization procedures are flowed across the network in a coordinated manner to facilitate collaboration between forces where joint decision-making can be effected through collaborating force structures.

Such integration between collaborating assets and networks should ensure the ability of geographically dispersed forces to create a high level of shared battlespace awareness. Potentially this cross network approach also creates an advantage by enabling the operations and motives of the network to remain ambiguous to the opposition, yet clearly understood by the home team.

NEC requirements have led to a significant step forward in a number of technologies that are critical to effective NEC operations, in particular:

- Near real-time dissemination of sensor data;
- Data fusion techniques;
- Automated target recognition and tracking;
- Accurate time and space reference frameworks;
- Geolocation technologies;
- Data transmission networks capable of handling large volumes of data;
- Self-forming and self-healing networks;
- Standardization of data transmission and processing formats;
- Decision support systems to distill data into information and the dissemination of knowledge critical to the decision-making process to the right place in the decision-making process at the right time.

Extending the reach of military power—exploiting network advantages

NCW and NEC philosophies center on the processes, networks and technologies that enable the achievement of information dominance by providing the capability to rapidly share information across the battlespace and to use communications systems to support distributed networks. By linking sensor and weapons systems (often referred to as effects systems) together over a network, NEC enables the synchronization and coordination of sensors and weapons to provide significant improvements in situational awareness and offensive and defensive capabilities. Such a step change in capabilities, where no one sensor or platform will operate as an independent entity, also provides doctrinal challenges to the orchestration, decision making, and deployment of assets, which may be globally dispersed and often directed by different "commands" in a traditional sense of the word.

The network topology, particularly related to the degree of integration of the network nodes, will determine the impact on collaboration and joint planning between the cooperating forces (see Table 2.2).

NEC technology brings with it the ability to allocate combat resources across the network in accordance with the priorities of the network rather than those of the individual commander. This is achieved through the ability to generate synergies through the selection of sensor and effects capabilities from across the network in accordance with the perceived threat facing the network as a whole. Sensors and effects systems mounted on the same platform can, in theory, be decoupled from each other across the network, providing significantly more sensor coverage and effects options. This

Network Type	Time Horizon	Network Characteristics
Joint composite tracking network	Immediate	Shared situational awareness, networked C2
Joint ISTAR network	Immediate: medium	Shared network priorities, collaborative engagement, dynamic replanning. Consideration of the network as an entity in its own right.
Joint planning network	Medium: long term	Access to GIG, integration of forward logistics requirements, generation of effects options.

Table 2.2 The Degree of Network Integration Determining the Impact on the Collaborative Warfighting Approach [3]

enables the weapons systems on one platform to be cued and tasked by the sensors and decision processes from another platform or part of the network.

Figure 2.13 illustrates the relatively limited sensor coverage that can be achieved by a single platform. Relying on platform-based sensors alone, the ship is not able to comprehend the larger situational awareness picture or effectively interpret its own situation within a broader context.

Figures 2.14 through 2.17 show the huge increase in situational awareness and weapons coverage that can be achieved through networking platforms collaborating together. The advantages come not only in the extended range of coverage of the sensors and weapons, but more particularly in the ability of the network to act as a single entity, prioritizing threats across the

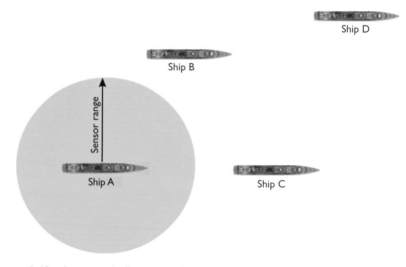

Figure 2.13 Sensor and effects range for a single stand-alone maritime platform.

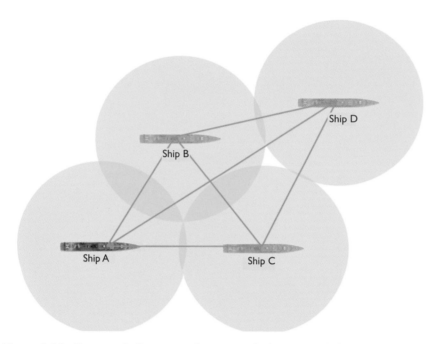

Figure 2.14 Sensor and effects range for a networked maritime platform, able to access a distributed sensor network.

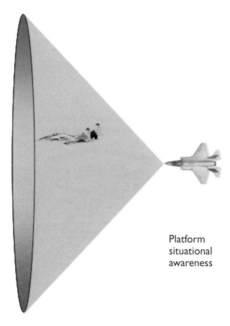

Platform situational awareness

Figure 2.15 Sensor and effects range for a single stand-alone airborne platform.

network and acting in the interests of the network as a whole rather than the individual platforms (Figure 2.18).

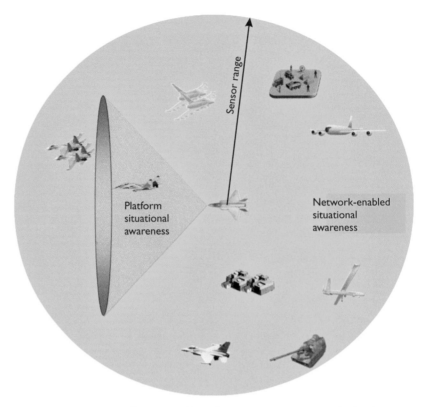

Figure 2.16   Sensor and effects range for airborne targets in a networked airborne platform, able to access a distributed sensor network.

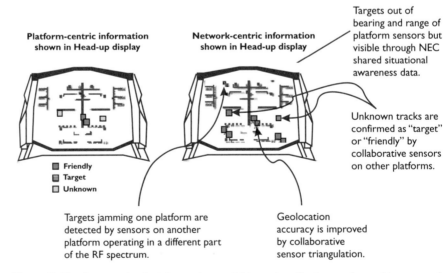

Figure 2.17   As a result, the information available to the pilot is transformed in terms of its accuracy and richness.

Figure 2.18 The head-up display on this F-15E is clearly visible in front of the pilot. When operating in a networked environment, the pilot can be sent details of priority targets, engagement sequences, and alerts that represent threats to the network rather than simply addressing threats to the platform itself.

Similarly, a fighter aircraft operating on its own will only be able to achieve a limited situational awareness, which may leave it exposed to fast moving threats that have been detected by other platforms. By sharing information between platforms, the fighter will be able to greatly increase its chance of being able to prioritize threats along with its chances of survivability.

A network-centric approach to information acquisition and the dissemination of battlespace information is critical to enable swift and decisive military decision making and the delivery of the desired effect. An NEC approach allows platforms and command and control (C2) capabilities to exploit shared awareness and collaborative planning, to communicate and coordinate responses, and to enable seamless synchronization across the battlespace. It underpins information dominance and the delivery of rapid, appropriate, and synchronized effects across the battlespace.

The ability to network sensors and communications systems has resulted in a huge increase in the amount of information that can be shared between collaborating forces. While advances in computing has enabled many processing systems to cope with this data with relative ease, the challenge of presenting huge volumes of information in a useable format to decision makers has increased significantly. Users need to be presented with timely, accurate, and relevant knowledge to enable effective decision making in those parts of the engagement process that require human decisions or assessment of the situational picture. Information superiority provides a relative military advantage only when it is effectively translated into superior knowledge and actionable decisions. Turning information into timely knowledge and getting that knowledge swiftly to the precise point in the network to enable rapid decision making is therefore also a key aspect of delivering an effective NEC system [10].

NEC operations envisage, and indeed enable, combat and C4ISTAR activities being undertaken by a large number of diverse but synchronized units, rather than by a small number of large structured units found in mechanized warfare. The diversity of assets that can be coordinated and employed in an NCW environment provides significantly enhanced capabilities through the ability to draw on a wide variety of specialist sensors and effects systems that would not be available to an isolated platform or group. In addition, distributed network operations benefit from the inherent redundancy of sensor or effects nodes and are therefore more robust when faced with an adversary intent on active or passive disruption.

The adoption of network-enabled concepts requires the integration of a large number of stand-alone sensor and dedicated communications links. This has involved both the integration of legacy systems, which were not originally planned to be part of a distributed network, and the adoption of common data interchange standards and message formats for new network-enabled capabilities. While significant progress has been realized in these areas, it is interesting to note that in the United States, for example, the adoption of NEC technologies and their concept of operations has generally remained the preserve of each of the armed forces rather than being fully coordinated at a central level, often creating problems that a true networked approach should naturally overcome.

In the past, planned and structured approaches to sharing information across multiple systems, typically on the same platform, have been adequate for the limited set of threats that platforms have had to face in previous decades, and the dedicated missions for which most platforms have been designed. Today, however, the use of extended sensor and effects networks has resulted in significantly improved flexibility and versatility to deal with threats that were not even envisaged in the recent past. No longer can platforms be dedicated to specific missions in an environment where unpredictability and agility are the order of the day.

In a networked environment, information about the threat disposition can be shared and accessed with relative ease between collaborating platforms, and the range of sensors that can be networked will typically give the battlespace commander a significantly improved perspective of the nature of the threat associated with individual targets and the prioritization of those targets in the battlespace as a whole. The ability to find, track, and destroy targets, through a process often described as the *kill chain,* increases disproportionately when operating in a networked environment. Speed of decision making across the *kill chain* improves as more accurate and accessible information from various sources is fused to a common data set, with a resulting decrease in the time between detecting and engaging targets.

The resulting improvements in situational awareness achieved through NEC also significantly enhance forward operational planning and rear logistics planning as adversary capabilities and intentions become clearer and the status of forward deployed forces are transmitted to rear echelons, allowing faster resupply and logistics planning. Improved situational awareness also ensures that friendly and neutral forces are clearly identified and the chances of accidental engagement rapidly diminish.

Network types, protocols, and characteristics

Unlike public domain Internet services, military networks tend to be highly organized and structured from an access and interface perspective, with a number of specific data handling, processing, and storage objectives shared by all its members. Within a distributed network, however, despite the formalities of the network structures, each member has his own specific role to play and own goal to achieve. Every member or node in a military network may be assigned a specific role within the organization, though the roles may vary with changing mission demands and priorities.

Within a networked environment, a node can consist of a single entity, a single instance of a physical thing such as an individual commander or other decision-maker, a sensor, a weapon, a vehicle, a server, a supply pallet, or even a mechanical component or a spare part—or a node may be a grouping of entities functioning as a single body, such as a staff, an analysis center, a combat unit, or an informal community of interest [4].

Networked organizations are characterized by standardized information sets enabling the widespread sharing of information and the prevalence of point-to-point relationships to enable information to be accessed directly by those who need it. Networked organizations operate significantly differently

We Must Fight the Net

"DoD is building an information-centric force. Networks are increasingly the operational center of gravity, and the Department must be prepared to 'fight the net.'"

—U.S. DoD Information Operations Roadmap, October 30, 2003.

to traditional military hierarchies typically found in the mechanized warfare era, as all networked nodes can in theory access the same level of information wherever they are rather than have access to information determined by their position in the hierarchy.

The structure of networks and how they interact with each other plays a determining role in the efficiency and operational protocols behind NEW. One way to define an organization's structure is to specify the nature of the interactions that take place among its members [11]. Interactions across networks are determined by the number of nodes (from which communications are sent and received) and their linkages (which represent two-way communications between nodes).

The interactions between the nodes, the ratio of nodes to linkages, and the patterns by which the nodes are connected determine the topology of the networks concerned. In the network of networks approach, multiple networks may interact with each other and networks with different configurations may therefore interact routinely (Figure 2.19).

Network nodes tend to fall into various functional roles—sensing, understanding, deciding, and communicating, each of which plays a role in supporting the OODA loop decision-making process:

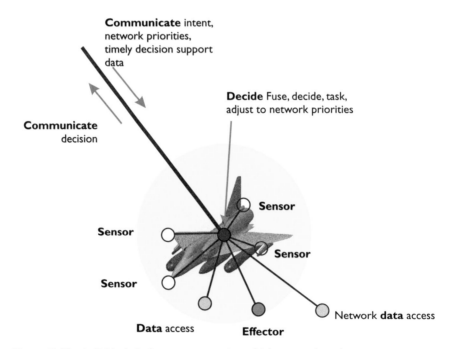

Figure 2.19   Individual platforms may contain multiple network nodes.

- **Communicate:** Actions and intent need to be communicated to other nodes on the network. Fused and processed data from multiple sensors or nodes may be communicated in the form of information. Integrated network applications may require long-range beyond-line-of-sight communications and some nodes may be used to relay communications over long distances to connect networks around the globe. These could, for instance, be devices mounted on UAVs or routed through satellite communications networks. Such transmissions will be used to connect larger areas of the network and to link up more distant nodes.

- **Sense (Observe):** Sense functions generate data about the environment. This may take into account data from other networked sensors (nodes) and/or other databases. The battlespace area that is being observed will be determined in accordance with the network priorities.

- **Understand (Orientate):** The function facilitates the generation and maintenance of a situational awareness picture, its implications relative to the position and capabilities of the platform and the threats and mission requirements. The understand function often incorporates inputs from sensors on local networks or other databases such as Global Information Grid (GIG) sources.

- **Decide (Orientate, Decide):** The decide function involves fusing sensor information from multiple sensors to take account of the commander's intent and priorities across the network in order to assist in the decision-making process. The decide function may require access to on-platform or off-platform databases along with real-time networked situational awareness information. Decisions can be automated, typically within the network, but decisions to engage targets are typically made by human decision makers.

- **Initiate Effects (Act):** In a network-enabled scenario, weapons systems (nodes) may be initiated from other platforms to reflect network rather than platform priorities (Figure 2.20). Increasingly, weapons themselves are becoming network-enabled allowing their targeting information to be updated during flight or even allowing the final target to be changed using en-route updates.

The determining and descriptive characteristics of a network can be summarized as follows [9]:

- **Types of nodes:** The utility of the military network will depend on the types of nodes connected. The variation in the functions of the nodes will greatly increase the inputs required for decision-making in the OODA loop and subsequently the number of options open to the commander for sensing, defending, and striking from across the network will be enhanced.

Figure 2.20   In a networked environment, aircraft such as the B1-B may act as a mobile weapons platform, or effects node in NEC parlance, with targeting information being uploaded to weapons immediately before launch, enabling the network to make priority targeting decisions and deliver an effect in a very short timescale.

- **Number of nodes:** The effectiveness of the network depends on the nodes within the network. The number and type of nodes needs to be optimized to ensure adequate geospatial functional coverage. A collection of nodes is referred to as a *cluster*. Efficient network design needs to balance the need for agility and multipath routing with the reduction in network capacity that rapidly becomes an impediment as the number of hops to the destination increases (Figure 2.21).

- **Number of linkages:** Communication linkages may involve the transmission of data over various media such as dedicated RF and microwave links, radar systems, optical links, or lasers. The number of linkages is the biggest contributing factor in determining the agility of the network to reconfigure to meet the unfolding operational requirements. In combat operations the number of linkages can also be useful where nodes are destroyed by enemy action or where parts of the network need to be isolated due to a computer network attack. Linkages that do not exist can also have as much of an impact on a network than linkages that do exist if the network is not capable of reconfiguration.

- **Routing protocols:** There are many ways that nodes and networks can connect to each other assuming that the messaging protocols are compatible across networks. There are also a number of ways that routing protocols can be characterized, and it is possible for a network to use several of the routing protocols described next at the same time. Some of the more common types of routing protocols are as follows [13]:

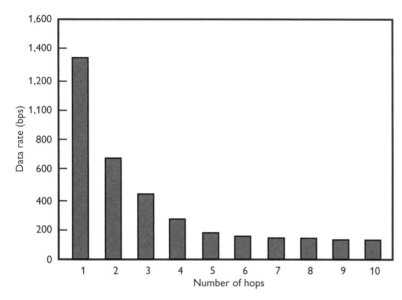

Figure 2.21 Network capacity diminishes exponentially based on the amount of hops to the destination [12].

- *Fixed routing:* This routing relies on a known and fixed pattern of nodes and connection protocols. It is often used for a fixed base infrastructure, but is not easily reconfigurable or adaptable to changing military requirements.

- *Hierarchical routing:* Similar to proactive/reactive routing, this type of routing depends on the depth of the nesting number of layers in the network, the hierarchical level where the node exists, and the depth of the preexisting knowledge of the network.

- *Adaptive routing:* Also known as *situation-aware routing*, this protocol takes into account the bandwidth available across various network routes to meet data transmission requirements.

- *Proactive routing:* Also known as *table-driven routing*, this type of protocol retains an up-to-date list of destinations and their routes by sporadically distributing routing tables throughout the network. The main disadvantages of such protocols are the amount of data management required and the inflexibility of the network if part of it is disabled or destroyed or needs to be isolated.

- *Reactive routing:* Also known as *on-demand routing*, this type of protocol finds a route on demand by flooding the network with route request packets. This approach, however, takes a relatively long time to establish a satisfactory route and can take up additional bandwidth or hold up other transmissions that could be used for more important purposes.

- *Flow-oriented routing:* This type of protocol finds a route on demand by following existing flows. However, it may take a long time to establish a new route without a prior knowledge of the network topology.

- *Hybrid routing:* Based on *proactive/reactive routing*, this routing is initially established through some proactively discovered routes and then provides additional routing to meet further demand through additionally activated nodes chosen through further exploration of the network. The routing capacity will depend on the additional number of nodes that can be activated and the speed with which those nodes can be incorporated into the routing.

- *Power conservation routing:* This approach finds the route through the network that uses the least amount of power for the data to reach its destination.

- **Addressing protocols:** Data transmissions connecting only two nodes may use point-to-point communication protocols (Figure 2.22), which can be relatively straightforward and provide for a high degree of flexibility in message format. Most network-enabled systems, however, require collaboration between groups of computers with the ability to handle simultaneous multiple inputs and outputs. This process is known generically as multipoint communications. There are three main types of multipoint communications addressing protocols:

  - *Unicast addressing:* A unicast address is designed to transmit a packet[1] of information to a single destination (Figure 2.23).

Figure 2.22 Point-to-point transmission.

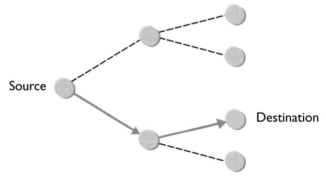

Figure 2.23 Unicast transmission.

- *Broadcast addressing:* A broadcast address is used to send a datagram[2] to an entire subnetwork (Figure 2.24).

- *Multicast addressing:* A multicast address is designed to enable the delivery of datagrams to a set of hosts that have been configured as members of a multicast group in various scattered subnetworks (Figure 2.25). Multicast protocols distribute data from a single sender to multiple receivers. Multicasting is not connection oriented and can be used to broadcast data across a broad range of network patterns.

- *Hypercast addressing:* A super-scalable, many-to-many multicast protocol for network applications, a hypercast address provides a socket-like interface to applications through which they can join and send data in a multicast group. Hypercast provides a means for applications to join multicast groups in order to send and receive data in those groups. For an application to join a multicast group, the application must specify the multicast group address when it creates a hypercast socket. Once it joins a multicast group, a node can send and receive data with simple commands [14].

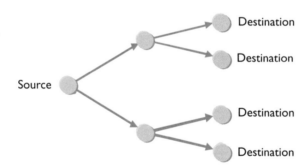

Figure 2.24   Broadcast transmission.

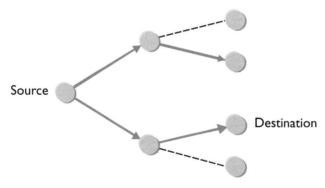

Figure 2.25   Multicast transmission.

- **Path length:** The length of a linkage is measured by the number of nodes through which the data must travel to reach its recipient. In a complex network, this may have a significant impact on data latency.
- **Number of connections at each node:** Distributed networks should have about two links per node as a maximum. Research suggests that even four or five connections per node can paralyze a network with too much feedback or feedforward [9]. Additionally, the greater the number of connections, the greater the bandwidth or transmission capacity needed at each node.
- **Clustering:** A clustering is a measure of the actual number of links between networks compared to the maximum number of links. Highly clustered networks tend to have pockets of connectivity, which can increase the connectivity and redundancy of the whole network [15].
- **Data latency:** Speed and currency of data transmitted across the network will have a significant bearing on the utility of the data at the nodes, as will the time taken to process data and use it in engagement decisions or pass it to other nodes for action. Data latency is influenced by the available transmission bandwidth, the number of nodes, the processing requirements, the path length, and the C2 structure where data may need to be collected and redistributed from a centralized nodal structure.

Data latency delay =

> Node processing delay +
> Queuing delay +
> Transmission delay +
> Propagation delay

- **Bandwidth:** The bandwidth is a measure of the volume of data (usually measured in bits per second) that can be transmitted between two points (Figure 2.26). The bandwidth between nodes in a network is a key determinant of the functionality that can be provided by each node. Bandwidth can be considered as the size of the pipe required to transmit the information between nodes.
- **Data transmission speed:** The bandwidth does not give a direct indication of data transfer speeds, as the data itself is often encoded in to a protocol that provides for strong encryption and for robust and clear transmission even in an environment where the signal is being jammed. Connection and synchronization protocols to ensure that no unintended users join the network also add to the overhead burden imposed on the original data.

Basic network building blocks

Traditional network theory has in the past principally been associated with Metcalf's law. Metcalfe's law states that the value of a network, expressed as

Figure 2.26 AWACS, Joint STARS, and E-2 sensor platforms were designed to carry a crew on board, trained and empowered to view the high-bandwidth sensor output, interpret that output in operational terms, and disseminate their interpretation over a low-bandwidth, dedicated link to selected users [8].

the number of connections, is proportional to the square of the number of users of the system (n^2), or in mathematical terms, the number of unique connections in a network of n nodes can be expressed as $n(n-1)/2$, which follows that the value is proportional to the square of the number of connected nodes (n^2). For example, 10 nodes would enable 100 possible connections, 100 nodes would enable 10,000 connections, and so on.

The utility of Metcalfe's law and the usefulness of the number of nodes that can be accessed through the application of its formula assumes that all members on the network are equally valuable. Generally, this is not the case in a dynamic environment where information from some nodes will be significantly more valuable than data from others. However, the basic tenet that n^2 connections are possible (even if they are not used) is still valid [16].

While Metcalfe's law explains many of the network effects of communication technologies and networks such as the Internet, social networking, and the World Wide Web, it is not directly applicable to combat networks other than explaining the distribution of raw data cross a network, because of the complexity of the functions performed by combat networks and nodes and the processes served by those nodes such as the OODA loop.

The capability gains achieved through increasing the nodes and linkages in a networked military system arise from the improved efficiencies brought about in the OODA decision-making process. Much depends on the way that

Network-Centric Enviornment

The U.S. Joint Forces Command defines a net-centric environment as:

"a Joint Force framework for full human and technical connectivity that allows all Department of Defense users and mission partners to share the information they need, when they need it, in a form they can understand and act on with confidence; and protexts information from those who should not have it."

—*U.S. Net-Centric Operational Environment Joint Integrating Concept*, Version 1.0, October 31, 2005, Joint Staff, Washington, D.C.

the incoming information is queued and processed at the decision nodes and how decisions are made. Decision processes may vary from the manual collection of relevant information through to automatic fusion and automatic decision-making. Typically, ISTAR processes may involve a significant amount of semiautomated or automated data processing. Decisions to deliver an effect to a target will always involve a human-in-the-loop unless automated defensive aids systems are involved.

In order for the OODA loop process to be improved, the entire process must be considered rather than any single element. The law governing the degree of improvement to an overall system when only part of the system is improved or when parallel processing is used to accelerate the processing of information is governed by Amdahl's law. For example, if 12% of a software program can be run by parallel processors while the remaining 88% of the operations are not parallelizable, Amdahl's law states that the maximum speed-up of the parallelized process is $1/(1 - 0.12) = 1.136$ times faster than the nonparallelized process.

Where information is queued in order for it to be processed, Little's law defines the variables that dictate how long it will take to complete a task within a process and is the key to understanding information processing efficiency. In this context, Little's law states that the number of packets of information in the system equals the average arrival rate, multiplied by the processing time. More importantly, Little's law implies that the throughput is independent of the arrival schedule of the packets of information, as information will queue until it can be processed.

In addition to consideration of the processing of information within a network, consideration must also be given to the alignment of physical assets to enable the options for the act stage of the OODA loop. The commander must wait for others to respond and must organize and position assets to engage. All of these events involve to greater or lesser degrees, one entity waiting for another, in effect, queuing up—issues that are not addressed simply by increasing the number of nodes or linkages within a network [17].

When connecting the basic network building blocks (Figure 2.27) into larger operational structures, there are a number of network patterns that can be adopted.

Hierarchical networks (Figure 2.28) generally require fixed nodes and linkages and have limited structural or dynamic flexibility. Data paths tend to be predefined, and such data flows are vulnerable to disruption by sections of the transmission network being taken out or by critical nodes being targeted or disrupted. Hierarchical networks are therefore best suited for stable

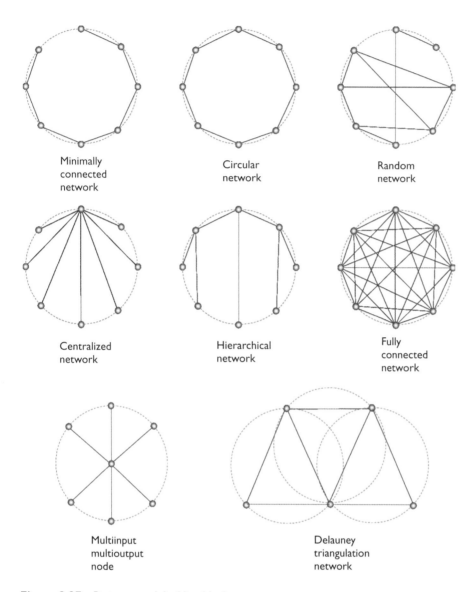

Figure 2.27   Basic network building blocks.

environments and predictable environments and are typically found away from front-line operations in support of reach-back C2 networks, logistics chains, and supporting military infrastructure requirements.

Hierarchical networks tend to organize communications through the use of building blocks of clustered nodes arranged through a family of systems (see Chapter 3) approach, with each building block a cluster under the control of a central data processing or C2 node. The central coordinating function that orchestrates and coordinates the activities of the lower-level system groups then integrates each building block into the larger network. As each node

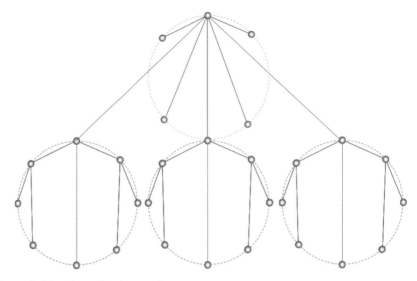

Figure 2.28   Hierarchical network.

typically communicates with only one or two other nodes in a predictable manner, bandwidth requirements are generally lower and can avoid higher bandwidth requirements associated with multiple access transmission protocols.

Processing requirements may also be modest as they can be structured to suit standardized inputs and outputs, which are in themselves typically limited in their variability, particularly as the networks increase in size and complexity with nodes farther away from the central hub. Hierarchical networks may also be used in a more dynamic environment where deployed units—often from different services or coalition partners—generate their own intelligence pictures in stovepipes, which are then reported back up the hierarchy for integration and redistribution. However, such a structure results in a network of stovepipes, each producing a separate "picture" of the battlespace. Only members of each stovepipe can access and integrate the information generated by the local network, and the real challenge becomes how to share the network information to other users.

General John Jumper of the U.S. Air Force likened this to tribal behavior where only members of that tribe can decipher and explain their hieroglyphics to members of other tribes [18].

In *time-sensitive targeting* (TST), where the rapid collection of data and decision making is required, hierarchical structures that need to collect all the sensor data centrally and distribute it to specific platforms can adversely extend the time line by introducing additional steps (collecting, processing, and distributing target data) and additional failure modes (allocation errors) [8].

In *fully connected networks* (Figure 2.29), all network nodes (typically at the system level) transmit and receive information to a common network, which can be accessed by other network nodes. In such an environment, even though the connectivity will be permanent between nodes, the bandwidth required will vary significantly depending on the functions being performed at each node; for example, a node acting as a C2 process will require a high bandwidth in its linkages to other nodes. An EW sensor simply transmitting data will require a relatively smaller bandwidth. All nodes participate equally. There is no hierarchy. In theory, all nodes can provide and access information to the same degree, although in practice the function of each node will determine the degree to which information is passed to and from the node. In a truly integrated network, there is no edge to the pattern as the distribution of the nodes and sensors can be adapted to meet the needs of the unfolding data collection and dissemination requirements.

Theoretically, and assuming full interoperability on technical, security, and cooperability levels, a fully connected network would provide the highest level of direct interaction by linking every node to every other node. Such an arrangement would require dedicated bandwidth availability between each node to ensure that data could be transmitted and each node would need to receive, process, and transmit high volumes of data and information for the network to operate effectively. Clearly, however, such an arrangement is impractical in a dynamic network environment where many different types of connections will exist on various platforms and participants.

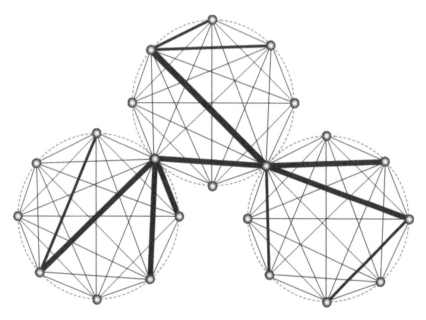

Figure 2.29 Fully connected fixed node network showing variations in bandwidth usage across the network.

For a fully connected network that consists of n nodes, every node that joins adds $n - 1$ linkages. This is costly to support, quickly overloads any available bandwidth, and means that the number of interactions possible for any node rapidly becomes overwhelming. Hence, even if such a system is constructed, the nodes within it must make an enormous number of decisions about when they will interact, with whom they will interact, and how much attention they will pay to any interaction or offer for an interaction. Hence, the endpoint on the spectrum of patterns of interaction— everyone with everyone else, all the time, and using the full range of media—is simply impractical as an approach for a large-scale military organization [19].

At a more practical level, enabling direct linkages between all nodes does not imply that direct connections permanently exist or that communications take place between all nodes. It does mean that, if required, any node can exchange information and interact with any other node as necessary.

A *Delaunay triangulation network* is a network type that provides connection with all nodes across the network using a triangular pattern between groups of nodes. Commonly used with hypercast addressing, the Delaunay triangulation topology can be constructed in a distributed fashion enabling robust networks with redundancy paths to be easily constructed. Such networks have been run with more than 10,000 connected nodes [20].

The richest, most efficient class of network construct is the *small world network* [15], which recognizes that moving information from one node to another requires only a small number of steps [21]. The distinguishing characteristic of small world networks is what is called a *large clustering coefficient*. This means that a link to any one node is readily connected to a number of other nodes through nodal clusters that can be thought of as "communities of interest" or "communities of practice." Each such community forms its own "small world," which may be linked to other small worlds by some individual nodes that act as connectors [19]. Such an arrangement is typical of network-centric interconnections and distributed networks where many discrete networks collaborate together for maximum effect, sharing information, synchronizing situational awareness, and working together to achieve the desired command intent.

As network connectivity evolves, future networks will comprise a mixture of hybrid structures that self-adapt to meet operational requirements and take into account the constraints in their environment (such as power, bandwidth, and latency). This flexibility will create patterns of interaction capable of forming complex adaptive structures.

Centralized versus decentralized network patterns

Centralized networks (Figure 2.30) comprise a central node connected to multiple subnodes, with the central node exercising clearly defined authority over the entire network. Centralized networks often tend to be used for unique systems or functionally specific roles or for fixed communications where requirements are at least reasonably permanent in nature, such as strategic communications to and from a headquarters unit.

Decentralized networks (Figure 2.31) can exhibit varying degrees of decentralization and mixed levels of control and authority across the network structure. Authority and control patterns may also vary across the network with time, depending on its functional role and relative proximity of each node to the operational requirements. Decentralized networks are often found in the family of systems architectures where systems with common operating interfaces are nested across a network to provide an extended systems capability.

Distributed networks (Figure 2.32) generate the most complex of network patterns. They exhibit no central control, no defined or clear authority and

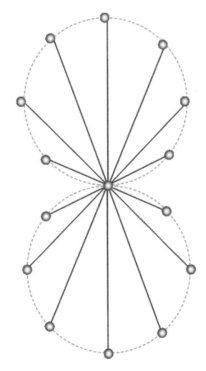

Figure 2.30 Centralized network.

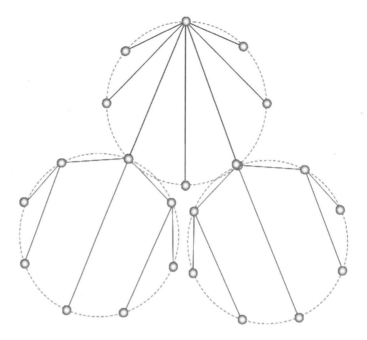

Figure 2.31 Decentralized network.

levels of control and authority will vary across the network, typically adapting to the complex and diverse needs of each node on the network. Such networks are commonly found in systems of systems networks where many systems and subnetworks are woven together to achieve the overall system of systems capability. Distributed networks may employ a fixed architecture as in the case of a system of systems network providing a strategic capability, or a transient architecture in the case of a battlefield sensor network, depending on their role and configuration.

There are three principal types of network connection types:

- *Fixed:* where structures are predefined and known to all nodes on the network;
- *Ephemeral:* where parts of the network join together and disband in a fixed pattern to meet mission requirements;
- *Ad hoc:* where random network patterns and routing structures are formed by nodes that join and depart from the network at will (Figure 2.33).

Fixed networks follow structured patterns and employ wired and wireless communication links between nodes. They are typically capable of high data rate transmissions and are often used as trunk routes for communications between other networks.

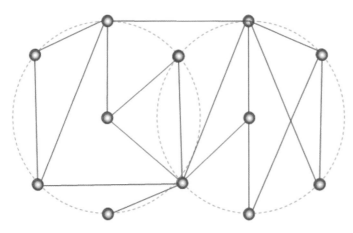

Figure 2.32 Distributed network.

Figure 2.33 The Raytheon Microlight JTRS radio automatically forms self-healing ad hoc networks and enables radios to communicate at 1 Mbps across the network with up to 8 hops with up to 2 km on a single hop. It transmits data using the Internet Protocol, allowing any PC-driven computer to accept and display information, images, and video.

Ephemeral networks are a collection of different networks and nodes, which are brought together for the purpose of accomplishing a specific task. The network persists for the duration required to provide the service, which could be anywhere from a few seconds (in the case of a weapons data link) to a few months (in the case of a deployed surveillance network).

Ephemeral networks consist of adaptive network structures, which form and disband using predefined rules to meet the needs of data collection, communications, or effects implementation. Ephemeral networks are ideally suited to meet changing threat scenarios where specific sensors, communications, or effects systems need to be added to a network in a fixed topology to achieve a specific task.

Ephemeral networks place additional demands on network routing, with point-to-point communications being delivered through a combination of dynamic routing, multihop networking, and fixed networks. Data bandwidth requirements are determined by the sensor and by the effects capabilities required, the role and position of the nodes within the network, and the nature of the threat or ISR requirement.

Ad hoc networks (Figure 2.34) are an autonomous collection of decentralized mobile users that communicate over relatively bandwidth-constrained wireless links. Since the nodes are mobile, the network topology may change rapidly and unpredictably over time. Network connectivity, including discovering the topology and delivering messages, must be executed by the

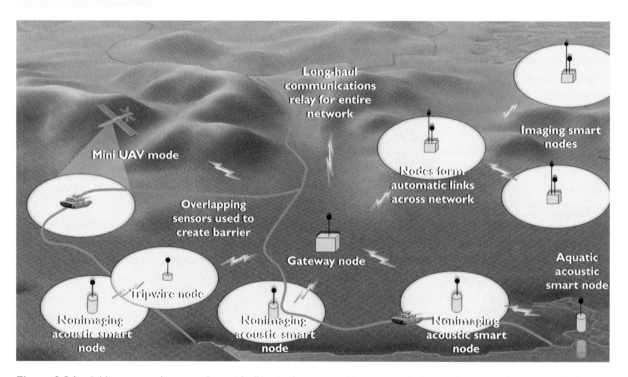

Figure 2.34   Ad hoc network protocols are ideally suited to networking unattended ground sensors. Data transmission rates for such sensors are typically up to 5 mbps depending on the power available.

nodes themselves with routing functions incorporated into mobile nodes [22]. Ad hoc networks that that may arrive and depart from a network are sometimes referred to as *Mobile Ad hoc Networks* (MANET).

Ad hoc networks are typically deployed in operational environments and are designed to maintain a low probability of intercept and/or a low probability of detection. Hence, nodes are designed to radiate as little power as necessary and transmit as infrequently as possible, thus decreasing the probability of detection or interception. A lapse in any of these requirements may degrade the performance and dependability of the network [22].

Ad hoc network addressing between collaborating nodes must be able to cope with a dynamic network topology, as sensors, effectors, and platforms enter and leave the network. In ad hoc networks, nodes and clusters do not have prior knowledge of the surrounding networks and associated topology and can use a variety of acceptable protocols to connect to available networks. Unlike ephemeral networks whose structures are predefined, the topology of ad hoc networks is almost infinitely variable. Nodes can dynamically join and leave the network without adversely interfering with the communication between other nodes. Nodes may also move physically relative to each other and still remain connected through changes to the network topology.

Ad hoc networks are ideally suited to meet changing threat scenarios or where the network may come under attack and needs to be rapidly reconfigured to isolate or bypass affected nodes or clusters of nodes. They provide a high degree of redundancy and can accommodate the addition or withdrawal of nodes and linkages as the need dictates. In addition, ad hoc networks are also more difficult for the adversary to identify and target as they are transient in nature and the number and type of participating nodes and linkage patterns will vary depending on the operational context.

Typically, new nodes wishing to connect to the network will announce their arrival and/or listen to activity from surrounding networks before initiating an appropriate connection protocol, which may vary depending on the type of network being accessed. If a new node can access multiple nodes within the network, it will choose the most appropriate node to connect to depending on the task required. It may also announce to other ad hoc nodes the list of nodes that it can access so that wider connectivity across the ad hoc network can be achieved. As nodes connect, each one will discover the topology and characteristics of all other accessible nodes and the protocols required to reach them. Such characteristics require the network configuration and management to be automatic and dynamic in nature, with multiple nodes joining and leaving the network, and the topology constantly changing.

The dynamic and self-organizing nature of ad hoc networks makes them particular useful in situations where rapid network deployments are required or it is prohibitively costly to deploy and manage fixed network infrastructure. Ad hoc nodes are typically optimized to accommodate multiple input, multiple output (MIMO) operations to provide data and information to meet multiple network requirements.

Optimized ad hoc networks can take into account a multitude of factors in their routing decisions. For example, they can optimize the available bandwidth by choosing a less busy part of the network or avoid nodes low on resources such as battery power or processing power. Additionally, routing paths can be optimized to use parts of the radio spectrum away from densely utilized frequencies. This ensures that lower transmission power can be used, as it does not have to compete with other signals, thereby reducing the likelihood of detection and reducing the reception error rates. As most network nodes will not have local access to the network, each node must also act as a router, meaning that it is responsible for passing traffic (voice, data, and video) from other nodes as well as managing local node traffic. Such an arrangement, of course, has significant implications for bandwidth management across the network where the number of networks nodes and scalability will clearly be determined by the internode bandwidth.

In ad hoc networks where data paths are established through multiple redundant data paths, resilience will be significantly determined by the number of concurrent connections to network nodes.

Ad hoc networks may employ multihop techniques. In a scenario, for example, where unattended ground sensors (UGSs), robotic systems, and UAVs are deployed by aircraft or ground forces as a networked system of systems, some nodes will be out of range of their destination. That final destination within the network may be another node or a communications relay for transmission to another network or a central C2 system. Using ad hoc and multihop techniques, sensors are able to interact with each other and use the onwards transmission capabilities of each node to forward data to the required destination, thereby creating a wireless self-organizing network.

In ad hoc network processes (Figure 2.35), the network nodes have the capability to determine how to best route the data to an appropriate node and self-organize the routing and interfaces within other networks. If the network topology changes through a sensor failing or enemy action, then the routing protocol will automatically find a new way to reach its destination. Sensors at the outer reaches of the network may need to be supplemented by the addition of further communications relay nodes if the distance between the initial nodes is too great for the network to be reliably formed.

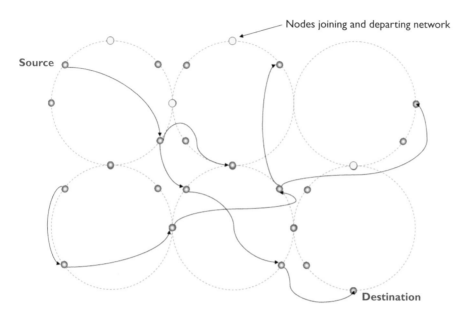

Figure 2.35 Ad hoc network showing self-routing network paths and transient nodes.

Multihop networks are another type of self-organizing network. In multihop networks, message packets are passed between nodes, which may be fixed, ad hoc, or ephemeral. Multihop networks using packet-switching protocols and multihop sequences can form an important part of reliable, large-scale, military ad hoc networks.

Typically information may be processed via various nodes on its journey to its ultimate destination. Like ad hoc networks, multihop networks provide a high degree of redundancy and may be designed to take into account network data transmission efficiency to consider factors such as power management, data latency, and, where wireless transmission is involved, the need to minimize the probability of detection (Figure 2.36).

Through a process called *packet switching*, involving the delivery of data by *store and forward* protocols, message packets in the form of datagrams are passed between nodes, which may be fixed or ephemeral. Typically information may be processed at various nodes for further onwards transmission, fusion, and processing. The routing protocol rules can be structured to meet network priorities to optimize the desirable characteristics of the routing path. For example, low data latency may be the primary concern within a targeting network, or alternatively, the use of the minimum hop count may be a priority to preserve the transmission energy required across a network of remotely deployed sensors.

Figure 2.36   Unattended ground sensor networks will use ad hoc routing principles to link to other networks and to minimize transmission power requirements.

If you are not on the network, you're not in the game.

Multihop networks (Figure 2.37) are particularly useful where transmission nodes are too far apart for a direct hop to be made. Accordingly, nodes in multihop systems are able to forward information packets on behalf of other users across multiple nodes in the network.

Given the potentially huge amounts of information that could be distributed to all nodes across the network, the challenge is to optimize information flows to minimize the bandwidth required to make the network operate efficiently. One model to address this issue is through a concept known as model-based communication networks [23] (MCNs), where each network node lets others know its constraints, and every node maintains dynamic models of itself and others so that nodes can alert each other when they

---

Future Standard Network Configuration Characteristics:

- **Self-configuration:** Automatic configuration of components;
- **Self-healing:** Automatic discovery and correction of faults;
- **Self-optimization:** Automatic monitoring and control of resources to ensure the optimal functioning with respect to the defined requirements;
- **Self-protection:** Proactive identification and protection from arbitrary attacks [7].

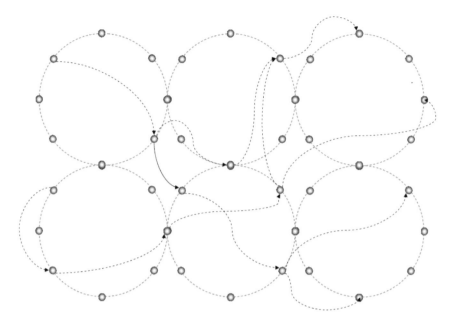

Figure 2.37 Multihop network showing self-routing network paths.

detect an event or acquire information that would be relevant to the needs of other nodes on the network.

Deployed network structures and challenges

With the myriad of network-enabled systems currently available, it is possible to construct many different types of complex network patterns using a variety of addressing protocols. As network protocols tend more towards Internet Protocol (IP)–based systems, interface standards are becoming increasingly standardized (although encryption techniques vary considerably) and the number of permutations and combinations of networks become almost limitless.

While assets on an ISTAR or effects network will need to be linked together, the degree to which they are connected, how they are connected, and the data rates and protocols that they require will depend on the roles that they play and the functions that they offer. Similarly, the networks formed to connect the nodes will be far from the utopian dream of universal connectivity where all nodes are connected to every other node and all have access to the same level of information. Rather, deployed network structures will comprise many point-to-point networks offering limited one-way data such as in the case of a distributed ground sensor, through to a hub providing data link connectivity for a Link-16 network.

Networking is not simply about connecting nodes or communications networks together. In addition to the technical challenges of optimizing network patterns and interfaces, operational challenges are now focusing on addressing how the data and information are processed and presented to the decision makers in an appropriate, timely, and relevant manner to assist in the OODA loop process (Figure 2.38).

Network designers and C2 planners need to consider that the volume of dynamic data introduced into OODA loop inputs may only serve to slow

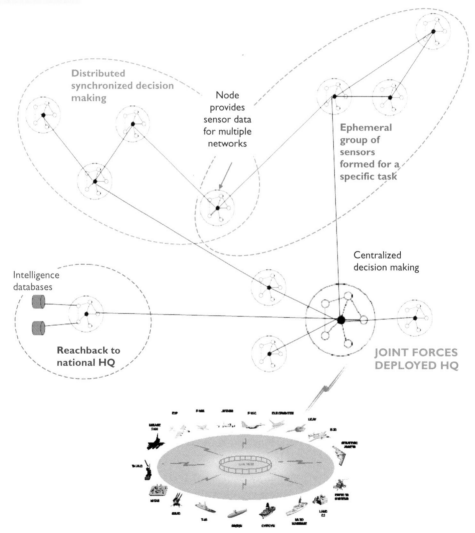

J-series datalink relays shared situational awareness picture to HQ via UCAV

Figure 2.38 Illustrative integrated network arrangements including centralized and distributed decision making, ephemeral networks, and reach-back.

down the decision-making process. Clearly, sharing huge amounts of data can confuse matters significantly if the data is not turned into structured, useable, and timely information and into a form where it can be exploited through the doctrine and concept of operations employed across the networked forces. With numerous sensors and intelligence sources being connected to increasingly large networks, the challenge is now more often one of information management rather than information acquisition.

With networks typically increasing in size and coverage, the issue of beyond-line-of-sight communications is now also an issue for platforms and units operating at the outer extremities of the network, and various solutions such as satellite communications networks are being used to extend networks across difficult terrain of over long distances.

> The challenge is now more often one of information management rather than information acquisition.

Networked operations are not a panacea for all C2 challenges, but if a networked approach is adopted across all aspects of military operations from concepts and doctrine to information management and training, the advantages can be significant. However, the greater the size of the network, the complexity of topology, and the network interfaces, the greater the challenge of achieving efficient networked operations. In particular, in complex networks, the challenge remains of balancing the interdependencies and requirements related to sensing, networking, processing, resource allocation, and management.

Among the many aspects that need to be considered in network design are the following:

1. Data latency:

    • Timing, especially for target engagement, is critical. How do you fuse information from across the network to assist with targeting that is a snapshot of what each sensor sees at a different moment in time?
    • How is bandwidth prioritized across a network? Which sensors and nodes take priority? What is the impact on the decision-making process of delayed data?

2. Data accessibility:

    • With the increasing volume data and information available, how do you know what information is available that will help make decisions?
    • How do you classify and tag data to ensure that security constraints re respected while not unduly impeding data accessibility?

- How do you provide timely access to a range of classified networks in coalition operations where potentially greater than 20 countries could be working together?

3. Data fusion:

- How do you integrate multiple application-specific networks from other C2 systems and across coalition networks where C2 and fusion protocols may vary across each partner?
- Where on the network is data fused and processed, and who has access to the resultant information?
- How do you effectively fuse sensor data from a diverse range of multispectrum sensors to form near real-time target tracks?
- How is target identification achieved from multisensor data fusion?
- How is nonlinear information such as intelligence reports fused into the shared battlespace picture?

4. Speed of decision making:

- When do you have enough information to make a decision that is adequate to achieve the commander's intent?
- How do you deal with constant updates of information coming in during a decision-making process?
- How much time do you permit to gather information from across the network?
- How can significantly increased information flows achievable through efficient networking be processed and presented to enable meaningful decision-making in a timely manner?

5. Command and control:

- Across increasingly large and complex networks, how do you deploy sensors and effects systems and prioritize the targets without disrupting the network?
- How will the command and control structure be set up to ensure a federated network rather than simply using the network to bring all decisions back to a central point?
- How will the balance between command and control be implemented across a network such that the commander's intent can be achieved, but so that there is sufficient devolved decision making to enable flexibility and speed of response at the edges of the network?
- Where across the network will decisions be taken? Which decisions will have priority over others? How will sensor and effect nodes be prioritized?

- What decisions need to be referred to human decision makers and what can be automated?

Although networked technologies are seen as the transforming facet of NEC, NEC evolution also needs to take into account the requirement to progress in parallel, the accompanying military operational doctrines. Net-centricity is a robust, globally interconnected network environment including infrastructure, systems, processes, and people [24], and as such it is not possible to network military assets without a significant change to the concept of operations and doctrinal infrastructure across a potentially complex military environment. In particular, doctrinal concepts around leadership, chain of command, organization, and materiel, for example, all need to be closely integrated to NEC processes to ensure that the decision making and kill chain decisions that need to be made with significantly increased tempo are not impeded through doctrinal concepts designed for platform-centric operations.

In the United Kingdom, the doctrinal approach is delivered through what are called defense lines of development (DLODs) (Table 2.3) and the introduction of new NEC capabilities are examined to ensure that the operational infrastructure can deliver the benefits from the improved capabilities.

The concept of lines of development recognizes that a network-enabled force structure needs to be supported by similar characteristics that support that structure, specifically by an agile and reconfigurable logistics supply chain that enables the rapid supply of men, materiel, and assets to meet the specific mission. In addition, supporting doctrine, leadership, and training also need to take account of the specific requirements of NCW in supporting this approach (Figure 2.39).

It is also important to realize that an effective network comprises not only the communications linkages themselves, but also the ability to integrate an array of subsystems into an effective *system of systems* approach. Such an approach to the management of NEC technologies and concepts is essential if full advantage is to be taken of the integration of the many capabilities found within modern military, government, and civilian infrastructures.

Command and control (C2)

Command and control (C2 or C²) and its associated systems are defined by the U.S. DoD.

Command and Control

C2 is defined as "the functions of commanders, staffs, and other Command and Control bodies in maintaining the combat readiness of their forces, preparing operations, and directing troops in the performance of their tasks. The concept embraces the continuous acquisition, fusion, review, representation, analysis and assessment of information on the situation; issuing the commander's plan; tasking of forces; operational planning; organizing and maintaining cooperation by all forces and all forms of support..."

—NATO Glossary of Terms

Training	The provision of the means to practice, develop, and validate, within constraints, the practical application of a common military doctrine to deliver a military capability.
Equipment	The provision of military platforms, systems, and weapons (expendable and nonexpendable, including updates to legacy systems) needed to outfit/equip an individual, group, or organization.
Personnel	The timely provision of sufficient, capable, and motivated personnel to deliver defense outputs, now and in the future.
Information	he provision of a coherent development of data, information, and knowledge requirements for capabilities and all processes designed to gather and handle data, information, and knowledge. Data is defined as raw facts, without inherent meaning, used by humans and systems. Information is defined as data placed in context. Knowledge is information applied to a particular situation.
Concepts and doctrine	A concept is an expression of the capabilities that are likely to be used to accomplish an activity in the future. Doctrine is an expression of the principles by which military forces guide their actions and is a codification of how activity is conducted today. It is authoritative, but requires judgment in application.
Organization	Relates to the operational and nonoperational organizational relationships of people. It typically includes military force structures, MOD civilian organizational structures, and defense contractors providing support.
Infrastructure	The acquisition, development, management, and disposal of all fixed, permanent buildings and structures, land, utilities, and facility management services [both hard and soft facility management (FM)] in support of defense capabilities. It includes estate development and structures that support military and civilian personnel.
Logistics	The science of planning and carrying out the operational movement and maintenance of forces. In its most comprehensive sense, it relates to the aspects of military operations which deal with: the design and development, acquisition, storage, transport, distribution, maintenance, evacuation, and disposition of materiel; the transport of personnel; the acquisition, construction, maintenance, operation, and disposition of facilities; and the acquisition or furnishing of services, medical, and health service support.

Table 2.3   U.K. MoD Defense Lines of Development

Effective C2 also involves the cognitive and organizational processes of C2, which are an important factor when considering the display, assimilation, and decision making of C2-related information.

The core of understanding C2 lies in its definition and the three key dimensions by which alternative C2 approaches can be characterized and differentiated [19]:

• The allocation of decision rights;
• The patterns of interaction that occur;
• The distribution of information.

C2 structures and data exchange requirements have varied enormously between the industrial age and the information age and can generally be classified by the form and interaction of these key parameters.

Command and Control Definitions [25]

- **Command:** The authority that a commander in the Armed Forces lawfully exercises over subordinates by virtue of rank or assignment. Command includes the authority and responsibility for effectively using available resources and for planning the employment of organizing, directing, coordinating, and controlling military forces for the accomplishment of assigned missions. It also includes responsibility for health, welfare, morale, and discipline of assigned personnel.
- **Control:** Authority that may be less than a full command exercised by a commander over part of the activities of subordinate or other organizations.
- **Command and control:** The exercise of authority and direction by a properly designated commander over assigned and attached forces in the accomplishment of the mission. Command and control functions are performed through an arrangement of personnel, equipment, communications, facilities, and procedures employed by a commander in planning, directing, coordinating, and controlling forces and operations in the accomplishment of the mission.
- **Command and control system:** The facilities, equipment, communications, procedures, and personnel essential to a commander for planning, directing, and controlling operations of assigned forces pursuant to the missions assigned.

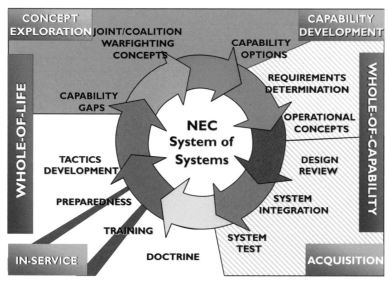

Figure 2.39 NEC doctrines impact across all doctrinal aspects of warfighting. (Courtesy of Thales Aerospace Division U.K.)

Each of the approaches listed in Figure 2.40 can be an effective way of discharging C2 requirements taking into account the quality and quantity of forces available and the structure of the opposing force—all factors that very much determine the most appropriate approach. Whatever approach is used,

Command structure	Characteristics	Examples
Autonomous	•Autonomous action authorized to support command intent. •Near real-time communications links across a network-enabled command structure. •Significantly fewer command layers. •Forces and capabilities available proportional to priority of outcome. •Increasingly small and diverse command units. •Less need for deconfliction of assets, as situational awareness is significantly enhanced through shared network datalinks.	On-call air support Special Forces operations Counterinsurgency warfare
Problem-driven	•Problem specified for each command unit. •Solution and timing to problem is determined by unit tasked. •Fewer milestones and constraints. •Collaboration across distributed command structure to acquire assets necessary to address problem.	Peacekeeping and stabilization operations
Goal-driven	•Specific time-bound goal specified for each command unit. •How goal is to be achieved is determined by unit tasked. •Assets allocated by central structure to achieve goal.	WW2 German "Blitzkrieg" operations
Task-driven	•Defined tasked solved by defined "text block." •Smaller units engaged in larger number of discrete tasks.	Air Tasking Order (ATO) structure Soviet "cold war" approach
Centralized	•Tactical situation continuously assessed against strategic goals. •Synchronized issue of command orders to all units from central control. •Multiple layers through which commands are flowed. •Limited speed and capacity of information flow between units. •Large military assets to be controlled with limited training and professionalism.	Soviet approach in WW2

Left margin: Increasing centralization (downward arrow)
Right margin: Increasing need for network enabled structure (upward arrow)

Figure 2.40   NEC doctrines impact across all doctrinal aspects of warfighting.

the advantages of good communications and networking across all of the assets involved are self-evident.

The organization of C2, as it has been employed for most of the twentieth century, has evolved to adapt to the accomplishment of military objectives despite the confusion generated by limited communications systems, data inaccuracy, and high information latency—all elements within what is often referred to as the *fog and friction of warfare*. In military operations, the speed, dynamic environment, and confusion that often accompany military operations can conspire to complicate and confuse the C2 process and reduce accuracy and understanding across the battlespace.

Within the C2 context, the evolution of Information Age technologies has dramatically changed the utility and accuracy of information, which in turn has forced the evolution of new forms of military organization and devolved approaches to C2 that seek to exploit the utility of information as a tool in its own right.

The Fog of War

"The general unreliability of all information presents a special problem: All action takes place, so to speak, in a kind of twilight…like fog. War is the realm of uncertainty; three quarters of the factors on which action in war is based are wrapped in a fog of greater or lesser uncertainty…The commander must work in a medium which his eyes cannot see, which his best deductive powers cannot always fathom; and which, because of constant changes, he can rarely be familiar."

—Carl von Clausewitz, *On War*

"War is a complex and chaotic human endeavor. Human frailties and irrationality shape war's nature. Uncertainty and unpredictability—what many call the 'fog of war'—combine with danger, physical stress, and human fallibility to produce 'friction,' a phenomenon that makes apparently simple operations unexpectedly and sometimes even insurmountably difficult. Sound doctrine, leadership, organization, moral values, and training can lessen the effects of uncertainty, unpredictability, and unreliability that are always present."

—USAF Air Force Doctrine, Document 1, November 17, 2003

The Commander's Intent

More often expressed as command intent, the *U.S. DoD Dictionary* provides a useful reference to summarize the military objectives at the heart of the C2 process:

"A concise expression of the purpose of the operation and the desired end state that serves as the initial impetus for the planning process. It may also include the commander's assessment of the adversary commander's intent and an assessment of where and how much risk is acceptable during the operation."

—Defense Technical Information Center, *DoD Dictionary of Military and Associated Terms*, Joint Publication 1-02

C2 typically involves a number of information-related functions and responsibilities that permeate through the organization originating from a focal point or node, which defines the desired outcome and strategic priorities for the engagement. C2 functions are performed through an arrangement of organizational structures, communications systems, facilities, and procedures employed in planning, directing, coordinating, and controlling an increasingly disparate range of forces and capabilities in the accomplishment of the mission.

The C2 process (Figure 2.41) involves identifying objectives, tasking the collection assets, using the information gathered to create an accurate and timely picture of situational awareness, assessing the implications of the situation, developing alternative courses of action, deciding on a course of action, transmitting orders, and ensuring that the desired outcome is achieved. This broad set of dynamic tasks requires a seamless, robust network linking all coalition forces and providing a shared awareness of the evolving situation.

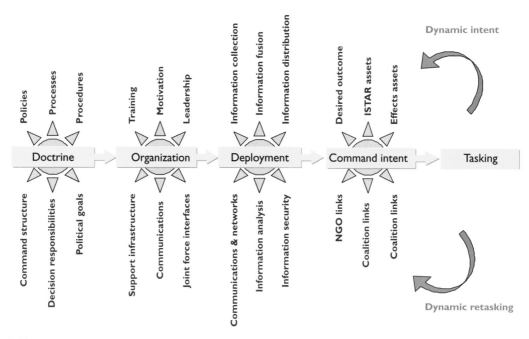

Figure 2.41   The command and control process.

Networked command and control

Networked command and control is not just about faster and more abundant connectivity. Networked C2 fundamentally changes the approach to C2 through an all-encompassing approach across a relatively unstructured network of sensors and effects systems to provide an environment where commanders can turn information into knowledge and knowledge into action in ways and with a tempo that was impossible to image only 10 years ago (Figure 2.42).

Networked C2 changes the way that assets are considered, with each asset within the battlespace being viewed as a node on a network offering a function into the network for the overall benefit of the collective goal rather than for a specific mission or purpose. It enables the commander to control all assets across the network to respond to the changing priorities and to achieve a coordinated response as appropriate.

Utilizing the full potential of NEC requires the confidence to move the C2 span of control to well beyond traditional—or human—organizational limits in terms of the tempo and volume of information that can be processed and the span of control that can be reached. Decisions that can be made as a result of self-synchronization and automation of decision-making require the

(a)

(b)

Figure 2.42 (a, b) A typical networked combat operations center.

confidence to adjust the doctrines and organization to optimize network-enabled outcomes. NEC philosophies theoretically enable information to be accessed and decision making to be effected at any connected node wherever it appears in the organization. This represents a significant contrast to traditional networks where command was effected from the center of the hierarchy and control was the preserve of the outer edges of the organization. The ability to move to a truly devolved C2 process where each node works towards the command intent represents a significant shift from traditional military thinking (Figure 2.43).

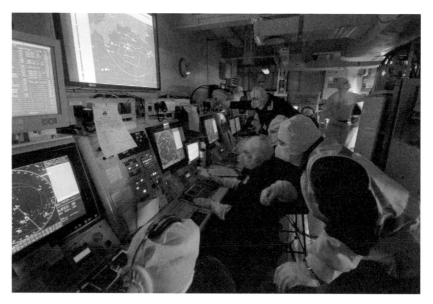

Figure 2.43 Networked C2 in the naval environment. Here in the operations room of HMS *Albion*, an Albion Class Landing Platform Dock (LPD) acts as a C2 center during exercises off Plymouth in the United Kingdom. (Photo courtesy of the U.K. MoD.)

Typically, C2 functions facilitated by the network include:

- **Command intent:** Defining and communicating the primary military objectives and associated timescales;
- **Force structure and employment:** Determining through the networked capabilities, how the force will be organized, the boundaries of authority and decision making within the force structure, and the orchestration of how each element will be employed to achieve the desired objectives;
- **Integration with military effects:** Defining how the C2 process will authorize the engagement of opposition forces with the appropriate military effect;
- **Information superiority and the organization of intelligence collection:** Achievement of these across the network to enable appropriate feedback to the decision-making process during the operation;
- **Facilitation:** Of the supply of physical materiel and timely information to the force structure to enable it to function effectively;
- **Effective decision making:** The ability to make effective decisions in a rapidly changing ambiguous environment against a background of irregular, uncertain and complex threat scenarios.

Across each of these functions, the need to share information in a timely manner is a key determinant of whether the outcome will match the command intent (Figure 2.44).

(a)

(b)

Figure 2.44 (a, b) C2 down to an individual soldier level is now quite feasible. Here a flexible display screen is shown in a personal digital assistant (PDA) device designed for the dismounted warfighter,

The C2 arrangements and organizational structure that may be employed in a networked environment are, to a large extent, determined by the degree of communications available between each unit or node in the battlespace. For example, the hierarchical network structure can be managed through much more limited communications capabilities than that required in a truly networked operation (Figure 2.45).

In turn, the degree of networking available will determine the ability of the enterprise to adapt to the most appropriate organization to achieve the

Figure 2.45   An EC-130E Airborne Battlefield Command & Control Center from the 42nd Airborne Command & Control Squadron, Davis-Monthan Air Force Base, Arizona, flies a training mission over Lake Mead, Arizona, in December 1999. The squadron's seven EC-130E Hercules aircraft have been modified to carry the Airborne Battlefield Command & Control Center capsule, which houses a 12-person battle staff. The battle staff includes operations, intelligence, and communications specialists to provide unified capabilities for combat operations.

The role of NEC in a C2 environment

"NEC is crucial to the rapid delivery of military effect... NEC will improve communication and understanding of strategic and military intent throughout the chain of command. Through NEC the command structure will improve its responsiveness to events on the ground and have the flexibility to respond in near real-time to fleeting targets, even where higher-level decision making is required prior to engagement."

—U.K. Secretary of State for Defence, *Delivering Security in a Changing World*, December 2003

command intent in the most effective manner. Greater information flow between the elements of the organization has, in turn, also driven the need to have greater degree of clarity in the scope and boundaries of dynamic decision making that a commander can exert towards a particular element within the organization and the communications capabilities across the military organization for appropriate employment of the command doctrine.

C2 in a NEC environment decentralizes the decision process and puts the decision making into the hands of the commanders in the field rather relying on a centralized command structure. It is a complex process that involves the collation and fusion of sensor and intelligence data into a meaningful knowledge set to determine and implement the appropriate decision and tasking requirements (Figure 2.46).

C2 structures in an NEC environment afford the option of employing centralized or decentralized C2, and exact structures can vary significantly depending on the doctrine and operational scenario. Appropriate decentralization of decision making, planning, and execution will significantly improve the tempo of operations, the speed of the OODA loop, and the ability to respond to the needs of the network rather than the requirements of the individual commander. (The merits of this are a hotly

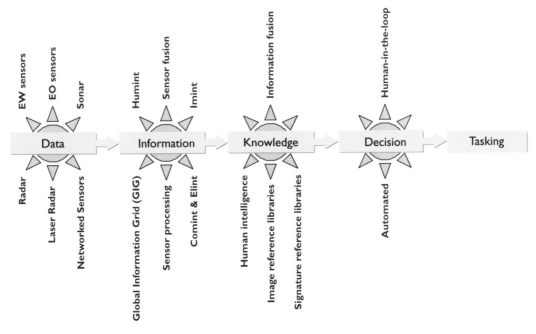

Figure 2.46 The transformation of data to decision is a complex process.

debated point in military circles, where the perception of the commander in the field can be that resources are often allocated to other, more important missions, with the result being that assets become scarcer and therefore less effective in a networked environment as they are allocated away from the "owner" to support missions that are of higher priority to the network.)

However, flexible and reconfigurable C2 structures should allow the commander to maintain effective C2 even when faced with rapidly changing situations. Inevitably C2 structures also evolve over time where experience in dealing with a particular threat is used to modify the topology of C2 structures. Typically, C2 structures can be classified by the degree of centralization exerted on the military organization, ranging from a centralized C2 system, where all decisions of any significance need to be referred to a central command point, through to an organization where command intent is expressed as a series of goals and the military organization adapts itself at a local level to achieve the desired outcome.

In a network-enabled environment, the term "control" within a command and control context takes on a subtly different meaning and shifts doctrine away from control in the form of assurance to a role associated more with ensuring that the military intent is achieved. In such an environment, individual nodes form an interactive part of the overall network and are therefore more involved in the decision and tasking process within the network to which they are connected (Figure 2.47).

Figure 2.47 The U.S. Army's Command Post of the Future (CPOF) will provide network connectivity to the outer edges of the network, enhanced decision making, and C2 functionality.

As well as being directed to undertake certain tasks, it is also possible for nodes to change the priorities of tasks undertaken across the network in an environment where the importance of each node is related to the activities being undertaken across the entity of the network, rather than being constrained by a predetermined hierarchical structure.

A network-enabled command structure pushes command decisions out across nodes, in line with the command intent, achieving a widely distributed, decentralized decision-making structure that is able to rapidly react to unfolding events rather than relying on decisions from a single node or commander as is the case with traditional C2 structures (Figure 2.48).

The organization of a networked C2 system and the functions that it provides must be designed to take account of the limitations of the system as determined by the capacity to collect data, process it, and share it around the network. These demands will vary depending on the C2 approach employed, but whichever is chosen, the C2 system needs to be adaptable enough to cope with the communications and feedback requirements of each model. Clearly, with the advent of networked communications around the battlespace, the C2 approaches open to the commander increase significantly, providing the ability to reduce the number of command structure layers, improve speed of communications, receive faster feedback from all participants, and adapt tactics to better suit the unfolding situation.

With the disappearance of both organizational rigidity and information stovepipes through a NEC approach to C2, barriers to information sharing

Figure 2.48 M-557 communication vehicles form one wall of a command center for the 2nd Armored Cavalry Regiment during Operation Desert Storm, February 1991.

and collaboration disappear, permitting a diverse range of unstructured networks that adhere to structured information exchanges instead of relying on structured organizational patterns. Rather than a rigid hierarchical structure, for example, an agile C2 system may permit a hub-and-spoke approach where spokes (or nodes) are tasked from the hub with a data set comprising military intent and desired local outcomes and self-organize themselves across their network to achieve the desired result.

The need for rapid and agile decision-making at each network node increases with networked C2 integration as local situations are better able to be monitored, their implications assessed for the entire network, and the use of assets connected to the network prioritized accordingly.

In order to achieve such adaptability, NCW requires C2 structures and processes that are highly flexible and able to integrate across varying network topologies to suit a range of missions. Generally, structures are designed to be modular, be able to rapidly and automatically incorporate new subnetworks, reconfigure themselves rapidly and efficiently, and employ reach back or reach forward. The network functionality correspondingly needs to match the network decision making processes as part of routine C2 functions to enable rapid decision making while retaining control of the pace of battle as reassessing priorities as new information and intelligence about the unfolding scenarios are fed through the OODA loop process (Figure 2.49).

The desired characteristics of an agile network-enabled C2 system generally converge around the principles of centralized control and decentralized execution. Typically desirable characteristics of such a structured but flexible C2 system include the following [5]:

- **Robustness:** The reliability and durability of communications systems and the ability to operate effectively even when the network is degraded through hostile action or network failure.

- **Resilience:** The ability to complete the required task and remain effective even when the network is damaged or degraded or tolerances are exceeded.

- **Responsiveness:** The ability to react to information and network requirements in a timely manner to ensure that the function performance of network nodes is not degraded.

- **Flexibility:** The ability to use the network topology to the advantage of the nodes employed within the network and to achieve the most efficient communications protocols.

- **Innovation:** The ability to use or create alternative C2 processes (Figure 2.50) to solve new problems and to solve existing problems or improve old processes using functionally equivalent approaches.

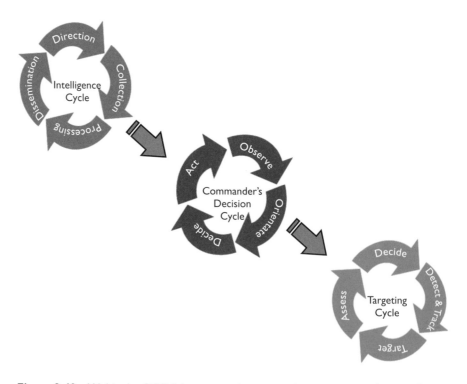

Figure 2.49 Within the OODA loop, several processes interact to contribute to the effectiveness of the overall C2 process.

Figure 2.50 The ability to change or reprioritize targets en-route for air, land, sea, and space assets greatly enhances the C2 options open to the commander. Here a B-2 Spirit bomber refuels from a KC-135 Stratotanker during a deployment to Andersen Air Force Base, Guam. The range and endurance of such platforms typically enable specific targets to be reprioritized during flight, with target engagement decisions being made late in the planning cycle. Upgrades have recently been announced to add to the B-2's capabilities to enable the engagement of moving targets using network-enabled targeting.

Complex networked C2 systems are evolving rapidly as low-cost, high-performance digital technology becomes universally available. Increasing use is also being made of sophisticated commercial technologies, including advanced secure transmission techniques and encryption to levels that in the past have been the preserve of the military (Figure 2.51).

Networked communications solutions can also use existing public telecommunications networks, including satellite communication systems, to establish a global high bandwidth communications network. This dual-use approach has the added advantage that it is difficult for an opponent to track military communications through a public network and consequently any increase or decrease in activity or type of activity will be difficult to detect. Of course, this advantage works equally well for the aggressor as the defender.

As demonstrated by recent conflicts in Afghanistan and Iraq, global communications networks can be established relatively quickly and cheaply. Given that most military systems are now migrating towards open system, IP-based protocols, it is a straightforward exercise to connect most C2 systems to a readily available global network. While civilian applications may be sufficient for basic connectivity requirements, a key challenge remains the

Figure 2.51   From a C2 perspective, the coordination of multiple platforms operating in the same battlespace is a long-standing and essential challenge [8]. Here USS *Ronald Reagan* (CVN 76) connects to Japan Maritime Self Defense Force (JMSDF) guided missile destroyer JS *Myoko* (DDG 175) during a fueling at sea (FAS) in March 2007.

huge and persistent data transmission capacities that are increasingly demanded by networked C4 and ISTAR systems, resulting in an ongoing need for military requirements to be supported by dedicated military networks.

While networked C2 provides many advantages, it also provides potential targets for the adversary. In an NEC C2 environment, adversaries will look to destroy the ability of the friendly forces by degrading or destroying C2 facilities and associated networks. Indeed, any node on a network will potentially become a target as opponents seek to destroy the adversary's means of coordinating a networked approach that delivers the flexibility and agility that is such a threat to the more traditional force. Nodes are more vulnerable to attack by electronic means than are traditional force structures as networks can be attacked via other networks or electronically often with little evidence of who is behind the action.

NEC does not answer all of the challenges of the C2 process, and it is important to ensure that one does not over-rely on the intelligence picture and the NEC-based decision-making processes to the exclusion of other inputs such as common sense and judgment of experience, which can often be obscured by the fog of war. Military victory achieved through NEC is, of course, not an end in itself, but rather a stepping stone on the way to a victory of a more strategic nature.

Command and control warfare (C2W)

Warfare involving the disruption of the opponent's C2 capabilities is referred to as *command and control warfare* (C2W) and involves actions taken to deny information by influencing, degrading, or destroying adversary C2 capabilities, while protecting friendly C2 capabilities against similar actions. Within a networked environment, where computer-computer and system-system data links are increasingly essential, more attention needs to be paid to the communications aspects of warfare, and the term C2W can and should be considered as a command, control, and communications warfare (C3W) challenge, rather than simply relying on a C2W approach.

C2W and C3W are based on the approach that if the adversary can be prevented from effectively using and communicating with his forces, the outcome may be as effective as physically destroying those forces, at least in the short term.

In order to pursue effective C2W strategies, operational domains are generally classified as follows:

1. **Operational security:** Prevents information about one's own C2 operations from being discovered and exploited by an adversary's signals intelligence systems.

2. **Military deception:** Feeds adversary forces and intelligence sources with information that may deliberately mislead them about one's own capabilities, operations, or intentions.

3. **Psychological operations:** Delivers information to emery forces and the environment and population in which they operate to influence

---

The Objectives of C2W and C3W

- Defeat the adversary by disrupting, disabling, or destroying its C2, C2W, and communications systems.
- Separate the adversary's command structure from its forces.
- Protect one's own C2 forces and communications systems from adverse adversary action.
- Ensure that the appropriate C2 approach can be securely implemented across the friendly force structure.

opinion and make it difficult for adversary leaders and commanders to adequately influence or control their forces or population.

4. **Electronic warfare:** Involves the exploitation of the electromagnetic spectrum and its interaction with electronic storage, processing, and distribution systems in order to gain intelligence and disrupt adversary operations by dominating the way in which the spectrum is exploited.

5. **Physical destruction:** Attacks on adversary C2 assets, typically including fixed and mobile command and communications centers designed to separate the command structure from the adversary's forces.

Decision making in an NEC environment

Networked warfare enables decisions to be taken right across the network, with decision-making power no longer confined to central nodes. As a result, many operational and doctrinal challenges exist as the desire for the center, seeking to influence the way the network is fought, is balanced with the need for delegated decision making disseminated to the outer reaches of the network. The speed of decision making and the fact that the most accurate information needed to take a decision lies at the point where it is gathered make it impossible for a central node to take all decisions in a timely manner. The aim, therefore, must be to ensure that the command intent is clear across the network to an extent that even an unpredictable shift in patterns and outcomes can be accommodated in local decision-making.

The military advantage derived from information dominance is referred to as *decision dominance* and aims to ensure that decisions are faster, are based on more accurate timely intelligence, and are therefore more effective than those made by opposing forces.

Networked C2 processes may take the form of any one of several manual or automated decision or communications processes (Figure 2.52). Such tasks may include the interactions associated with voice communications, decision aids and expert systems, databases, data processing, and fusion algorithms, with each function autonomously performing selected tasks with varying degrees of manual intervention.

The level of automation within a NEC environment depends on the criticality of the function that will be affected by the decision. Even within an automated NEC environment, human intervention is still required for critical decisions and authorization of effects on to a target.

Among the tasks that C2 functions are likely to perform are:

Figure 2.52 Airmen from the USAF 25th Air Support Operations Squadron at Joint Base Pearl Harbor-Hickam, Hawaii provide Air C2 coordination during Exercise Balikatan on March 20, 2010, in the Philippines. (Image courtesy of the U.S. Air Force.)

- Translating a commander's intent into messages or instructions such as digitally encrypted voice messages or digitally encrypted engagement tasks;
- Linking the OODA loop steps between collaborating assets;
- Requesting additional situational awareness information;
- Tasking ISTAR sensors;
- Tasking effects systems;
- Requesting information from sensors to fuse into an end product;
- Notifying decision entities of issues that require immediate attention;
- Reporting back to the command structure.

Effective C2 in a networked environment involves the allocation of command intent objectives into key centers of gravity, allocation of functional ISTAR and effects elements that support these centers of gravity, and the integration into the network of the battlespace nodes or assets that form each functional element [26] (Figure 2.53).

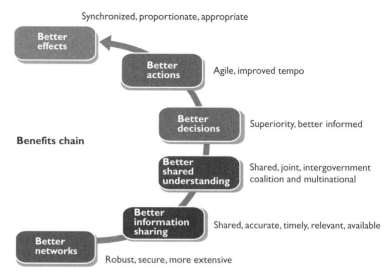

Figure 2.53   A shared information chain significantly contributes to better decision making and improved outcome [27].

Within any well-networked shared data environment, the information available to a commander from fixed and live sources can be overwhelming. Much of the future challenges of NCW will focus on how this information is managed so that it optimizes the decision effectiveness. Such an approach centers on *contextual decision making* where processing algorithms will sort fixed and dynamic data into packages that are relevant to the situational context concerned and categorize the decision that needs to be taken.

Network-based decisions also permit the commander to specify the desired outcome and then have the network organize itself to achieve that goal, a contrast to the era of mechanized warfare, where command decisions were very much focused on achieving a specific mission and were heavily reliant on the immediate resources available within relatively short distances and slow reaction times across the information chain.

Intelligence, surveillance, and reconnaissance (ISR)

ISR describes those activities related to the tasking, synchronization, collection, processing, and exploitation of sensors and assets, including the processing and communication systems required to gain information and knowledge concerning a target (adversary). ISR inputs provide the commander with the situational awareness to enable him or her to plan the required offensive or defensive operations. The focus of ISR is strictly on target information and information systems gained through sensor capabilities [28]. Commanders use information derived from ISR sensors and systems to maximize their own forces' effectiveness while ensuring a full exploitation of an adversary's weaknesses to effectively counter their opposition's strengths.

The provision of ISR data for C2 purposes needs to be balanced between the need for accuracy and quality versus the criticality of the information being gathered and the speed with which it is needed for decision-making purposes or additional processing.

Intelligence, surveillance, target acquisition, and reconnaissance (ISTAR)

ISTAR integrates sensor and intelligence capabilities and contributes to the overall situational awareness picture by providing an insight into the opponent's disposition, intentions, capabilities, and vulnerabilities (Figure 2.54).

ISTAR capabilities (Table 2.4) required in an effective network range from long-range, long-endurance (persistent) surveillance across air, land, and sea domains, through to the identification and targeting of individuals to enable their engagement at short notice by network-based effects systems. The networks that support ISTAR systems therefore need to reflect the data collection, processing, and dissemination requirements for specific operational applications (Figure 2.55).

Integrating ISTAR in the command and control environment

ISTAR is a true system of systems concept where information is acquired, processed, and shared across a network in a timely manner to enable a shared situational awareness view to be accessible to all members of the network.

Functions required to deliver ISTAR capability

Figure 2.54 The practical application of data collection can be seen in the identification, surveillance, target acquisition, and reconnaissance (ISTAR) chain, which is at the heart of understanding how information dominance is achieved in the battlespace.

	Now	**Next**	**Future**
Architecture	Common ground stations	Cooperative networks	Adaptability
	Smart networks	Assisted decision making	Ephemeral networks
	Near real-time operation	Interoperability	Unlimited bandwidth
ISTAR of the Air	E-scan radar	Persistent ISTAR	Multiplatform
	ESM geolocation	Difficult targets	Multistatic
	NCCT/CEC	Improved NCTR	Multispectral
	Networked sensors		
ISTAR of the Land	UAV systems	Improved ID	Persistent
	All weather	Difficult targets (urban, counter ISTAR)	Counterdeception
	NCCT		Multispectral
	Networked sensors		
ISTAR of the Sea	ESM performance	ESM architecture	Multiplatform
	Bistatic radar	Improved ID	Multistatic
	CEC	Autotarget recognition	Multispectral
	Networked sensors		

Source: Barry Trimmer, Technical Director, Thales Aerospace Division, U.K.

Table 2.4 ISTAR Capabilities Evolution

Figure 2.55 Platforms such as the Dutch Navy ship HrMs *De Zeven Provincien* (F 802) are capable of acting as advanced ISTAR sensors, collecting information using sensors covering multiple bands within the RF spectrum and staying on station for long periods of time to provide persistent surveillance.

ISTAR systems integrate with C4 systems to create a common tactical picture and ensure that the communications and data networks form information highways for the transmission of sensor and effects data to and from C2 systems (Figure 2.56).

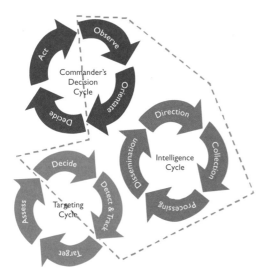

Intelligence, Surveillance, Target Acquisition, and Reconnaissance (ISTAR) links surveillance, reconnaissance, and target acquisition systems and sensors to cue maneuver and offensive strike assets, with particular emphasis on the timely passage of critical and targeting information. It encompasses the collection and management of information and intelligence to provide situational awareness for commanders and their staffs to exploit opportunities by directing maneuvers and supporting the tasking process. **(JWP 0.0-1)**

Figure 2.56 Integrating C4 and ISTAR in the OODA loop: intelligence and targeting cycles.

ISTAR is fundamental to the successful conduct of military operations.

Networked ISTAR enables the integration of sensors across the kill chain (see Appendix E) (find, fix, assess, track, target, and engage) to be completed in minutes rather than hours through a rapid assimilation of sensor information into a shared network environment, which can be used to rapidly increase the speed of decision making processes within the OODA loop. Increased speed of decision making also enables a wider range of effects to be exploited through tasking them earlier in the kill-chain sequence, thereby enabling a more effective engagement of a wider range of targets with a greater certainty of correct identification and optimal selection of the most suitable effects system.

C4ISTAR in a network-enabled environment

In order to acquire and process the data needed to achieve information superiority, it is clear that intelligence gathering and processing systems must combine the requirements of command, control, communications, and computers (C4 or C⁴), intelligence, surveillance, target acquisition, and reconnaissance, often shortened to C4ISTAR (Figure 2.57). Supporting commanders in their decision-making processes is the primary function of C4ISTAR, and these processes extend far and wide across the battlespace and across all phases of an engagement from information operations to effects targeting and from situational awareness to longer-term military planning.

In an increasingly collaborative military environment, networked C4ISTAR (Figure 2.58) must support a wide range of user requirements from single

Figure 2.57 C4ISTAR is the term used to describe the information and decision support systems to assist commanders at all levels to plan, direct, and control their activities. It provides timely input of the situational awareness picture into the commander's decision-making process and provides the connectivity necessary to employ national and multinational forces effectively and efficiently as part of the expanded network.

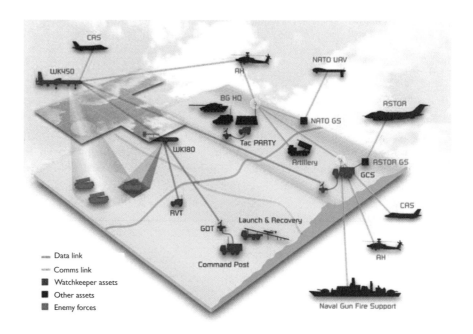

Figure 2.58 The modern battlespace is dominated by C4ISTAR networks. As illustrated here, a distributed sensor network comprises a set of sensor nodes, a set of processing nodes, and a communication network connecting the nodes together. (Courtesy of Thales Aerospace Division UK.)

The demand for C4ISTAR information will always exceed the capability to provide it.

service users to joint operations by forces from the same country of multinational coalition operations often involving many countries.

Future improvements in battlespace capabilities will certainly be centered on NEC principles and their ability to deliver an enhanced ISTAR capability, supported by capable C4, linked to precision effects and attack systems. This places a high premium on network capability and capacity, specifically to provide timely and accurate information for targeting and putting effective kill-chain links and processes in place.

Effective integration of C4 and ISTAR systems provides flexible interoperability between joint forces, the ability to deliver precision effects without the need for huge force structures, and the ability to create and sustain a faster tempo of operations.

Often the term C4ISTAR is divided into a number of more specific abbreviations (Figure 2.59):

- Such as C4 with an emphasis on communications aspects of the network;
- or C4I with an emphasis on i for intelligence;
- or ISR with an emphasis on situational awareness;
- or STAR with an emphasis on network sensor capabilities.

In an NCW environment, key elements of the force mix can be described within the context of networked C4 and networked ISTAR.

Networked C4

The four Cs equate to:

- **Command:** The exercise of military authority to attain a specific objective or outcome.
- **Control:** The process of verifying and adjusting military activity so that the commander's military intent is accomplished.
- **Communications:** The ability to exercise the necessary liaison to exercise effective command between tactical or strategic units to command.

$$C^4 \quad ISTAR$$

$$C^4ISTAR$$

$$C^4I \quad STAR$$

Figure 2.59   C4ISTAR.

"Networked C4ISR is a critical prerequisite for transforming our forces, providing for an increasingly transparent battle space, swift and effective decision making, and rapid, parallel, effects-based operation."

—*U. S. DoD, Information Operations Roadmap,* October 30, 2003

- **Computers:** The computer systems and compatibility of computer systems. Also includes data processing.

Networked ISTAR

NEC and NCW themes are often described in the context of intelligence, surveillance, target acquisition, and reconnaissance (ISTAR) and their associated effects systems:

- **Intelligence:** Timely knowledge of the disposition and intentions of friendly and enemy forces and neutral parties. Includes collection and analysis activities from intelligence sources available across the network and distribution of intelligence to appropriate users in a timely manner across the network.
- **Surveillance:** 24/7 observation of enemy capabilities and known targets of interest in all-weather environments to produce a shared situational awareness picture.
- **Target acquisition:** To support highly accurate target identification and engagement decisions. Includes the need to acquire fixed targets and mobile targets and to acquire targets that are hidden or obscured.
- **Reconnaissance:** 24/7 all-weather observation of the battlespace and individual kill boxes and preeffects and posteffects planning and assessment.
- **Effects systems:** A diverse range of weapons systems ranging from lethal to nonlethal, whose deployment and use can be closely matched to operational requirements that achieve the desired disruption or defeat of the detected threat.

Intelligence warfare and the intelligence cycle

Often referred to as I2W, intelligence warfare is not just about information; it is about converting timely relevant information into meaningful intelligence. I2W requires the analysis and fusion of information from multiple sources specific to the particular objective or mission requirements to provide a picture of not only what the enemy is currently doing, but to predict what he will most likely do over time.

The need for intelligence gathering and intelligence warfare will depend on the type of warfare being undertaken and the phase of the engagement underway. The complexity of the intelligence task has certainly become significantly more complex with forms of modern warfare where the enemy is harder to identify, where the timescales involved become much shorter and where the diversity of intelligence gathering means and targets is almost exponential.

Use of Intelligence

"Commanders use intelligence to anticipate the battle, visualize and understand the full spectrum of the battlespace, and influence the outcome of operations."

—JP 2-01 Joint and National Intelligence Support to Military Operations, October 2004

Particular challenges for intelligence warfare exist in the various forms of irregular warfare, where the provision of *actionable intelligence*[3] is especially difficult. The ability for targeted individuals to hide among the population, the tactics employed, and the distributed nature of insurgent organizations make finding, identifying, and engaging such targets difficult. Intelligence efforts (Table 2.5) may focus on nontraditional areas such as cultural, social, political, and economic issues rather than military capabilities and key leaders. Fusing information obtained from multiple sources, methods, and levels is required to provide timely, accurate, and relevant intelligence to all levels [29].

There are three types of military intelligence:

1. **Strategic intelligence:** Strategic intelligence is used at the government level to direct national policy decisions. It supports and informs plans and policies at the national and international levels.

2. **Operational intelligence:** Operational intelligence supports joint-force theater-wide plans and operations within a region or theater to direct and inform planning related to the employment of forces.

3. **Tactical intelligence:** Tactical intelligence is used for tactical operations at the component or unit level to support immediate action.

The focus of intelligence operations is the analysis and exploitation of information from multiple sources to meet the needs of various planning cycles. The speed with which the information is required and the degree of processing and analysis that can be afforded are very much dependent upon the use to which the information will be put.

The process by which information is converted into intelligence and made available to users is described by the U.S. DoD as the *intelligence cycle.* There are six phases in this cycle (Figure 2.60):

1. **Planning and direction:** Determination of intelligence requirements, development of intelligence architecture, preparation of a collection

Anticipatory	Complete
Timely	Relevant
Accurate	Objective
Usable	Available

Source: [30].

Table 2.5   Attributes of Good Intelligence

Joint Intelligence Preparation of the Battlespace

"The analytical process used by joint intelligence organizations to produce intelligence assessments, estimates and other intelligence products in support of the joint force commander's decision making process."

—*U.S. DoD Joint Command & Control Doctrine Study,* February 1, 1999 [25]

Intelligence

The U.S. DoD defines intelligence as:

"The product resulting from the collection, processing, integration, analysis, evaluation, and interpretation of available information concerning foreign countries or areas. Also, information and knowledge about an adversary obtained through observation, investigation, analysis, or understanding."

—U.S. Joint Forces Command

Counterintelligence

The US DoD defines *counterintelligence* as:

"Information gathered and activities conducted to protect against espionage, other intelligence activities, sabotage, or assassinations conducted for or on behalf of foreign powers, organizations or persons, or international terrorist activities, but not including personnel, physical, document or communications security programs."

—U.S. Government Executive Order 12333 as amended

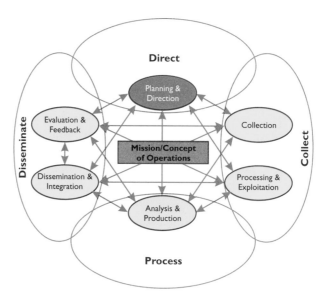

Figure 2.60   The intelligence cycle is sometimes abbreviated to the four steps of direct, collect, process, and disseminate.

plan, and transmission of information collection requests to appropriate collection sources and agencies.

2. **Collection:** Acquisition of information, preferably from multiple sources, and the provision of this information to processing elements.

3. **Processing and exploitation:** Conversion of collected information into forms suitable for the production of intelligence (Figure 2.61).

4. **Analysis and production:** Conversion of processed information into intelligence through the integration, analysis, evaluation, and interpretation of source data and the preparation of intelligence products in support of known or anticipated user requirements.

5. **Dissemination and integration:** Delivery of intelligence to users in a suitable form and the application of the intelligence to appropriate missions, tasks, and functions.

6. **Evaluation and feedback:** Continuous assessment of intelligence operations during each phase of the intelligence cycle to ensure that the commander's intelligence requirements are being met and adjustment of intelligence collection requirements to suit.

A range of information gathering techniques can perform the assimilation of situational awareness and intelligence information and contribute to the compilation of an accurate picture of the adversary's activities and intentions.

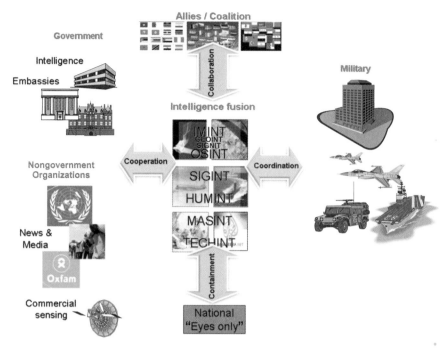

Figure 2.61   The intelligence process [31].

The feeds for these intelligence requirements come from multiple military, governmental, and nongovernmental sources. Military intelligence sources are typically summarized as follows:

- **GEOINT:** Geophysical intelligence.
- **IMINT:** Image intelligence.
- **SIGINT:** Signal intelligence, the collection of signals intelligence, comprising:
    - *ELINT:* Electronic intelligence, the collection of electronic signals, typically radar and electronic sensor signals.
    - *COMINT:* Communications intelligence.

Dynamic tensions in intelligence delivery

- Share versus secure;
- Velocity versus vulnerability;
- Context versus content;
- Collect versus connect.
- Infer versus confirm.

- **HUMINT:** Human intelligence; the collection of intelligence by interpersonal means rather than by electronic systems, that is, information reported from the field, from agents, military police, refugees, civilians, and NGOs. An accurate interpretation of intelligence from HUMINT sources requires an understanding of culture, politics, attitudes, and decision-making processes of the opposing side.
- **MASINT:** Measurement and signature intelligence relies on intelligence collected from multiple sources to describe the signatures (distinctive multispectral characteristics) of fixed or dynamic targets and results in intelligence that detects, identifies, and tracks these targets in environments in which they are likely to be encountered [32, 33].
- **TECHINT:** Technical intelligence relies on the analysis of scientific and technical intelligence of the characteristics and capabilities of enemy weapons and sensors and to improve the knowledge of the opposition's military capabilities.
- **OSINT:** Open source intelligence gathered from publicly available data such as Internet sources, press agencies, and declassified government papers.

GEOINT

Sometimes also considered a subset of MASINT, GEOINT, or geophysical intelligence, refers to the exploitation and analysis of information to describe, assess, and visually depict physical features and geographically related activities on the earth's surface. GEOINT sources include imagery and mapping data, seismic, thermal, and other multispectral measurements that may be collected from sensors placed directly on the earth's surface or by satellite or other means, such as ships, UAVs, or special-purpose aircraft. As described later, GEOINT makes extensive use of multispectral and hyperspectral imaging techniques.

IMINT

Image intelligence, or IMINT, is obtained from photographic or multispectral sensors including high-resolution synthetic aperture radar (SAR) imagery (Figures 2.62 and 2.63). It can be analyzed as a discrete image and interpreted to detect features and military facilities and capabilities (Figure 2.64). When images are compared over time, an approach called *coherent or incoherent change detection* can be utilized where images at a pixel level can be compared over time to see what has changed. Multiple static images may also be merged from different sensors and angles to enhance image features and understanding (Figure 2.65).

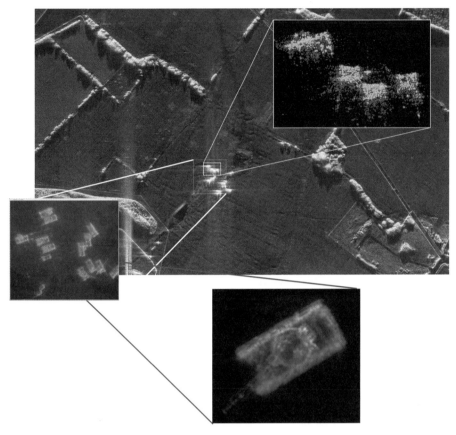

Figure 2.62   IMINT from SAR radar trials of a T72 tank formation showing high- and low-resolution images. (Images courtesy of QinetiQ.)

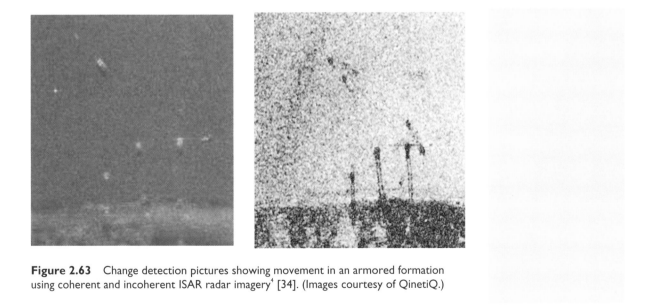

Figure 2.63   Change detection pictures showing movement in an armored formation using coherent and incoherent ISAR radar imagery[4] [34]. (Images courtesy of QinetiQ.)

Figure 2.64   Image intelligence is not limited to that from terrestrial or airborne sources.

(a)

(b)

Figure 2.65   (a) An intelligence specialist reviews aerial reconnaissance imagery on a light table in the carrier intelligence center (CVIC) aboard the aircraft carrier USS *John C. Stennis* (CVN 74). The imagery came from the "Tomcatters" of Fighter Squadron 31 (VF-31) during their final Tactical Air Reconnaissance Pod System (TARPS) mission. (b) The TARPS pod is shown being removed from an F-14D Tomcat following the sortie.

SIGINT, ELINT, and COMINT

The analysis and interpretation of signals across the electromagnetic spectrum are generally referred to as signals intelligence (SIGINT). This, in turn, comprises communications intelligence (COMINT) and electronic intelligence (ELINT), which is concerned with radar and active sensor emissions (Figure 2.66).

Electromagnetic devices are used by both civilian and military organizations for communications, navigation, sensing, information storage, data processing, and command and control functions. As such, the analysis and exploitation of these systems form an integral part of the intelligence that feeds in to the military decision-making processes.

Effective signals intelligence can help develop a better understanding of adversary capabilities, threats, and intentions through long-term, in-depth intelligence collection and exploitation. In turn, this can help the understanding of adversary vulnerabilities and their disposition and during hostilities enable faster prioritization of targeting opportunities to attack short-term transient targets.

COMINT comprises the interception and analysis of communications signals to provide intelligence on and the disruption of communications patterns and information transmission. This is typically achieved by intelligence gathering through the analysis of enemy communication patterns and the decryption and analysis of those signals. COMINT systems need to be able to intercept and analyze a wide range of civil and military communications systems. COMINT techniques use advanced interception and decryption methods to acquire anything from faxes to mobile phone communications, e-mails, and satellite communications. The very nature of the technology used means that these communications can often be intercepted a long distance from the battlefield itself (Figure 2.67).

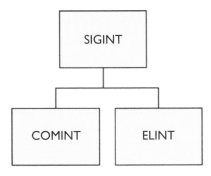

Figure 2.66   Signals intelligence (SIGINT) comprises communications intelligence (COMINT) and electronic intelligence (ELINT).

Figure 2.67　Specialist signals interception aircraft such as the BAE Nimrod R1 (illustrated) and the USAF Rivet Joint aircraft forma a key asset in electronic intelligence collection missions and the interception of voice communications, decryption, and analysis from virtually all type of communications networks. (Photo courtesy of Steven Hadlow.)

Given the huge volume of communications traffic that needs to be intercepted to find the elements that are important to the intelligence process, much COMINT analysis is heavily automated with computer programs searching for key words, relationships, and patterns in intercepted text or in speech through automated voice recognition techniques. More detailed analysis of leads or real-time analysis is usually undertaken by linguists working quickly in real time to decode cryptic messages passed in a foreign language (Figure 2.68). Once identified, intelligence from COMINT sources is carefully analyzed and quickly processed into a report and communicated to the operational commanders [35].

COMINT has become particularly challenging in recent decades. The proliferation of digital technology has resulted in the widespread availability of low-cost digital personal communication devices such as GSM phones and

Figure 2.68 Cryptologic technicians monitor electronic emissions in the electronic warfare module aboard the Nimitz-class aircraft carrier USS *Ronald Reagan* (CVN 76).

handheld satellite communications that are highly mobile, difficult to intercept, and resistant to conventional electronic attacks. In space, civilian satellites are often used for military communications and data traffic, and at a terrestrial level, optical fiber and microwave networks, linked to increasingly more powerful and secure computers, constitute the basis for powerful information systems that are equally capable of supporting military and civilian communications and data networks. The multiplicity and unpredictability of communications networks and transmission methods constitute a formidable challenge for communications intelligence systems, which require increasingly sophisticated decryption algorithms and high-speed computers.

Deception and disruption of communications activities can be achieved, for example, through broadcasting false signals, and the rebroadcast of previous signals and various types of jamming can also be employed to disrupt the use of communications systems. In order to effectively achieve this, intelligence sources need the frequency transmission characteristics associated with each emitter source to rebroadcast spoof signals using the same frequencies and codes as the emitter they are copying.

ELINT comprises the interception and analysis of noncommunications signals across the electromagnetic spectrum to provide information on the opponent's activities and sensor systems, and, as with COMINT, good intelligence information will permit signals to be altered so that they provide false information about the characteristics of the platform providing the return.

ELINT includes the analysis of radar intelligence (RADINT), which deals with intelligence gained by the analysis of radar returns and from the study of radar signals and patterns.

During times of open hostility, jamming and active deception of signals will be commonplace, but typically COMINT and ELINT activities are passive activities undertaken over long periods of time so that patterns, techniques, and operations may be understood and countermeasures developed. Information gathered is used to develop threat libraries to better understand the operational use and usage patterns of the opponent's equipment and to gain a picture of deployment patterns of opposing forces (Figures 2.69 through 2.71).

HUMINT

HUMINT specialists work in a systematic and methodical way to assimilate crucial pieces of intelligence from people such as refugees, prisoners of war, or local people who live in the area where the army is operating or who have

Figure 2.69   The analysis of signals intelligence will often reveal valuable information about communications patterns, frequencies, and locations. In advanced systems where decryption of the intercepted signal can be achieved, an analysis of communications content can also prove invaluable to the intelligence analysis and decision superiority process.

Figure 2.70   The sensor suite on the USAF's RC-135 Rivet Joint aircraft allows the crew to detect, identify, analyze, and geolocate signals across all common communications bands. The mission crew can then forward gathered information in a variety of formats to a wide range of consumers via Rivet Joint's extensive communications suite. The crew comprises 34 people, including the aircrew, electronic warfare officers, intelligence operators, and airborne systems engineers.

Figure 2.71   The RC-135U Combat Sent provides strategic electronic reconnaissance information to the president, secretary of defense, Department of Defense leaders, and theater commanders. Locating and identifying foreign military land, naval, and airborne radar signals, the Combat Sent collects and minutely examines each system, providing strategic analysis for warfighters. Collected data is also stored for further analysis by the joint warfighting and intelligence communities. The Combat Sent deploys worldwide and is employed in peacetime and contingency operations.

access to people are engaged in or who will influence activities in the area of operations (Figure 2.72). HUMINT provides commanders with timely, accurate, and often unique intelligence on an enemy's intentions, capabilities, and way of working [35].

Of all the various intelligence sources, HUMINT from a trusted source or corroborated from multiple sources can be particularly reliable, as other forms of intelligence rely on the integrity of electronic systems and are therefore susceptible to electronic deception. (Of course, HUMINT can also be open to misinterpretation or deception, and care needs to be taken accordingly.)

The collection of HUMINT, especially in urban operations, does not necessarily involve use of specialized operators or electronic sensors. Troops and patrols operating in urban areas can provide timely and focused HUMINT if they understand the importance of intelligence collection and the techniques involved. In a HUMINT context, the concept that "every soldier is a sensor" can be a powerful source of highly relevant intelligence.

Thanks to the accessibility of modern communications, HUMINT can be easily collected by making use of the Internet, mobile phones, landline phones, cameras, and short message services (SMS) messaging and does not always require direct contact with the local population. These accessible communications channels enable HUMINT operators to recruit informers

"When I took a decision, or adopted an alternative, it was after studying every relevant—and many an irrelevant—factor. Geography, tribal structure, religion, social customs, language, appetites, standards—all were at my finger-ends. The enemy I knew almost like my own side. I risked myself among them a hundred times, to learn."

—Colonel T. E. Lawrence
Letter to Liddell Hart, 26 June 1933

Figure 2.72 A soldier from the 3rd Royal Australian Regiment Battalion Group chats to a local in the town of Gleno during Operation Astute to assist the Government of Timor Leste to restore peace to the nation.

over the Internet, while on patrol, or even via leaflets distributed over an area of interest. HUMINT is, however, also susceptible to being considered as a factual source and must be corroborated ideally from other (often non-HUMINT) sources before decisions that rely on its integrity can be made.

In this regard, all intelligence sources need to be interpreted in the context of the environment in which the intelligence is being collected. Data from a single source cannot be interpreted in isolation or without taking into account intelligence from additional sources or, indeed, the social, political, and cultural environment in which the data is being gathered.

Intelligence analysis is about more than processing information and identifying patterns. It looks beyond what is happening and what may happen to use intelligence to generate scenarios, responses, and options. This involves generating insights into how the adversary thinks, interprets information, and may respond to particular courses of action, information that can be obtained particularly effectively through HUMINT sources.

MASINT

Measurement and signature intelligence (MASINT) provides technically derived intelligence to detect, locate, track, identify, and describe the specific characteristics of fixed and dynamic target objects and sources [36]. MASINT is a relatively slow and time-consuming process used to build a reference library of multispectral and multicharacteristic target signatures so that the information can be fed into ISTAR and targeting systems for operational references and identification purposes. MASINT is particularly useful in countering deception as it enables multiple signatures to be correlated and compared with known target signatures through the use of a wide range of signature measurement techniques.

The electromagnetic signatures and traits by which a target may be identified cover a wide range of physical and electromagnetic characteristics. Processing by the use of MASINT analysis of these characteristics can provide a useful reference library of target characteristics. Typically, those characteristics cover features such as [37]:

- Chemical, biological, radiological, and nuclear activities and associated signatures;
- Emitted energy (e.g., nuclear, thermal, and electromagnetic);
- Radiated, reradiated, or reflected energy (e.g., electromagnetic and sound);
- Mechanical sound (e.g., acoustic signatures from engines, propellers, or machinery);

Scanning Arrays, Staring Arrays, Staring Plane Arrays, Focal Plane Arrays (FPAs), and Charge-Coupled Devices (CCDs)

Staring arrays and **focal plane arrays** are types of image sensing devices consisting of an array of frequency-sensitive pixels arranged at the focal plane of the lens.

Scanning arrays are arrays that have to perform multiple scans of the area of interest to compile a useable picture.

Staring arrays (Figure 2.73) are different from scanning arrays in that they compile the desired picture in one look, without having to build up the picture through a series of scans using a narrow beam. Staring arrays are commonly found in surveillance sensors, particularly space-based satellites, where large areas of territory can be imaged without having to risk the unreliability found in mechanically scanned systems.

Focal plane arrays (FPAs) are commonly found in missile guidance sensors where their ability to image a target can provide highly accurate guidance cues for the final stages of the missile's flight. Unlike a traditional lens, which tends to concentrate the image on a focal point, focal plane devices ensure that the entire array is in focus. These arrays are most commonly used for producing images at a particular wavelength and in military applications are often used at infrared frequencies to detect heat from targets or exhaust plumes from missiles—often using FPAs operating at different parts of the IR spectrum to distinguish the target from decoys.

Charge-coupled devices (CCDs) are arrays that are sensitive across the visual spectrum frequency band.

- Magnetic properties (e.g., magnetic flux, geomagnetic properties and anomalies);
- Motion (e.g., vibration, or movement);
- Material composition;
- Change detection (measurement in changes of any of these characteristics over time).

MASINT covers a broad area of intelligence gathering and analysis and encompasses many scientific disciplines, but can broadly be considered to comprise the following subsets.

Electro-optical MASINT

Electro-optical MASINT covers the analysis of electro-optical signatures of military equipment and capabilities in visual or near-visual frequency ranges and is complementary to the field of IMINT. While IMINT is primarily focused on creating a visual picture that can be interpreted for visual clues, electro-optical MASINT supports the more detailed interpretation of the picture to distinguish, for example, features between real targets and decoys. While IMINT relies largely on human interpretation of a visual picture,

Figure 2.73   Defense Support Program (DSP) satellites use a staring array of 6,000 infrared sensors to detect heat from missile and booster plumes against the earth's background to provide early warning of attack from ICBMs and smaller missiles. This geosynchronous satellite system proved useful in 1990–1991 during the first Gulf War to detect the launch of Iraqi Scud missiles in time to give early warning to potential targets.

electro-optical MASINT can provide clues based on a quantitative assessment of the specific electro-optical frequency patterns and intensities and produce a reference library of the spectral characteristics to assist with image interpretation.

Analysis of precise frequencies within the electromagnetic spectrum can reveal specific signatures associated with the characteristics of military capabilities ranging from the ultraviolet flash of a missile launch to imaging lasers to the hyperspectral signature associated with spillages of particular chemicals at manufacturing plants.

Infrared MASINT

Sensors operating at specific frequencies are often employed to search for specific military characteristics such as the telltale heat (or ultraviolet) signatures associated with missiles, artillery, and indirect fire projectiles. Sensors can be ground or space-based depending on what is being searched for, with activities over large areas of land or sea typically being observed by space-based staring infrared sensors. Some countries, most notably the United States, when implementing wide area infrared anomaly detection and missile early warning capabilities, have placed in orbit a series of space-based staring array sensors that detect and locate infrared signatures from heat sources such as missile rocket motors but also from other intense heat sources such as nuclear or conventional explosions and industrial activities (Figure 2.74). Such signatures, which are associated with measurement of thermal energy and its duration and location, do not provide pictures in the IMINT sense, but suggest the presence of activities that would merit further investigation.

Spectroscopic MASINT

Spectroscopy involves the analysis of radiated energy and its associated wavelengths from a target or area of interest. Spectroscopy can be applied either to targets that are already radiating energy, such as engine exhausts or hot structures, or alternatively targets can be stimulated with a laser or other energy source to release energy, enabling the material characteristics to be measured. Unlike IMINT, MASINT is not a true imaging technique, although it can be used to build a synthetic picture of the target using measurements of energy, wavelength, and geometrical coordinates (Figure 2.75).

Figure 2.74 The electro-optical sensor package for the RC-135 Cobra Ball surveillance aircraft includes long-range visible and medium wave infrared surveillance telescopes, and long-range laser ranging systems enabling ballistic-missile tests to be monitored at long range.

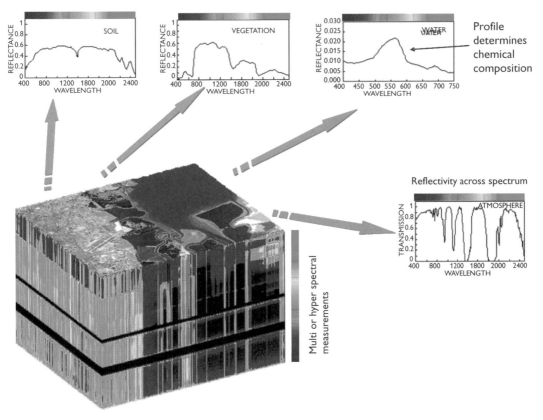

Figure 2.75   Multispectral and hyperspectral imaging measures the intensity, absorption, and reflectivity of materials at various frequencies to determine their characteristics [38].

Spectroscopy typically covers a range of frequencies and typically extends into the IR and UV ranges, permitting a much wider discrimination of target characteristics than would be possible at optical wavelengths detectable by humans. Spectroscopic measurements can reveal whether a building is made of concrete, wood, or brick, whether a road is paved or earth, and whether liquids on their surface are water or other chemicals, for example.

Unlike optical cameras that only look at visual red, green, and blue frequencies, multispectral and hyperspectral imaging looks at electromagnetic frequencies above and below visual wavelengths and measures the intensity, absorption, and reflectivity of materials at various frequencies.

Multispectral imaging is a color-based image discrimination technique exploiting the principle that objects are not typically blackbodies—they emit or reflect some wavelengths preferentially to others. Multispectral systems typically sense up to 10 adjacent channels with each channel equating to a discrete frequency band. Multispectral systems usually merge the individual images from up to 10 of these channels to form a composite image.

Hyperspectral imaging is similar to multispectral imaging but employs up to several hundred channels to capture an object's signature across many bands of the electromagnetic spectrum, resulting in an image using tens to hundreds of colors in each pixel as a multidimensional discriminant. This information is used to generate a "data cube" that can reveal information about objects (that narrowband scanners cannot detect) through exploiting the ability to determine the chemical composition of materials and gases in the environment. The ability to identify individual chemicals through the spectral emissions measured by hyperspectral imaging enables, for example, the discrimination of painted vehicles from foliage and can even identify and classify the composition of emission from factories or the chemicals within a gas cloud.

Spectroscopic measurements plot energy measured by radiant intensity against the frequency of the emission. The frequency range covered by spectroscopic systems is a function of the number of discrete spectral bands in the sensor system, and the range of frequencies covered will determine the amount and of detail and type of characteristics that can be obtained about the target.

Sensor systems cover different ranges of frequency bands:

- Monospectral (covering 1 band);
- Multispectral (covering 2 to 100 bands);
- Hyperspectral (covering 100 to 1,000 bands);
- Ultraspectral (covering more than 1,000 bands).

Multiband detection can be optimized by employing many different segments of the electromagnetic spectrum. Targets appear different across the spectrum because of their chemical composition. By selecting individual sensors optimized for their performance in a particular spectral range, combined multiband sensors can exploit multiple portions of the spectrum, greatly increasing detection capabilities.

In *wideband* the resolution of the sensor image and accuracy of detection characteristics is driven by its operating bandwidth. Wideband is a relative term used to describe a broader range of operating frequencies enabled by the use of improved component sensitivities and wider spectral coverage.

Ultrawideband yields even resolution of detection characteristics by exploiting a significantly broader frequency range, octaves or even decades wider in operating frequencies.

The greater the spectral frequency coverage (Figure 2.76), the more discrete the information or greater the resolution that can be provided. The emission and absorption characteristics clearly identify the type of material being analyzed and serve to fingerprint the properties of the target being observed. The ability of such techniques to identify material is considered particularly valuable by the intelligence community when identifying signatures of material used in the production or testing of weapons of mass destruction (WMD) such as nuclear or chemical warfare products (Figure 2.77).

Nuclear MASINT

Nuclear MASINT covers the detection of nuclear energy and materials and the analysis of the effects of ionizing nuclear radiation on materials, people, equipment, and the environment. Nuclear MASINT sensors are often space-based and are usually combined with other scientific fields such as seismology and atmospheric sampling to detect nuclear activities.

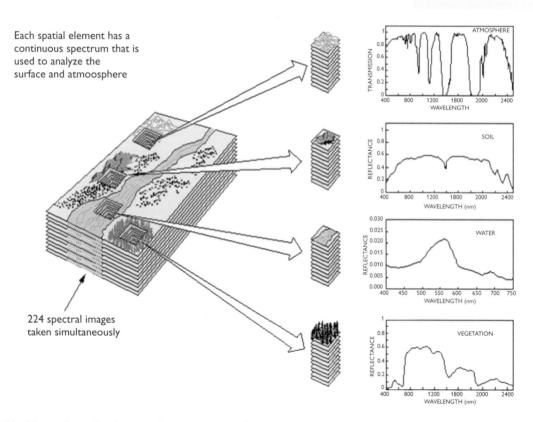

Figure 2.76 The analysis of multispectral images across different wavelengths can be used to distinguish landscape features and anomalies to assist with intelligence gathering.

Figure 2.77   The Morenci open-pit copper mine in southeast Arizona is North America's leading producer of copper. This Advanced Spaceborne Thermal Emission and Reflection Radiometer (ASTER) image uses short wavelength infrared bands to highlight in bright pink the altered rocks in the Morenci pit associated with copper mineralization.

Geophysical MASINT

Geophysical MASINT focuses on the analysis of the physical environment to understand the impact of geophysical characteristics on military operations and to detect changes in those characteristics that may have military implications. Analysis of geophysical characteristics over time enables the detection of changes in the environment that may indicate enemy activity or the presence of enemy facilities including evidence of tunneling, roads, and other activities that disrupt the natural environment. Geophysical intelligence includes the study of properties related to the weather, acoustic, seismic, gravimetric, and magnetic environments.

Radar MASINT

Radar MASINT is related to SIGINT in that it measures the characteristics of electromagnetic radiation, but from radar sources rather than from communications signals (Figure 2.78). Radar MASINT uses processing and classification techniques to measure the radar characteristics of targets under a variety of conditions and from different angles to build a reference library of these characteristics to assist with target identification. Radar MASINT data, for example, is particularly applicable to noncooperative target recognition (NCTR), where the characteristics of radar reflections from a target are analyzed in detail to classify the type of platform being illuminated.

In addition, the study and classification of radar emission characteristics enable a library to be built to help identify types of radars detected by threat warning systems. Every radar has its own distinctive emission patterns, which will vary depending on the mode in which the radar is operating and can alert a platform, for example, to whether it has been detected and whether a radar has locked on in preparation for a missile launch.

Since all radars have slight variations in their build characteristics (Figures 2.79 and 2.80), through taking very accurate measurements, it is possible to identify the emissions from individual radars—not just the radar type. Such fingerprinting is known as specific emitter identification (SEI).

Figure 2.78 The Electronic Support Measures System (ESM) on a Royal Navy Type 23 Frigate is capable of taking accurate measurements of enemy radar characteristics to enable a library to be built of hostile emitter characteristics.

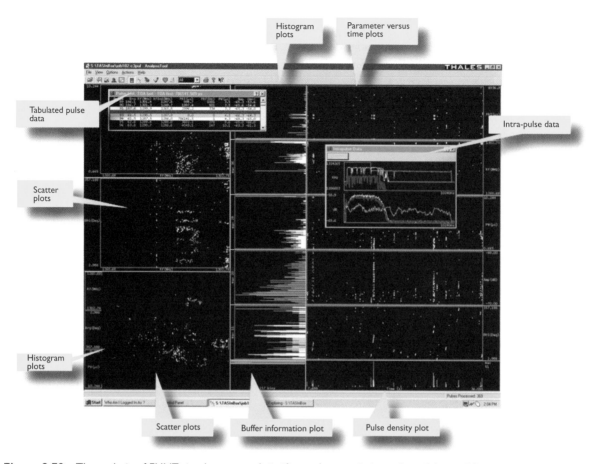

Figure 2.79 The analysis of ELINT signals can reveal significant characteristics and capabilities of foreign equipment and help with the automated identification of emitters in electronic warfare system. The illustration shows the automated analysis and classification of radar signals. (Image courtesy of Thales.)

Detailed knowledge of radar characteristics can also be used to rebroadcast false signals to give the impression that the platform emitting the signals is totally different to its real identity (Figures 2.81 and 2.82).

Materials MASINT

Materials MASINT involves the detection, collection, processing, and analysis of gas, liquid, or solid samples and their analysis in the better understanding the capabilities or methods of employment of an opponent's military capabilities. MASINT is of particular use in planning responses to nuclear-biological-chemical (NBC) or chemical, biological, radiological, and nuclear threats (CBRN) as it is often referred to, both in civilian and military environments (Figure 2.83).

Figure 2.80 The Indian ship Rajput class destroyer INS *Rana* (D 52) follows behind the Nimitz-class aircraft carrier USS *Ronald Reagan* (CVN 76) during Malabar exercise 08. Radar characteristics of such ships are analyzed and stored in libraries for future cross reference of radar signals (U.S. Navy).

Figure 2.81 An active jamming pod on a German Luftwaffe Tornado. Such pods are capable of emitting signals to give the impression that they are in fact another type of aircraft, so a Tornado, for example, could emit the electronic characteristics of a Boeing 737 civil airliner.

Figure 2.82 Information from RADINT sources is also used to program antiradar missiles such as the MBDA Alarm missile shown on an RAF GR4 Tornado. The ability for such missiles to detect and classify hostile radars is at the heart of their capability.

Figure 2.83 The Fuchs tactical NBC reconnaissance vehicle, which can keep up with deployed forces, detecting, identifying, sampling, and reporting CBRN hazards. These vehicles are equipped with radiation survey, meteorological, chemical, and biological sensors and can collect atmospheric, soil, water, and vegetation samples, immediately analyzing and reporting their findings. Its computer systems can predict likely propagation patterns and report the results on multiple networks using NATO standard ATP45(B) tactical symbols. (Photo courtesy of Rheinmetall AG.)

Radio frequency MASINT

Radio frequency MASINT concentrates on characterizing the signatures of military systems and equipment, based on their intentional and unintentional RF emissions. While intentional RF emissions will have been analyzed by ELINT sensors, RF MASINT also considers the unintentional radiation emitted from radar and radio systems (e.g., sidelobes) and other radio frequency emission characteristics that will help identify specific platforms that are often equipped with a multitude of RF transmission systems. RF MASINT also includes the analysis of electromagnetic pulses that emanate from electrical power generation systems, nuclear explosions, large conventional explosions, and EMP weapons.

TECHINT

Technical intelligence seeks to acquire a better knowledge of the characteristics and capabilities of enemy weapon systems and sensors so that they can be detected and identified more easily. This also enables effective countermeasures to be developed and deployed against them. TECHINT is more formally defined as [39]: "Intelligence derived from the exploitation of foreign material, produced for strategic, operational, and tactical level commanders. Technical intelligence begins when an individual service member finds something new on the battlefield and takes the proper steps to report it. The item is then exploited at successively higher levels until a countermeasure is produced to neutralize the adversary's technological advantage."

The effectiveness of TECHINT is dependent on the timely collection, exploitation, and analysis of foreign warfighting equipment and associated materiel (Figure 2.84). The interpretation and exploitation of TECHINT such as foreign weapon and sensor systems, materiel, and technologies are referred to as scientific and technical intelligence (S&TI), which involves the detailed assessment of research and development capabilities, the application of those technologies to weapons and sensor systems, and the analysis of the enemy's capabilities "by all appropriate and available means."

OSINT

Open-source intelligence (OSINT) is obtained from commercial radio and television broadcasts, newspapers, magazines [36], and other written publications and from analyzing and interpreting that information, often cross matching it with other intelligence sources to make conclusions about military capabilities. OSINT has the advantage that it can be gathered relatively cheaply and safely from worldwide, easily accessible sources.

(a)

(b)

Figure 2.84   (a, b) A search team discovers an MiG-25 Foxbat buried beneath the sands in Iraq during Operation Iraqi Freedom. Several MiG-25 interceptors and Su-25 ground attack jets have been found buried at the Al-Taqqadum air field west of Baghdad.

Scientific and technical intelligence

Scientific and technical intelligence (S&TI) looks at foreign scientific and technical developments that have or indicate a warfare potential. This includes medical capabilities and weapon system characteristics, capabilities, vulnerabilities, limitations, and effectiveness, research and development activities related to those systems, and related manufacturing information. S&TI collects information regarding adversary equipment and information that is needed to preclude scientific and technological surprises and advantages by an adversary that could be detrimental to friendly personnel and operations [36].

Information warfare and information superiority

Information warfare and information superiority concepts lie at the heart of NEC and NEW. The concepts associated with NEC and NEW describe actions that are taken to influence the interpretation of information and the subsequent decision-making process by the opponent, while protecting the integrity of information and the quality of decision making by one's own forces. The goal of information warfare is to achieve control of the accuracy of information available to each side and centers on the concept and importance of achieving information superiority early on in any engagement.

Achieving information superiority requires relative rather than absolute information dominance over your opponent. The degree to which the dominance can be achieved will influence the speed of the outcome, but the outcome itself will be largely determined by the fact that relative information dominance has been achieved rather than the degree to which it has been achieved. Achieving information superiority requires a combination of offensive and defensive information operations to ensure superior collection of accurate and timely information to one's own forces while denying the same to the opponent.

A relative information advantage can [40]:

- Be persistent or transitory;
- Exist in some areas of the battlespace but not others;
- Be measured in the context of a task or set of tasks;
- Be created by taking actions to reduce the need for information and/or increase the information needs of an adversary;
- Be achieved through the synergistic conduct of information operations, information assurance, and information gain and exploitation.

Information warfare is likely to be increasingly seen as a means of inflicting nonovert damage on an adversary, which may or may not precede an overt military response. Such activities have the advantage that they leave little direct evidence from which to identify an attacker and can be faster, cheaper, and just as effective as other more direct forms of attack. Military focus in the run-up to any overt conflict would most likely be through a computer

Information Superiority

The U.S. DoD defines information superiority as:

"The operational advantage derived from the ability to collect, process, and disseminate an uninterrupted flow of information while exploiting or denying an adversary's ability to do the same."

—*U.S. Army Information Operations: Doctrine, Tactics, Techniques, and Procedures Field Manual*, No. 3-13, November 2003

Information Dominance

The U. S. JCS reflects that information superiority is:

"Degree of dominance in the information domain that permits the conduct of operations without effective opposition."

—U.S. Joint Chiefs of Staff (JCS)

Information Technologies [41]

What Is Different

- There are few geographical boundaries in the information infrastructure.
- Target access points are changing and may not be in geographical proximity to the target.
- The military can no longer create and control the battlespace.

network attack aimed at disrupting and destroying an adversary's capability to acquire, process, and distribute data related to military, national infrastructure, or economic assets (Figure 2.85).

Dominating the information domain is as essential today as controlling land, sea, or air space was during the era of mechanized warfare. The military advantage gained through information warfare is today as significant as the military advantages of massed armor formations or bomber formations in the era of mechanized warfare. It enables forces with lower numerical strength to concentrate their effect in the right place and at the right time to make a disproportionate impact on their relative military strength.

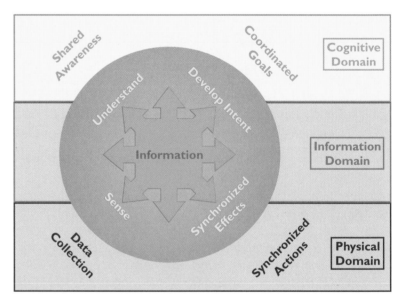

Figure 2.85 Information transfer across all domains is key to ensuring effective situational awareness and decision making [61].

During overt hostilities, information networks remain key assets for both sides, with each trying to disruption information gathering, assessment, and dissemination, while at the same time trying to defend their own systems and protect from military strikes against important information nodes such as computer centers, satellite ground stations, and so on.

Some analysts, for example, believe that there is no more important vulnerability to modern societies than their heavy reliance on information networks [42], with the implication clearly being that disruption and destruction of those information networks would take a high priority in any future conflict. The same vulnerabilities will be true for other nations who are heavily reliant not only on military network capabilities but also on civilian information infrastructures, and recent attacks on national information networks are well documented.

To achieve information superiority, commanders must focus efforts to improve their own situational and operational awareness while influencing adversary perceptions in a way that leads the adversary to make decisions that are suboptimal for their situation and of benefit to the tactical requirements of friendly forces. Such an approach to information management demands decisive judgment and operations in narrow windows of opportunity at times and places that are advantageous to the friendly forces and which enable the exploitation of weaknesses in the tactics of the opposing forces. A relentless approach to information management should provide sustained information superiority and the upper hand for friendly forces. The speed with which the enemy reacts to actions often determines the length of time that friendly forces can maintain an information superiority advantage, and the use of information warfare–based approaches to the OODA loop also dramatically improves the speed of decision making for friendly forces.

Conversely, preventing adversary commanders from exercising effective C2 is the goal of friendly forces operating in the information environment. Friendly forces seek to achieve information superiority by destroying or interfering with C2 systems or the data they contain and transmit.

Adversaries may apply a variety of synchronized offensive and defensive information operations to produce complementary and reinforcing effects in order to protect their C2 systems from attack. C2-related information operations target specific areas of the C2 infrastructure for exploitation, aiming to destroy or disrupt the command chain and its personnel, disconnect the various C2 links from operational forces, and disrupt and corrupt the data stored in decision-critical information systems. Effective information operations are resource intensive and time critical and it is difficult to sustain information superiority across all operational settings for long periods of time (Figure 2.86). Thus, commanders do not seek to sustain

In NCW, information is itself both a force multiplier and a target in its own right.

"The mission of destroying missile sites and arms depots is almost the easy part. The critical task ... is to continue to reshape the information environment and target points of fracture in the opposition."

—P. W. Singer, *Foreign Policy Studies*

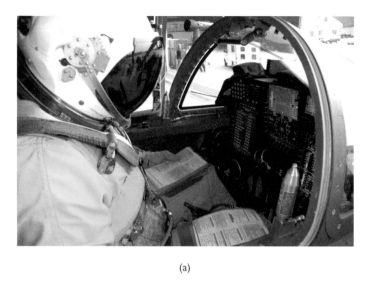

(a)

(b)

Figure 2.86 (a, b) Information superiority requires persistent and enduring surveillance and intelligence activities. Specifically, spectral surveillance using sea, land, air, and space-based sensors, immediate identification and geolocation of targets of interest, shared data and a common situational awareness shared across secure networks. Aircraft such as the U-2S Dragon Lady are essential in achieving multispectral persistent surveillance.

information superiority over an extended period (although they will want to retain broad information superiority advantages for enduring periods), but will rather look to achieve localized information superiority for specific periods in order to produce decisive and timely operational results. Such an approach provides a capability within a window of opportunity to enable decisive operations at times and places that the commander chooses.

An adversary, depending on the level of technology available may employ approaches to information warfare similar to those of the friendly force or possibly of a more asymmetric nature—either way the opposing forces will look to create a relative advantage in the quality and timeliness of information available to them over that available to their opponents. The characteristics of information that make it of significance to the commander in the field are many, and interference with any element can degrade the information input and subsequent quality of output of the OODA loop.

Technological advances in recent years have greatly increased our ability to collect, process, disseminate, and exploit information. The challenge today is not one of collection but of the integration of huge amounts of sensor data that can now be collected, and the timely distribution of quality information (for a definition of information quality, see Chapter 3) around the battlespace in near real time to aid the OODA process. Overcoming this challenge is particularly important for the rapid achievement of information superiority.

For every intelligence collection or effects engagement, the degree to which information characteristics will play a determining role in the outcome will vary. For example, in a missile engagement timeliness and accuracy of data will be key determinants of the outcome. For a carrier battle group making a decision on course and speed, accessibility, consistency, and degree of synthesis of the information used in that decision will be key.

In U.S. doctrine, information operations are divided into five core capabilities aimed at achieving decisive military effects or preventing the enemy from doing the same [43].

- **Electronic warfare (EW):** Any military action involving the use of electromagnetic or directed energy to control or exploit the electromagnetic spectrum or to attack the enemy.
- **Psychological operations (PSYOP):** Planned operations to convey selected information and indicators to audiences to influence their emotions, motives, objective reasoning, and ultimately the behavior of foreign governments, organizations, groups, and individuals. The purpose of psychological operations is to induce or reinforce foreign attitudes and behavior favorable to the originator's objectives.
- **Operations security (OPSEC):** A process of identifying critical information and subsequently analyzing friendly actions attendant to military operations and other activities to:
    - Identify those actions that can be observed by adversary intelligence systems;

- Determine indicators that adversary intelligence systems might obtain that could be interpreted or pieced together to derive critical information in time to be useful to adversaries;
- Select and execute measures that eliminate or reduce to an acceptable level the vulnerabilities of friendly actions to adversary exploitation.

- **Military deception (MILDEC):** Consists of actions executed to deliberately mislead adversary military decision makers regarding friendly military capabilities, intentions, and operations, thereby causing the adversary to take specific actions (or inactions) that will contribute to the accomplishment of the friendly mission.

- **Computer network operations (CNO):** Comprised of computer network attack, computer network defense, and related computer network exploitation enabling operations [44].

Information Operations

"The integrated employment of the core capabilities of Electronic Warfare, Computer Network Operations, Psychological Operations, Military Deceptions and Operations Security, in concert with specified supporting and related capabilities, to influence, disrupt, corrupt or usurp adversarial human and automated decision-making while protecting our own."

—U.S. DoD, *Information Operations Roadmap*, October 2003, declassified

Information operations are focused on disrupting or controlling the opponents command and communications processes while preserving the integrity of one's own C2 networks, and in U.S. doctrine [45] are described in terms of:

- Deterring, discouraging, and dissuading an adversary by disrupting his chain of command while preserving the integrity of friendly C2 operations;
- Protecting friendly plans and misdirect those of the opponent;
- Controlling and disrupting the opponent's communication and sensor networks while protecting friendly networks.

In summary, effective information operations allows the commanders to understand more clearly the situational picture unfolding on the battlespace, permits them to take advantage of opportunities more rapidly and more effectively, while denying adversary commanders the information needed to make timely and accurate decisions or leading them to make suboptimal decisions that are more favorable to friendly forces.

As explained in *Understanding Information Age Warfare* [45], the *information position* of an organization is the degree to which it is able to fulfill its need for information in order to function effectively. A force with relatively simple information requirements, for example, a terrorist cell, may have a relatively strong information position, although it possesses relatively little information. At the same time, a coalition force seeking to employ sophisticated weapons and conduct successful counterinsurgency operations may possess a great deal of information, but still have a relatively weak information position because of the massive amount of information it requires and the quality of the information it needs to operate effectively.

Clearly, a well-informed networked force can move quickly to initiate effective actions and is able to engage all assets linked by the network in the prosecution of its military objectives.

Offensive information warfare

With modern society's almost total dependence on information technology (IT), the potential impact of offensive information warfare represents a significant risk. IT systems permeate every aspect of modern life in both civilian and military domains and much disruption can potentially be caused by hostile attacks on network-dependent systems. Even opponents with limited traditional military capabilities are readily able to make use of information warfare strategies with minimum cost and preparation, and an effective attack on a critical network node of IT system can cause significant disruption to an opponent's infrastructure. While military C2 and network systems might be generally better protected than civilian networks, the potential for disruption and distraction of governments and populations is significant as a result.

Attacks on computer networks and associated infrastructure may either be overt, where the attackers are not concerned about covering their tracks but rather set out to maximize the damage and disruption inflicted, or covert in nature where information is not touched, but activity related to data and networks is collected for subsequent analysis and use in future lethal attacks or related intelligence activities.

Conducting effective information warfare is not a random process. Maximizing the effectiveness of information warfare attacks requires an extensive investment in understanding the other side's architectures, the criticality of their network nodes, operating procedures and patterns, access techniques, and the likely impact of attacks on the target. Such knowledge is best obtained over lengthy periods of time and involves probing attacks and intelligence gathering operations over several years to prepare for a time when offensive operations may be required.

Information networks, especially those that are linked to the Global Information Grid are vulnerable to attack from a variety of different sources.

Offensive information operations

Offensive information operations or offensive information warfare seeks to destroy, degrade, disrupt, deny, deceive, exploit, and influence adversary decision makers, military forces, supporting populations and those who can

DoD Information Operations

"The integrated environment of the core capabilities of electronic warfare, computer network operations, psychological operations, military deception, and operations security, in concert with specified supporting and related capabilities, to influence, disrupt, corrupt, or usurp adversarial human and automated decision-making, while protecting our own."

—U.S. Army War College, Information Operations Primer Department of Military Strategy, Planning, and Operations, November 2006, AY07 Edition

affect the outcome of friendly operations. Offensive IO also targets the information and information systems (INFOSYS) used in adversary's decision-making processes to disrupt the associated processes and to prevent their use against friendly forces or associated infrastructure [46].

Offensive information warfare can be applied in a number of different domains, which can generally be classified under the following headings:

- Computer network penetration (firewall attack, hacking);
- Computer network exploitation (viruses, Trojans);
- Network deception (within the network);
- Network denial (denial of service);
- External deception and denial (active EW);
- Electronic attack (EMP, directed energy, or physical attack).

Offensive information operations assist in the achievement of the information superiority advantage by creating a disparity between the quality and timeliness of information available to friendly forces compared to that available to adversaries. Types of active attack may vary in their severity and duration, but the coordination of attacks is essential to maintain a convincing approach to the enemy's understanding of the aspirations, strengths, and disposition of the adversary it faces (Table 2.6).

The following information operations effects assist in the creation of relative information superiority information advantage. Many of them will overlap, and typically offensive information operations will comprise of several objectives [46]:

- **Destroy:** To damage a combat, sensor, or C2 system so badly by kinetic or nonkinetic means that it cannot perform any function or be restored to a usable condition [47].

Core Capabilities	
Electronic warfare	Military deception
Computer network operations	Psychological operations
Operations security	
Supporting Capabilities	**Related Capabilities**
Information assurance	Public affairs
Physical security	Civil-military operations
Counterintelligence	Defense support to public diplomacy
Physical attack	
Combat camera	

Table 2.6 Joint IO Definition

- **Disrupt:** To break or interrupt the flow of information across C2 networks.

- **Degrade:** To use nonlethal or temporary means to reduce the latency, effectiveness, or efficiency of the data associated with adversary combat, sensor, or C2 systems. The degree of degradation can reduce the target's value and the need to employ physical destruction to achieve a suitable level of impairment. Degradation of these systems can also reduce the quality of adversary decisions and actions, especially where alternative information sources are not available and where the enemy is not aware that the input or output of its systems has been impaired.

- **Deny:** Involves protecting access to information concerning friendly force capabilities, dispositions, and intentions that adversaries may need for effective and timely decision making.

- **Deceive:** To seek to lead the opponents to incorrect conclusions about the friendly forces' strengths, disposition, and intentions by deliberately providing false information to the enemy. The provision of such information needs to be carefully managed to ensure that it is consistent with other information sources and that the means in which it needs to be obtained by the enemy gives no clues as to its real nature.

- **Exploit:** To use the enemy's own C2 systems to the advantage of friendly forces by gaining access to enemy networks and C2 systems to collect information or to introduce erroneous information upon which the enemy will make command decisions.

- **Influence:** To manipulate information to influence the opponent's emotions, motives, and reasoning and subsequently their perception and decision-making processes. The hope is that such influence will then cause opponents or those who support the opponent's aspirations in a more indirect manner, to behave in a manner more advantageous to friendly forces. Perception management also seeks to influence the opponent's current operations and forward plans through reducing their desire to oppose friendly forces.

Defensive information operations

Defensive information warfare or defensive information operations is primarily concerned with ensuring the protection of one's own information assets, information sources, and the processes that go into the compilation of the information-based intelligence picture. It seeks to ensure timely and accurate information access for friendly forces operating across many networks and security levels while denying opponents access to information systems, data, or the transmission of data that may be used to undermine friendly C2 operations.

Defensive information warfare can be applied in a number of different domains, which can generally be classified under the following headings:

- Computer network defense;
- Encryption;
- Deception;
- Information assurance.

Defensive IO seeks to limit the vulnerability of C2 systems to adversary action and to prevent enemy interference with friendly information and information systems. Timely identification of adversaries, their intent, and their capabilities are the cornerstone of an effective response to an adversary's offensive IO. Defensive IO phases include [46]:

- **Protection:** Actions taken to ensure that opponents cannot access valuable information or that equipment does not fall into the hands of those who could use the knowledge gained against friendly forces (Figure 2.87). Protection of information systems requires that access to those systems, by both humans and other computers is strictly controlled and that security protocols are regularly tested and changed to avoid any enduring damage should a network be penetrated.
- **Detection:** To discover or determine the existence of an intrusion or an attempted intrusion into friendly information systems.
- **Restoration:** To bring information systems back to their original state and to reintroduce their functionality back into the network.

Figure 2.87   Staff at the U.S. Air Force Cyber Command, Barksdale Air Force Base, Los Angeles, update antivirus software for Air Force units to assist in the prevention of cyberspace hackers.

Information Warfare Aircraft

EC-130H Hercules Compass Call

The EC-130H is a specially modified Lockheed Hercules aircraft designed to undertake sophisticated stand-off information operations and electronic warfare in conjunction with other airborne and land-based EW assets [48] (Figure 2.88).

Figure 2.88   Compass Call is the designation for a modified version of Lockheed corporation's EC-130H Hercules aircraft configured to perform tactical command, control, and communications countermeasures. Modifications to the aircraft include an electronic countermeasures system, air refueling capability, and associated navigation and communications systems.

The system is configured to disrupt enemy C2-related communications and limit the ability to coordinate dispersed military assets through attacking the quality, reliability, and timeliness of information that those assets do receive. Specifically, the aircraft uses various jamming techniques to prevent or disrupt enemy communications and degrade the transfer of information essential to C2 of weapon systems and other resources. It primarily supports tactical air operations, but also can provide jamming support to ground force operations. In particular, it is understood that the EC-130H can target and disrupt enemy sensor and communications systems by introducing computer algorithms such as viruses into sensors and communications networks. Planned upgrades will add an electronic attack capability to defeat early warning and acquisition radars to support offensive joint force operations, along with more agile interception, digital signal processing, jamming, and deception capabilities focused on enemy C2 and navigation systems.

The EC-130H aircraft carries a crew of 13 people comprising four flight crew and nine mission systems specialists. The mission crew includes the mission commander (electronic warfare officer), weapon system officer (electronic warfare officer), mission crew supervisor (covering cryptology and linguistics), four analysis operators (linguists), one acquisition operator, and an airborne maintenance specialist to support the mission equipment permanently located in the aircraft's cargo bay.

The Compass Call (Figure 2.89) forms a powerful asset in prosecuting offensive information operations and has demonstrated the ability to disrupt and deceive enemy C2 networks in many recent military operations including Kosovo, Haiti, Panama, Iraq, Serbia, and Afghanistan.

Figure 2.89　A crew member studies a system operating guide before performing tactical command, control, and communications countermeasures in an EC-130H Compass Call over Afghanistan.

EC-130J Commando Solo

The EC-130J Commando Solo (Figure 2.90), based on the Lockheed Hercules airframe, is specially modified to conduct offensive information operations, including psychological operations (PSYOPS) and broadcasts to civilian and military audiences in AM, FM, HF, and TV communications bands [48]. A typical mission consists of a single aircraft operating in secured airspace flying an orbit that can reach the desired civilian or military target audience. With built-in air-to-air refueling capability, the aircraft is able to remain on station for long periods of time.

Figure 2.90   An EC-130J Commando Solo operated by the USAF Special Operations Command's 193rd Special Operations Wing prepares to land at Kandahar, Afghanistan. The EC-130J conducts information operations, psychological operations and civil affairs broadcasts in the AM, FM, HF, TV, and military communications bands.

Operationally, PSYOPS provides an effective and adaptable means of exploiting the psychological vulnerabilities of hostile forces to create fear, confusion, and paralysis, thus undermining their morale and fighting spirit [49]. Many modifications have been made to Commando Solo aircraft including improved navigation systems, active and passive self-protection equipment, and the capability to broadcast radio and color TV signals on all worldwide standards (Figure 2.91).

Figure 2.91   A USAF crew member adjusts the output of a transmission during an EC-130E Commando Solo II training mission.

In 1990–1991 during Operations Desert Shield and Desert Storm, Commando Solo was deployed to broadcast the Voice of the Gulf, a radio station set up by the U.S. military to convince Iraqi soldiers to surrender.

In 1994, Commando Solo was used to broadcast radio and TV messages to the citizens and leaders of Haiti during Operation Uphold Democracy. TV and radio broadcasts from the Commando Solo aircraft featured President Jean-Bertrand Aristide and helped to contribute to the orderly transition from military rule to civil democracy.

Continuing its tradition, in 1997 Commando Solo operations supported the United Nations' Operation Joint Guard with radio and TV broadcasts over Bosnia-Herzegovina in support of UN Stabilization Forces operations. In 1998, the unit and its aircraft participated in Operation Desert Thunder, a deployment to Southwest Asia to persuade the Iraq leadership to comply with UN Security Council resolutions. The Commando Solo was again sent into action in 1999 in support of Operation Allied Force. The aircraft was tasked to broadcast radio and television to Kosovo civilians to prevent ethnic cleansing and assist in the expulsion of the Serbs from the region.

More recently, the Commando Solo aircraft have been deployed to Afghanistan to broadcast messages to the local population and Taliban fighters during Operation Enduring Freedom and to the Middle East in support of Operation Iraqi Freedom [48].

- **Response:** In information operations, to react rapidly and effectively to an adversary's information operations attack or intrusion.

The techniques used in defensive and offensive information warfare are combined in an appropriate manner at different stages of conflict to produce information dominance. Other offensive and defensive information warfare operations unrelated to electronic combat systems can also be employed such as perception management and psychological operations that are outside the scope of this book.

The impact of NEC on military doctrine and capabilities

The military value of NEC is directly related to the impact that such technologies have on determining the outcome of any engagement, and the recognition that the information provided by those capabilities is of immense value in its own right. The information generated by NEC allows commanders to consider how their assets can be used to achieve the most effective operational outcome, enabling the emphasis to be shifted from a focus on the number of platforms and weapons systems to a mission emphasis, where precision effects can be targeted at precisely the right time and right place to maximize the impact on the adversary.

As information warfare becomes increasingly important as a means of accurately understanding the disposition of friendly, enemy, and neutral forces, the emphasis is increasingly switching towards attacking the enemy's information systems, thereby preventing him from understanding the situation he faces and disrupting the ability to coordinate an effective response. If one considers the OODA loop, for example, it is easy to understand how the weaknesses in the information-dependent links and inputs of the decision-making processes can be exploited. Equally, the importance of protecting the OODA loop from enemy attack is self-evident.

Targeting priorities also change in an NEC environment as the threats associated with traditional targets are better understood, and the need to disrupt or destroy information-rich targets such as networks and communications centers move higher up the priority list; indeed, traditional targets of the mechanized warfare era such as tanks, aircraft, and ships can often be much more effectively disabled by attacking their links into C2 systems rather than relying on actual physical destruction.

Additionally, for targets traditionally termed *time-critical*, the concept of network-enabled warfare brings with it the dimension of persistent surveillance and where a target has been classified as such in the past, persistent surveillance removes or greatly extends the timeline within which a target must be engaged before a sensor loses track of its location.

NEC brings with it a fundamentally different approach to warfighting that enables assets to be linked together as a single entity, rather than deployed and fought as a collection of semi-independent assets. In a networked environment, the intelligence picture becomes the integrated and shared picture compiled from information gathered from across the network, and the military priorities become those of the network rather than those of the constituent elements. The key to a networked organization's power is therefore dependent upon how the capabilities of its constituent parts are organized and integrated together to develop synergies and the ease with which the network can self-synchronize to form a cohesive entity.

"It takes a network to defeat a network" [50].

Military options open to commanders in dealing with the engagement of threats will also increase as decision makers have at their disposal the additional abilities to attack networks and communication systems without having to launch overt military operations. This capability is known as a *soft kill* rather than a *hard kill* (see Appendix F), with hard kill referring to the physical destruction of targets. In an NEC environment the priorities will therefore switch to focus on networks, C2-related information systems, and communications nodes, rather than weapons systems or more visible forms of military power. The increased target set and means for disabling or exploiting it add a new dimension to the complexity of the military planning operation and the task of developing an appropriate threat response.

Network-Enabled Doctrine

Doctrine consists of the fundamental principles by which military forces guide their actions in support of national objectives.

- Land, sea, air, and space are separate domains requiring exploitation of different sets of physical laws to operate in, but are linked by the effects they can produce together.
- Doctrine is about effects, not platforms. This focuses on the desired outcome of a particular action, not on the system or weapon itself that provides the effect.
- Doctrine is about using mediums, not owning mediums. This illustrates the importance of properly using a medium to obtain the best warfighting effects, not of carving up the battlespace based on service or functional parochialism.
- Doctrine is about organization, not organizations. Modern warfare demands that disparate parts of different services, different nations, and even differing functions within a single service be brought together intelligently to achieve unity of effort and unity of command.
- Doctrine is about synergy, not segregation. True integration of effort cannot be achieved by merely carving up the battlespace.
- Doctrine is about integration, not just synchronization. Synchronization is "the arrangement of military actions in time, space, and purpose to produce maximum relative combat power at a decisive place and time." Integration, by comparison, is "the arrangement of military forces and their actions to create a force that operates by engaging as a whole." Synchronization is, in essence, deconfliction in time and space between different units.
- Doctrine is about what's important, not who's important.
- Doctrine is about the right force, not just equal shares of the force.
- Doctrine is about the need to share, not the need to know.

—Adapted from USAF Air Force Doctrine Document 1, November 17, 2003, and U.S. DoD Joint Publication (JP) 0-2

In an NEC environment, the destruction, disruption, or disabling of information-based assets provides an immediate and tangible operational payoff. The destruction of C2 networks separates forces from both receiving instructions and reporting on progress, leaving them as autonomous units at best and as ineffective and removed from the fight at worst. Targeting dedicated ISTAR assets and associated sensors removes the eyes and ears of the enemy, and targeting data links removes the ability for platforms to share and receive an accurate and timely situational awareness picture.

Other opportunities exist for degrading specific information-based assets for example jamming or deceiving navigation systems such as GPS (Figure 2.92) and Glonass, thereby feeding inaccurate information to ISTAR systems and platforms, many of which rely on accurate positioning information for functions such as targeting, reporting, and automatic takeoff and landing.

Other dedicated sensor platforms are also particularly vulnerable to attack and would be targeted early on in any conflict by an opponent who had such a capability. Such platforms increasingly act as airborne C2 nodes to manage the fight, process multiple sensor inputs from collaborating networks, and prioritize and allocate targets. Shooting down an Airborne Warning And Control System (AWACS) early in the fight, for example, would not only deny a wide area surveillance capability, but would at the same time remove a key C2 node from the fight. Soviet planners have long recognized the

Figure 2.92   A GPS-guided Joint Precision Air Drop System (JPADS) bundle, known as Screamer 2K, floats to the ground after being dropped from the back of a C-130 Hercules over Afghanistan in 2008. In this mission, over four bundles were dropped from an Alaska Air National Guard C-130 to resupply U.S. Army troops on the ground with ammunition and water.

strategic importance of these assets and believed that attacking them directly would limit NATO's ability to maintain an extended and coordinated air campaign. The Su-27 Flanker (Figure 2.93), as well as being a formidable air-superiority fighter, retains the role of a long-range interceptor being assigned to carry the Novator KS-172 AAM-L long-range anti-AWACS missile specifically for this role [17]

Similarly, SIGINT aircraft such as the RC-135 Rivet Joint (Figure 2.94) and Nimrod R1 and ground surveillance aircraft such as the JSTARS and ASTOR would represent high value information assets and are likely to be targeted early on in any major hostilities. Eliminating such important nodes from an information-dependent environment early on in hostilities could provide a decisive advantage to hostile forces, especially if they are able to preserve their own ISTAR capabilities from hostile attack or interference.

In a networked environment many different platforms will play important roles in establishing networks, processing and fusing ISR data, and allocating targets. Depending on the operational requirements and the platforms available, these assets may be airborne, land-based, or seaborne platforms. Networks specific to air, land, or sea operations frequently operate in parallel, sharing critical information on threats, priorities, and targeting, calling on assets from collaborating networks for intelligence updates or support for strike operations. For example, naval task forces may well coordinate expeditionary air operations by hosting a combined air operations

Figure 2.93   The Su-27 Flanker would be used early on in any conflict to target key airborne ISTAR assets and C2 nodes. (Photo courtesy of Sukhoi.)

Figure 2.94 An RC-135 Rivet Joint flies over Afghanistan, January 2007. The intelligence reconnaissance aircraft carries a crew of 34 and is configured to intercept and analyze a wide range of analog and digital communications signals.

center (CAOC) prior to a more permanent land-based facility being established, while air platforms such as JSTARS will frequently be used to direct ground forces to engage with enemy mobile targets (Figure 2.95).

For offensive engagements, dynamic retasking of sensors and effects systems across the network permits multiple engagements of a target using the most appropriate effects systems selected from across the network rather than reliance on the concept of a single platform and a single engagement (Figure 2.96).

NCW is essentially a three-step process often involving multiple interacting networks:

1. Using self-synchronization to achieve a shared situational awareness;

2. Using shared awareness to achieve a common understanding of the threats and disposition of forces;

3. The subsequent prioritization across the network of threats and allocation of assets to deal with those threats and to pursue the command intent. Such an approach leads to a dramatic increase in both agility and effectiveness for all participants on the network.

Figure 2.95 The E-8C Joint Surveillance Target Attack Radar System is the only airborne platform in operation that can maintain real-time surveillance over a corps-sized area of the battlefield. A joint U. S. Air Force–Army program, the Joint STARS uses a multimode side looking radar to detect, track, and classify moving ground vehicles in all conditions deep behind enemy lines.

The introduction of NEC capabilities demands a new approach to military tactics. While radios and data links have in the past 30 years or so connected limited groups of collaborating assets with varying degrees of effectiveness, it is only in recent years that true integrated networking has become possible, enabling timely and accurate shared situational awareness and the tasking of sensors and effects systems to reflect the unfolding priorities across the battlespace. Unlike traditional mechanized warfare, where asset capabilities, strengths, and dispositions are broadly known, modern warfare assumes much more complex patterns and dynamics involving multiple diverse assets and threats. The distribution of these assets takes on a much more random

Figure 2.96 Ships such as the U.S. Littoral Combat ship are increasingly capable as acting as central nodes in complex networked environments. Here the littoral combat ship *Independence* (LCS 2) is shown underway during builders' trials in July 2009.

and distributed pattern with decision making taking place across a network rather than being directed from a central point and radiating out to the outer edges of the formation. Where such engagements take place between opposing forces, the interactions between each network become increasingly complex as the independence of nodal actions results in a high level of unpredictability and increases to an almost infinite level the evolving patterns of the engagement.

As Von Moltke pointed out [51], "no plan survives first contact with the enemy." The rigid organizational structures of conventional warfare are ill-suited to today's dynamic engagement patterns where flexible and agile networks metamorphasize around each other as the engagement unfolds.

Within a large-scale networked environment, the options open to the commander are also almost infinite; a situation that creates opportunities and challenges in its own right. However, the ability to accurately understand the unfolding battle, to rapidly penetrate and disrupt C4I systems, and to disrupt the opposing force's supporting infrastructure also creates a potent combination of speed, accuracy, and agility. Network-enabled capabilities enable not only a more precise targeting of threats, but also a better understanding and identification of the nature of the threats (Figure 2.97). In urban warfare scenarios, for example, networked sensors are able to analyze information about patterns of movement of individuals and vehicles to form a *pattern of life* in order to compare normal patterns with activities that are out of the ordinary (Figures 2.98 through 2.100). Pattern of life analysis is based on the presence of the abnormal and the absence of the normal.

Such analysis requires persistent observation over a period of time that will provide meaningful patterns of life from which any deviation can be observed. ISTAR information provided for such purposes needs a standardized time and space reference to ensure that patterns can be accurately matched. Among the ISTAR data that can be used for such analysis is the following:

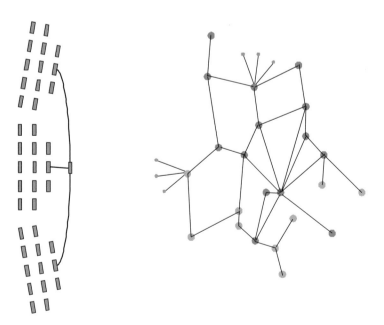

Figure 2.97   Conventional warfare versus network-enabled warfare.

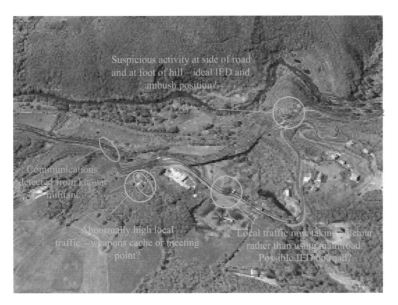

Figure 2.98 Pattern of life analysis can provide extremely useful intelligence activity; persistent surveillance and networked sensors are key to providing such information.

Figure 2.99 Pattern of life analysis is greatly enhanced by advances in automated mapping of ground moving targets. In this screenshot, an E8-C JSTARS aircraft maps moving targets using its GMTI radar overlaid on a background geographical map. The challenge now switches to processing the vast amount of data captured on such missions and identifying areas of abnormal activity for further investigation.

Figure 2.100   An A-10 Thunderbolt II ground attack aircraft performs a low-level strafing run with its 30-mm cannon during a combat search and rescue demonstration. The digitally upgraded A-10 has been equipped with satellite-guided precision weaponry and advanced network-based data links for sharing information with ground-based warfighters.

- GMTI: pattern analysis at various times of the day/week/season;
- SAR imagery: to show any changes to terrain or ground obstacles;
- Video imagery: visual reference to local activities;
- Still imagery: pixel-by-pixel analysis of accurately matched updates of air-to ground photos to show where the ground has been disturbed;
- HUMINT: intelligence reports from local sources;
- COMINT: pattern matching of communications transmissions;
- ELINT: radio communications patterns.

Networking opens up significant flexibility in strategic and tactical operations, allowing both proactive planning and faster reactive responses to unfolding events. The broad reach of intelligence information provided by a network of many C2 and ISTAR nodes can provide an essential framework for coordinated action across the battlespace, with the ability to rapidly adapt to new threats and priorities as the battle unfolds. Information gathering assets can be brought into the network or dropped as the battle evolves, providing resilience and a degree of immunity to hostile action taken by opposing forces to disrupt the network. Networking enables many more inputs to be considered in the decision-making process and significantly enhances the quality of the decision-making process, the quality of the decision itself, and the quality of the execution process (Figure 2.100).

Whereas in the past the quality of information was traded for speed, networking also permits more accurate and faster collection and distribution of data without the need to sacrifice quality. This is particularly important as the tempo of modern operations is significantly faster than in previous generations of warfare and decisions need to be taken based on near real-time information; in modern warfare, fractions of a second count.

Network-enabled concepts bring with them significant practical advantages to the warfighter. The ability to achieve seamless integration of data, by fusing the data from multiple diverse sensors, intelligence sources, and databases, provides a real revolution in situational awareness, which is at the heart of any effective military response.

As the quality of information and situational awareness improves, the time needed to recognize that a change in operational tactics is required reduces, while the opportunity to anticipate requirements ahead of time increases. The ability to process more information more rapidly enables specific command intent decisions to be made more frequently—translating into significantly improved agility and faster deployment of effects systems. Faster decision making, however, also requires greater integrity of information quality accompanied by a high degree of information assurance to guarantee that the decisions made at the end of the process have a high degree of integrity to ensure delivery of the desired outcome.

As more sensors and different sensor types become available across the network, the power of networked warfare and the ability to produce a coherent view of the battlespace will only be fully realized if a means is found to effectively integrate an order of magnitude more information than is the case today. Once the decision-making process has benefited from enhanced quality of information within the decision-making process, the challenge turns towards selecting and optimizing the appropriate response across a distributed networked environment.

In order to ensure coordination of purpose across the network, the commander's intent must be established and disseminated early in the engagement and updated to all nodes in the C2 network in a timely manner (Figure 2.101). Once the command intent has been established, all participants in the network share a joint understanding of the desired outcome, the key stages in the achievement of the outcome, and an ability to collaborate efficiently and effectively either individually or in collaboration with other sensor and effects systems.

Where a distinct network is engaged in achieving a discrete military objective, the ISR sensor commander needs to have the operational capability to maneuver, task, and prioritize the networked sensors to provide specific intelligence information at a time and location to suit the mission requirements.

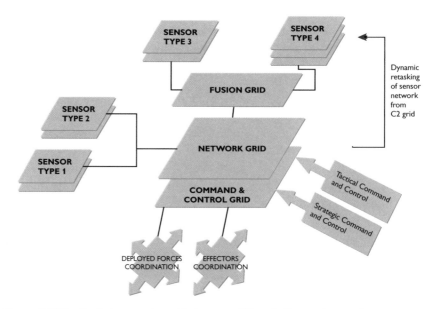

Figure 2.101   Typical network-enabled sensor, C2, and effectors integration.

Where military forces are working in a joint and networked environment, tasking of sensors should shift towards an emphasis on requesting specific intelligence, and the principle node determining ISTAR priorities should then prioritize—for the benefit of the entire network—the intelligence gathering or effects tasking sequence and subsequent dissemination of the intelligence picture (Figure 2.102).

Networked battlespace

The networked battlespace provides a step change in capability to the military commander—both at a strategic and operational level. It challenges the traditional hierarchical C2 structures and replaces them with a network where the span of control can be almost infinite, where all nodes have the ability to shape the priorities and outcomes in the direction of the commanders' intent rather than simply being tasked to achieve a specific objective, and where all can play a decisive role in an engagement. It supports rapid and better informed decision making, enabling C2 to be exercised without the constraints of organizational hierarchies, distance, terrain,or weather.

Networking of sensors, databases, command structures, decision makers, and effects systems will lead not only to improved shared and near real-time situational awareness but also to increased cooperation and unity of purpose between military units, coalition units, and their supporting networked

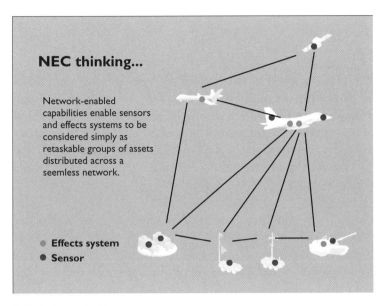

Figure 2.102   NEC thinking.

infrastructure, all aspects of which will contribute to much improved information superiority and subsequently superior decision superiority. The advantages of implementing a network-enabled force include a much improved perspective on the unfolding battle through improved control and timely feedback as to the progress and impact of command decisions, speed in the identification and targeting process, precision in the application of force and its associated timing, improved choice and suitability of effect, and improved force awareness and protection, resulting in improved combat effectiveness.

Such capabilities support the increasing need to conduct high-tempo, dispersed, and varied operations in order to gain the advantage over increasingly diverse threats and to counter the force disposition and tactics of opposing forces. The agility associated with such an approach is important in that it forces the opponents to engage at a time and place and on terms determined by friendly forces rather than to control the unfolding actions and the timing themselves. The ability to coordinate dispersed operations rapidly and effectively through the use of advanced information technology is therefore likely to be a central aspect of future combat operations [52]. In this task, networked operations play an essential role.

In summary, the benefits of NEC, which need to be addressed in operational planning and across all aspects of doctrine and training, can be summarized as follows.

Every sensor, effector, soldier, and platform can be considered as a node on the network

In traditional mechanized warfare, platforms were considered as the basic unit of deployment, and military capabilities were very much measured by the number and type of platforms available. In an NEC environment where information collection abilities are as important as the effects systems they support, the force strength becomes a more complex picture and capabilities can be measured by their value to the network rather than their value as a deployed platform.

Network-enabled technology permits every sensor and effector located on each soldier or platform to be considered as a node on the network, available for the benefit of the network as a whole. As military networking evolves to more open Internet Protocol (IP)–based approaches, every node will be capable of being assigned its own IP address and, in the case of electronically scanned radar, for example, may undertake several tasks in parallel, tracking one target while simultaneously scanning a sector for hostile forces (Figure 2.103).

Sensor and data fusion

Sensors and data are linked together to provide data fusion from multiple sources from sensors operating in different parts of the spectrum to provide multispectral threat recognition, leading to a higher degree of certainty in threat assessments and engagement decisions.

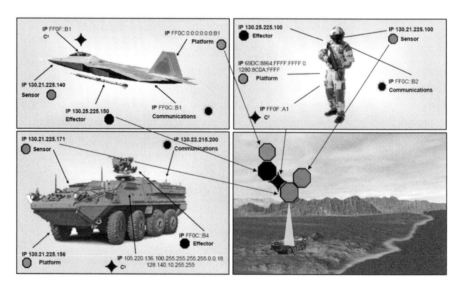

Figure 2.103 Internet analogy: every weapon, sensor, and platform has a new IP address. (Image courtesy of Raytheon.)

Increased tempo and speed of decision making

Linked with the improvement in accuracy of near real-time situational awareness and an integrated C2 network is the ability to make significant improvements in the speed of decision-making. Speed enables the ability to curtail enemy intentions well in advance of them being able to achieve their military objectives. The increased speed with which information can be shared across the battlespace means that battle management becomes significantly more dynamic in nature, with replanning during a mission reflecting the unfolding changes in the shared situational awareness picture and the ability to rapidly and effectively disseminate direction including the commander's intent (Figure 2.104).

Shared situational awareness

With improved and shared common situational awareness through near real-time information sharing and collaboration, users can relate the information to their particular situations and perspectives, draw common conclusions, make compatible decisions, and take appropriate action related to the overall situation [53]. Shared near real-time situational awareness (Figure 2.105) also greatly enhances survivability through knowledge of the disposition of enemy/friendly and neutral forces and reduces the chances of "blue-on-blue" fratricide (also called friendly fire) (Figure 2.106).

Figure 2.104 Controllers in the combined air operations center (CAOC) at an air base in Qatar monitor the status of ongoing missions supporting Operation Iraqi Freedom. The CAOC was the nerve center for all U.S. Central Command air operations when the first air strike occurred in Operation Iraqi Freedom on March 20, 2003. Cruise-missile attacks and the start of massive air operations with thousands of sorties a day followed this opening strike. By May 2, 2003, major combat was over and the stabilization phase of the operation began.

Figure 2.105   DARPA's TIGR (Tactical Ground Reporting system) is a multimedia reporting system for soldiers at the patrol level, allowing users to collect, log, and share information to improve situational awareness and to facilitate collaboration and information analysis. TIGR uses before and after photographs and updated imagery to manage the changing tactical landscape and to provide the most current views of the battlespace. The Google Earth and Wikipedia forum type of user interface allows multimedia data such as voice recordings, digital photos, and Global Positioning System (GPS) tracks to be easily collected and searched, enabling soldiers on patrol to map out routes, taking note of possible obstacles or high risk areas.

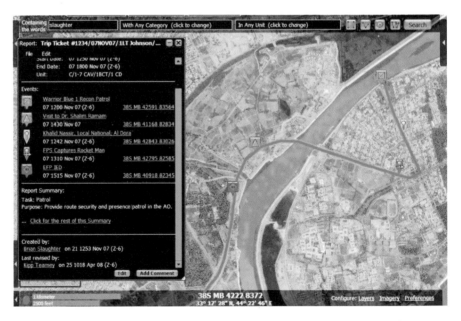

Figure 2.106   TIGR enables soldiers to see what routes they and other patrols have taken in the past and to record any areas of note or suspicious activity or to enable them to point a computer mouse at a point or building on a map and open a window to show the history of incidents at that particular location.

Decision superiority

Information superiority achieved through NEC capabilities delivers
significantly improved quality of information into the OODA loop
decision-making process. The improvements into the input of the process
result in significantly improved outputs measured by quality and
effectiveness of engagement or tasking decisions (Figure 2.107).

Collaborative engagement and crosscueing

A principal goal enabled by NEC is the near real-time synchronization of
both sensors and less dynamic data sources in a shared situational awareness
environment. Such capabilities enable a collaborative approach across a
network in detecting targets, prioritizing threats, and deploying the best
placed and most appropriate effects systems to deal with the threats. This
collaborative process dramatically improves the accuracy of target
identification, geolocation, and timeliness of information provided across the
network, with cumulative improvements in all of these areas often as high as
10 times the efficiency and accuracy provided by stand-alone systems. Given
the increasing collaboration in military environments with coalition forces, it
is increasingly important to be able to operate over common data exchange
standards. In this environment, for example, there are additional challenges
of multilevel security required to operate with coalition and allied forces,
some of whom will invariably be permitted access to more sensitive
information than others.

Figure 2.107   An MQ-9 Reaper unmanned aerial vehicle descends into an airfield in
Afghanistan in November 2007 after a mission. The Reaper is able to carry both
precision-guided bombs and Hellfire missiles. Integrating Reaper UAVs within a
network-enabled environment has allowed rapid action to be taken against time-critical
targets of opportunity.

Cross-cueing is an important part of a networked engagement process and involves one sensor handing over the track or engagement sequence to another sensor to maintain continuity across the kill chain (Figure 2.108).

Deployment of effects

Accurate situational awareness enables networked sensors to focus on improving threat prioritization and to deploy the effects systems available across the network to best match the appropriate response to the threat (Figure 2.109). Multiple and varied effects systems can then be put at the disposition of the network (rather than individual commanders) to support the engagement across a variety of effects requirements and timeliness. The lethality of effects is also enhanced through the speed with which the appropriate effect can be selected and brought to bear on the target and, accordingly, the chances of collateral damage minimized. Effects systems or netted weapons need to be designed from the outset to be integrated into a networked environment with many features being able to be selected up until the point of impact or delivery of the effect. The big jump in capability comes from turning current dedicated-control, single-purpose weapons into flexible service providers, limited only by the weapon's capabilities and by policy restrictions such as doctrine and rules of engagement [54]. Future effects systems will be instantly adaptable to meet the specific targeting requirements of the network and, as is increasingly the case, will be delivered from a platform that loiters within the target area awaiting a call for fire from the network commander.

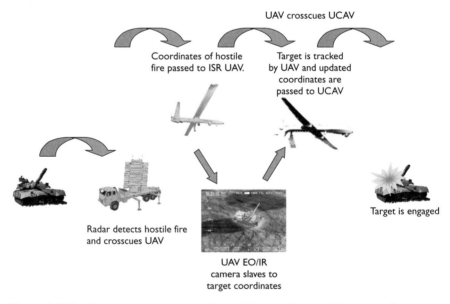

Figure 2.108 Crosscueing using network-enabled principles can deliver a significant improvement in combat effectiveness and speed of response.

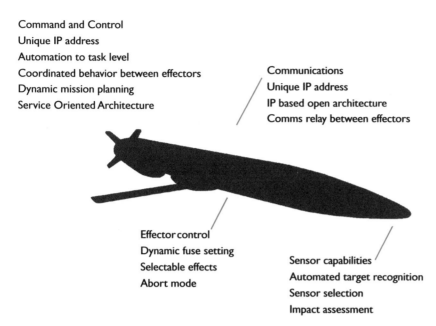

Command and Control
Unique IP address
Automation to task level
Coordinated behavior between effectors
Dynamic mission planning
Service Oriented Architecture

Communications
Unique IP address
IP based open architecture
Comms relay between effectors

Effector control
Dynamic fuse setting
Selectable effects
Abort mode

Sensor capabilities
Automated target recognition
Sensor selection
Impact assessment

Figure 2.109 Networked effects play an important role in capability delivery through the provision of net-ready weapons, acting as independent, versatile battlespace assets or as addressable elements of a control or launcher system.

NEC permits the network to be fought rather than individual elements

Because NEC links assets together across the battlespace as a connected network, the network itself becomes an entity, and information shared across the network enables priorities for the network to be addressed rather than those of the individual assets. Associated with this concept is the need to establish effective C2 principles to allocate ISR and effects assets.

A coordinated network, a shared situational awareness picture, and an embedded precision effects capability can offset many disadvantages of having a smaller or less well-equipped force by deploying and utilizing its own forces more effectively. Fighting the network as an entity enables a more rapid and effective domination of the information superiority landscape to deny the enemy knowledge of the intentions and actions of the smaller force and to strike with precision at the opponent's most vulnerable information-related nodes (Figure 2.110).

It is worth noting, however, that despite the advantages provided by NEC, sheer numbers will still generally defeat a networked force where the network cannot provide sufficient physical assets and effects systems to overcome the numbers disadvantage.

Figure 2.110   With the advent of NEC, virtually every asset can become a node on the network. Here a support vehicle for the Army Brigade Combat Team Modernization program is fitted with a situational awareness display.

The Networked Battlespace in Action—The U.S. Army Land Warrior Program and the Army Brigade Combat Team Modernization Program

The key enabler in effectively executing missions is soldier connectivity to and integration with the network and C2 structure. This connectivity provides the soldier with superior situational awareness, both mounted and dismounted, enabling the soldier to effectively perform battle command functions while maximizing soldier and force lethality and survivability.

Such a combination of network technologies enables the modern soldier to see first, understand first, acquire once, and target once, in effect delivering a combination of decision superiority and effects superiority.

Future soldiers will be equipped with hand-carried devices capable of enabling remote network interface to operate various unmanned systems, including UAVs Class I, all UGV, NLOS-LS, and T/U-UGS (Figure 2.111). Such devices will enable remote control of critical functions, and provide the dismounted soldier with key battle command functions to optimize his dismounted situational awareness.

Helmet subsystem
• Color SVGA helmet-mounted display
• Audio headset & microphone

Backplate
• Communications subsystem
• Enhanced computer subsystem
• Battery

Weapons subsystem
Integrates weapon-mounted sensors
such as Multifunction Laser, Daylight
Video Sight, Thermal Weapon Sight.
Weapon user interface device provides
control of weapon-mounted sensor
functions and voice communication
from the weapon.

Navigation subsystem
• Selective Availability/Antispoofing
  Module (SAASM) GPS & antenna
• Dead-reckoning module

Enhanced soldier control unit
• Embedded keyboard
• Integrated MBITR push-to-talk
• Analog audio interface

Figure 2.111 U.S. Land Warrior configuration. (Photo courtesy of U.S. Army IPEO Soldier.)

Rifleman's Radio

The latest U.S. Army network radio (Figure 2.112) can carry voice and data signals up to 2 km in open ground and 1 km in urban terrain, uses the high-bandwidth JTRS Soldier Radio Waveform (SRW), and transmits GPS location information. It will transmit only voice and limited data, not images and video, and will automatically pass location information so a soldier's position will show up on vehicle-mounted blue-force tracking displays. The rifleman's radio will be a smaller cousin to the JTRS HMS variant that will transmit images and video from forward sensors such as the UGS and SUGV.

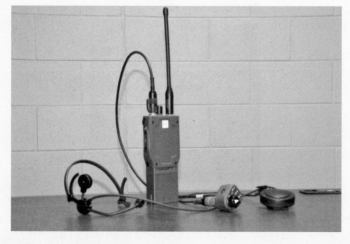

Figure 2.112 The AN/PRC-154 Rifleman Radio.

Unattended Ground Sensors (UGS)

The Unattended Ground Sensors (UGS) program (Figure 2.113) has, to date, been divided into two major subgroups of sensing systems:

- Tactical-UGS (T-UGS) (Figure 2.114), which includes: Intelligence, Surveillance and Reconnaissance (ISR)-UGS; Chemical, Biological, Radiological, and Nuclear (CBRN)-UGS; and Urban-UGS (U-UGS)[5].
- The unattended ground sensors (UGS) field will include multimode sensors for target detection, location, and classification; and an imaging capability for target identification.

(a)

(b)

Figure 2.113   (a, b) UGS systems can be hand-employed by soldiers or robotic vehicles either inside or outside of structures. In urban operations these systems are particularly suitable for monitoring choke points and high-risk areas such as corridors, stairwells, sewers, and tunnels.

Figure 2.114   A tactical unattended ground sensor.

The UGS system will also include a gateway node to provide sensor fusion and a long-haul interoperable communications capability for transmitting target or situational awareness information to a remote operator or to the common operating picture through the JTRS network. The UGS can be used to perform mission tasks such as perimeter defense, surveillance, target acquisition, and situational awareness (SA), including chemical, biological, radiological, nuclear, and high-yield explosive (CBRN) early warning

Class I Unmanned Aerial Vehicle (UAV)

The Class I Unmanned Aerial Vehicle (UAV) provides the dismounted soldier with reconnaissance, surveillance, and target acquisition (RSTA) (Figure 2.115). Estimated to weigh less than 20 kg, the air vehicle operates in complex urban and wooded terrains with a vertical takeoff and landing capability. It is interoperable with selected ground and air platforms and controlled by dismounted soldiers. The air vehicle also features an EO/IR/LD/LRF capability to perform the RSTA mission and uses autonomous flight and navigation systems to ease operator workload. It will also perform limited communications relay in restricted terrain, a tremendous deficit in current operations.

(a)

(b)

Figure 2.115 A Class 1 UAV takes off on a reconnaissance flight to search for a simulated improvised explosive device (IED).

Class IV Unmanned Aerial Vehicle (UAV)

The Class IV Unmanned Aerial Vehicle (UAV) supports the Brigade Combat Team and U.S. Naval Forces with communications relay, long endurance persistent stare, and wide area surveillance (Figure 2.116).

Figure 2.116 An RQ-8A Fire Scout Vertical Takeoff and Landing Tactical Unmanned Aerial Vehicle (VTUAV) System prepares to land aboard the amphibious transport dock ship USS *Nashville* (LPD 13). This is the first autonomous landing of the Fire Scout aboard a Navy vessel at sea. With an on-station endurance of over 4 hours, the Fire Scout system is capable of continuous operations, providing coverage at 110 nautical miles from the launch site. Utilizing a baseline payload that includes electro-optical/infrared sensors and a laser rangefinder/designator, Fire Scout can find and identify tactical targets, track and designate targets, accurately provide targeting data to strike platforms, employ precision weapons, and perform battle damage assessment.

Unique missions include dedicated manned and unmanned teaming (MUM) with manned aviation, wideband communications relay, and standoff chemical, biological, radiological, and nuclear (CBRN) detection with on-board processing. Additionally, it has the payload to enhance the reconnaissance, surveillance, and target acquisition (RSTA) capability by crosscueing multiple sensors.

Small Unmanned Ground Vehicle (SUGV)

The small unmanned ground vehicle (SUGV) (Figure 2.117) is a small, lightweight, man-portable UGV capable of conducting military operations in urban terrain, tunnels, sewers, and caves.

The SUGV and its multiple payloads are an aid in intensive or high-risk functions such as urban intelligence, surveillance, and reconnaissance (ISR) missions or investigating chemical and toxic materials without exposing soldiers directly to the hazard.

(a)

(b)

Figure 2.117   (a, b) The SUGV can be carried as a backpack and is ideal for identification, surveillance, target acquisition, and reconnaisance missions.

Networked logistics support

NEC permits the status of units to be shared across the network, enabling more efficient logistics planning and resupply of operational units.

Network-enabled coalition warfare

NEC is particularly important in the context of international coalition operations where shared interoperability is essential if joint operations are to be undertaken. A good example of this was the first Gulf War, where American and British forces were able to operate together in the opening stages of the war and during the period of active offensive operations, while other nationalities, due primarily to their limited ability to interface with U.S. and U.K. C2 systems, were reduced to support roles or subsequent peacekeeping roles, where NEC requirements were of a lesser priority.

Interoperability comprises of more than just networked technologies; national policies, tactics, techniques, procedures, and service culture must also be appropriately aligned. In this context interoperability can be defined in a number of ways: at an operational level, where interoperability is about national policy, operational doctrine, releasability, and procedures, and at a technology level, where interoperability is about common or compatible systems, standards, hardware, and software [55].

Coalition leadership is often determined by political rather than military considerations. This in itself makes coalition operations more of challenge, as various nations need to work together despite widespread differences in interoperability standards and equipment capabilities required to support their common political and defense aspirations.

Within a coalition context, planning interoperability needs to consider whether systems will coexist, whether they will exchange data between themselves, or whether they will form part of an integrated network system [56].

Given the diversity of capabilities associated with coalitions, specific coalition interoperability issues need to be mapped against these considerations and may include [57]:

1. Maintaining secure interoperability.

2. Command system interfaces:

    • Information, data and service description languages;

- Information and service exchange mechanisms;
- Structures for information and system management;
- Management interoperability gateways.

3. Coalition common operating picture.

4. Communications interoperability.

5. Command system adaptability and management with low performance communications.

While there are no standard groupings that comprise coalition forces, there are several frameworks that typically bring together coalitions of the willing. The more common Western coalitions are shown in Appendix A.

At a more tactical level, collaborating forces will aim to establish a common mission network to ensure that units in the field are able to effectively collaborate for a precise aim or time period often to ensure the delivery of a specific mission or capability.

Now that the direct conventional military threat to Europe and the United States has receded, and overt state versus state conflict tends to be limited to relatively short periods of high intensity activity, there is now an increasing emphasis on the need to support ever more collaborative and wide-ranging operations from counterasymmetric warfare to peacekeeping and humanitarian relief. As a result of these diverse requirements, operational scenarios are becoming more complex and unpredictable and can generally be classified into one of three categories: deliberate intervention, asymmetric warfare, including counterterrorist action, and peacekeeping, including stabilization operations (Figure 2.118).

Recent global conflicts such as those in Kosovo, Iraq, and Afghanistan have been characterized by the participation of a wide range of multinational forces. Each of those forces will be equipped with C2 systems and networks that are specific to the perceived needs of those countries both from a technological perspective and a security perspective (Figures 2.119 and 2.120).

Although it may technically be possible to connect many of the C2 systems and networks together or even make the information available to participating nations, many systems have been tailored to contain classified data only available to a particular nation or coalition partner. As multiple standards and data exchange formats move towards IP-based open standards, the problem is becoming largely one of trust and politics between nations rather than incompatibility between systems. This challenge is particularly

Figure 2.118 An F/A-18C Hornet assigned to the Stingers of Strike Fighter Squadron (VFA) 113 refuels with a British Royal Air Force L-1011 aircraft in southern Afghanistan.

Figure 2.119 July 2007: A French Rafale M combat aircraft performs a catapult-assisted launch from the flight deck of the nuclear-powered aircraft carrier USS *Enterprise* (CVN 65).

acute where key intelligence systems containing information ideally required by all participants are hosted by a nation using a system which at worst is restricted to use by its own nationals and which at best security constraints do not permit information to be shared outside a group of favored nations,

Figure 2.120 Three F-111s from RAAF Base Amberley fly alongside a Republic of Singapore Air Force KC135 during refueling operations of the coast of Darwin. As part of air operations during Exercise Pitch Black 04, the Republic of Singapore Air Force KC-135 air-to-air refueler supplies vital aviation fuel to participating Air Forces off the coast of Darwin, Australia. Coalition operations not just involve the physical integration of assets, but increasingly require that those assets form part of the coalition sensor and effector network.

often thereby excluding full contributions to be made by many other willing participants, including nonmilitary forces such as nongovernmental organizations (NGOs).

Politics aside, the wide range of coalition missions can better be supported by NEC technology providing a secure and reconfigurable networked structure rather than by traditional rigid military structures. In these scenarios, electronic combat systems and their associated networks need to operate in support of achieving the objectives associated with each coalition outcome, recognizing the significant variations in intensity of conflict, the degree of information dominance required, and the roles that those systems must play.

In coalition operations, networked interoperability is particularly important for:

- Rapidly achieving a shared situational awareness picture;
- Sharing intelligence from diverse sources about an adversary's capabilities, systems, methods, and decision-making processes;
- Achieving unity of command intent through the use of interoperable C2 systems to effectively coordinate deployed forces and to ensure reachback to national intelligence and governmental networks;

- Avoidance of mistaking coalition forces for the enemy or misidentifying enemy forces as friendly.

The ability to develop and maintain a common command intent, a shared situational awareness to understand and exploit the available resources, and the ability to coordinate diverse coalition forces require a level of information exchange that can interface effectively with multiple systems and networks, a coalition-based planning process where all partners may participate, a common concept of operations, and a set of compatible C2 procedures to implement and report on joint-force operations. Such interaction across multiple networks demands significant advances in interoperability, security protocols, and organization to deliver the right data and information at the right time to coalition commanders across extended networks, to use that data and information to achieve information and decision superiority, and to ensure coordinated planning across the network and efficient allocation of assets to ISR and targeting operations (Figures 2.121 and 2.122).

Coalition information networks

Joint operations involving allied and coalition forces have driven the development of various programs to support the exchange of classified information between the United States and its allies using high, medium, and low bandwidth data rates.

Figure 2.121 Japan Maritime Self Defense Force (JMSDF) ship JS *Haruna* (DDH 141), USS *Lake Champlain* (CG 57), and USS *Russell* (DDG 59) steam in formation between the Ronald Reagan Carrier Strike Group and JMSDF. (U.S. Navy photo.)

Figure 2.122   Four Royal Danish Air Force pilots fly F-16 Fighting Falcons in formation behind a USAF 100th Air Refueling Wing KC-135 Stratotanker during a refuelling mission over Denmark, December 2009.

For the controlled sharing of information, two fundamental principles are generally established:

1. **Definition of rules:** Specifying who can access what information under which conditions.

2. **Enforcement of rules:** Ensuring that information dissemination conforms to the rules.

Information access rules are derived from a combination of factors including mission type, nationality, and situation. In environments where coalition forces (as well as NGOs and civil authorities) are working together, the definition of these rules (or policies) in a way that can be rapidly adapted to ongoing situations is difficult. Issues include inconsistent definitions of impact levels and protective markings, variations in clearance levels, and how to dynamically assess the risks and benefits associated with the sharing of a piece of information.

Currently, static rule sets need to be augmented with dynamic rule sets that can be identified and assessed against risks as they emerge. Two key innovations for this style of operation are semantic labeling[6] of data and policies that define how labeled data can be combined and transformed. For example, downgrading an image resolution might mean that the image can be shared with an NGO, or providing less precise location information about a sighting of insurgents might allow this information to be shared more widely [58].

High-bandwidth communications are generally provided either by microwave or fixed ground stations, or for mobile platforms via satellite communications, proving access to the Non-Secure Internet Protocol Router Network (NIPRNET)/Secret Internet Protocol Router Network (SIPRNET). Access is facilitated through high assurance firewalls and multilevel Web servers. Message traffic is directed through the supporting U.S. network operations center (NOC) where it is processed and transmitted to an Allied communications technical control facility for transmission into a coalition nation's secure networks (Figure 2.123).

Classified information can be exchanged at a medium data rate using networks such as the coalition-wide area network (CWAN) dedicated for 5-eyes[7] data exchange, with sensitive data being communicated over the

Figure 2.123   HMAS ANZAC conducts Officer of the Watch Maneuvers with South African Ship, SAS *SPIOENKOP* off the Cape of Good Hope. Pictured is SAS *SPIOENKOP* with HMAS *ANZAC*'s Seahawk Helicopter flying above.

NATO Secret WAN (NATO SWAN), which supports NATO activities, such as exercises, operations, and logistics planning (Figure 2.124).

Encryption standards supporting classified information exchange must also be common to coalition networks with standard encryption keys shared between partners enabling access to the appropriate networks and levels of information sharing.

Low-bandwidth communications generally involve access to the SIPRNET/NIPRNET via a U.S. NOC, which provides information assurance before transmission over a coalition, a tactical wide area network. Tactical networks generally rely on relatively low-bandwidth bearers such as HF and UHF radio networks and may also access higher-bandwidth communications nodes across an extended communications infrastructure.

In addition to these fixed communications networks, most ISTAR systems exchange data through a series of NATO standard interoperable data links such as Link 11 and Link 16 to share a near real-time situational awareness picture of friendly and opposing forces.

Increasingly interoperability between coalition forces is being made easier through the exploitation of software-defined radios and communication

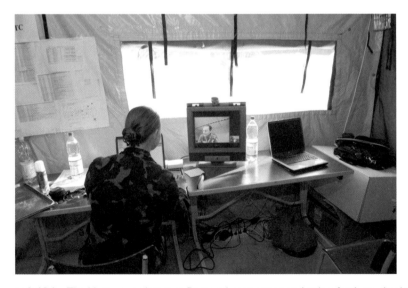

Figure 2.124 The Voice-over-Internet Protocol connection and video feed are checked during the testing phase of exercise Combined Endeavor 2007 in Lager Aulenbach, Baumholder, Germany. The exercise brings the United States and other nations together, especially those in NATO or Partnership for Peace, to plan and execute interoperability testing of command, control, communications, and computer systems in preparation for future combined humanitarian, peacekeeping, and disaster relief operations.

systems where encrypted waveform interoperability can be achieved through programmable software algorithms embedded in software defined radios (SDRs) that allow adaptation and interoperability with most military and civil radio waveforms.

In terms of military capability and effectiveness, it is also important to remember that different allies will bring with them dissimilar capabilities that, when combined across a network, may provide a disproportionate benefit to the overall capabilities of the network due to their diversity and unique capabilities.

Although network protocols may be common between a group of coalition partners, there is also a need to standardize the definition and interpretation of the information shared across the networks. For example, if one partner defines a track as unknown using its own definition and passes it to another partner on the network, it may subsequently be displayed as hostile or friendly if that partner does not have the same definition of the same track characteristics.

In summary, military advantage in coalition operations can only be achieved by an efficient fusion of information systems, command structures, and intent across participating forces. Mission objectives that exploit and integrate the diverse ISR and targeting capabilities from across the coalition partners are at the heart of a successful military intervention if speed, confusion, and disruption of key information-related systems are to be used to achieve information superiority and ensure an early tactical advantage. Electronic combat systems are indispensable across all stages of the information superiority battle, in particular for gathering intelligence information, prioritizing offensive and defensive operations, and actively disrupting enemy activity. In coalition operations it is therefore essential to achieve interoperability between those systems if early effect and maximum advantage is to be achieved.

The impact of NEC on sensor performance

Improvements in stand-alone sensor performance have traditionally concentrated on improving sensitivity, range and resolution. A different dimension to sensor performance can be achieved through a NEC approach, delivering a step change in the quality of the output through the integration of sensors into a networked environment (Figure 2.125). Significant improvements in the quality of sensor-derived information such as detection thresholds, accuracy, and certainty can be achieved by overcoming the limitations of individual sensors through the use of data fusion—and at relatively little cost compared to the sensors themselves.

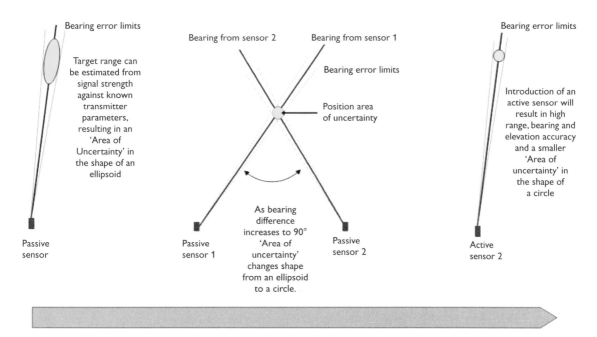

Figure 2.125 Use of passive networked sensors and the introduction of active sensors rapidly improve geolocation accuracy.

The performance advantage realized by networking sensors is determined by the function of the type of sensors being employed (e.g., active or passive, imaging or sensing, frequency sensitivity), the type of objects or target being detected or analyzed (e.g., missiles, aircraft, vehicles, ships, terrain, and buildings), and the environment in which the sensors are operating (e.g., weather, terrain, and background noise).

The challenges in integrating sensor performance are those associated with the need to translate sensor data into a common data set, irrespective of the sensor from which the data was collected and the need to integrate the data into an accurate space/time framework shared between all contributing sensor sources. The integration of data from dissimilar sensors such as the need to match infrared or visual imagery from a target with radar returns and the correlation of that information to assist in target identification can be technically challenging.

Despite these challenges, networked sensors can generate significant improvements in the accuracy and timeliness of the situational awareness picture across the battlespace, especially where additional sensors can be tasked from across the network to support target detection and identification.

The operational performance of a sensor network (a collection of networked sensor nodes) in generating improved battlespace awareness depends upon a number of factors, including [2]:

- The performance of individual sensors;
- The translation of sensor data into a form that can be integrated with other sensors;
- The ability of different sensors to image the same target;
- Suitable sensor geometry, including the locations of the sensors with respect to each other and the objects of interest;
- The velocity of information output from the sensor;
- Fusion algorithms;
- The operation of sensors within accurate space and time reference grids;
- The ability of the network to connect and task sensors and to transmit and receive potentially large volumes of information.

In a networked environment the ability to fuse data from multiple sensor sources and distribute the fused output to collaborating forces greatly enhances the decision-making processes and effective combat power across the network (Figure 2.126).

Fusion of active sensors can provide an improved probability of detection through comparison of weak signals at detection thresholds. Improvements in track and bearing accuracy can also be achieved, while the characteristics of the target itself viewed from a different angle will help establish early on its positive identity (Figure 2.127).

Fusion of multispectral data from different sensor types can also provide an improved probability of detection and identification through comparison of target characteristics at different frequencies (radio frequencies, millimeter wave, infrared, visible regions, ultraviolet); multisensor data fusion in particular can greatly assist with the location, identification, and tracking with sufficient accuracy to support engagement decisions.[8]

In order to benefit from the integration of multispectral data into a common framework, it is essential that relative positions of targets are referenced against a common space/time framework (Figure 2.128), which in turn means that the geolocation and time data of the platform against which the time data, range, and three-dimensional bearing of the target is being measured must be extremely accurate.

Dynamic tasking and coordination of networked sensors

Network ISTAR performance can be enhanced by dynamically tasking available sensors in accordance with network priorities to suit either ISR

Battlefield targets can be detected using multiple sensor techniques—the more that are used, the greater degree of certainty of identification of the target and its intentions. Networked sensor fusion enables sensors across multiple platforms to corroborate inputs to aid in the speed and accuracy of this process.

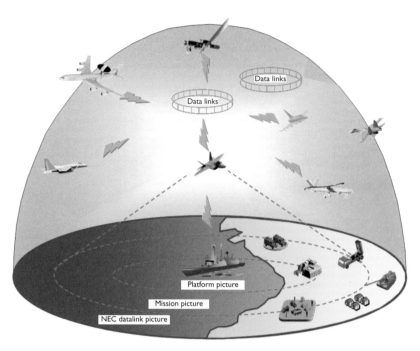

Figure 2.126 A network-enabled approach uses multiple networked data links to share sensor information, to cue effects systems, and to build a shared real-time situational awareness picture. The fighting entity becomes the extended network rather than the individual platforms or systems. The sensor data collection priorities and targeting priorities are those of the extended network rather than the platforms on which the sensors and effects systems are housed, and many functionally different networks may interact to provide this capability.

Information Quality Needed to Engage a Target

A high degree of certainty is needed to engage a target and requires accuracy and certainty in the following areas:

- Positive identification of target if targeting is preplanned: matching target data against multispectral libraries;
- Positive identification as "hostile" from at least 2 sources if targeting is unplanned: visual, IR, radar, NCTR, IFF, NCCT/CEC identification, ELINT library matching, human intelligence;
- Geolocation: to pass precise target location to effects system; greater accuracy required than for tracking;
- Track and velocity vector: to pass precise target engagement vectors to effects system;
- Safe engagement zone to avoid collateral damage: known disposition of blue/red/neutral forces;
- Confirmation of command intent and engagement authorization.

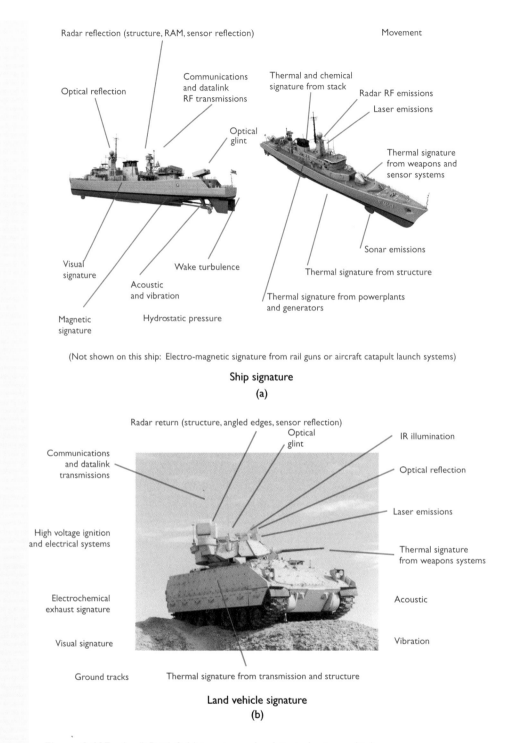

Figure 2.127 (a–d) Battlefield targets can be detected using multiple sensor techniques—the more that are used, the greater degree of certainty of identification of the target and its intentions. Networked sensor fusion enables sensors across mulitple platforms to corroborate inputs to aid in the speed and accuracy this process.

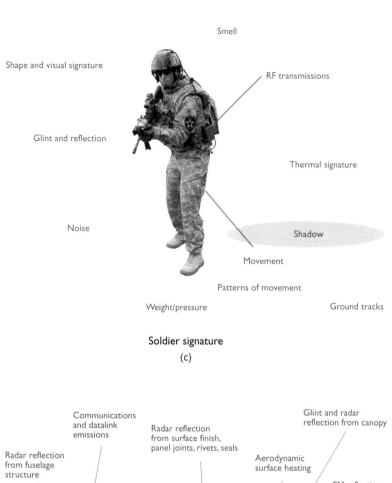

Smell

Shape and visual signature

RF transmissions

Glint and reflection

Thermal signature

Noise

Shadow

Movement

Patterns of movement

Weight/pressure

Ground tracks

Soldier signature
(c)

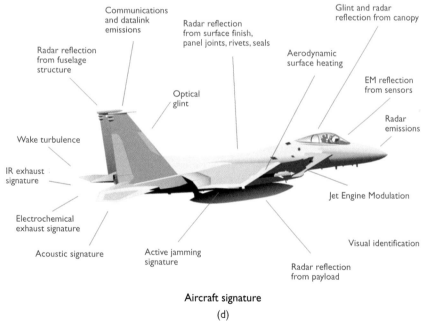

Communications
and datalink
emissions

Radar reflection
from surface finish,
panel joints, rivets, seals

Glint and radar
reflection from canopy

Radar reflection
from fuselage
structure

Aerodynamic
surface heating

Optical
glint

EM reflection
from sensors

Wake turbulence

Radar
emissions

IR exhaust
signature

Electrochemical
exhaust signature

Jet Engine Modulation

Acoustic signature

Active jamming
signature

Visual identification

Radar reflection
from payload

Aircraft signature
(d)

Figure 127 (continued)

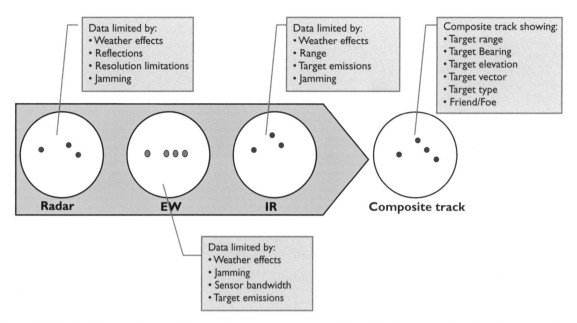

Figure 2.128   Combining multispectral sensor data can greatly assist with the identification and tracking of a range of targets, especially at low detection thresholds and in adverse conditions.

missions or to support targeting operations. Sectors within the battlespace that represent high areas or interest or sectors from which enemy forces may appear can also be monitored more effectively through a dynamic tasking approach. This enables limited sensor resources to support many assets with the needs of the network taking priority over the needs of individual assets.

Using networked sensors for target identification

Where sensors operate on their own their performance is limited by their inherent design limitations such as the target detection characteristics of the specific frequency in which they operate, detection ranges, their spatial coverage, and scan rates. By networking sensors together, particularly sensors of different types, these limitations can be significantly reduced, particularly in challenging environments where unintentional interference or intentional jamming on sensor frequencies interferes with the requisite performance.

Using networked sensors for target engagement

The quality of information provided by the ISTAR network is particularly important when it comes to engagement decisions. Confirming an engagement decision requires a significantly better quality of information than establishing the presence of a target or its associated track data.

Fusion of multiple passive sensors can enhance the bearing and geolocation accuracy together with extending the spatial coverage available to those connected to the network. Passive sensors will typically only provide bearing and elevation data about a target. However, by using fusion techniques relying on the triangulation of bearing data from multiple sensors, the target range can also be determined and a track established. Where sensors are used for targeting as well as intelligence collection, increasing the numbers of sensors that are used for triangulation much reduces the error ellipsoids associated with triangulation plots to a level that can permit positive identification and accurate target location data to be generated. This in turn permits engagement decisions from passive sensors to be made more quickly and with greater confidence than is possible from stand-alone sensors.

Active radar sensors and passive SIGINT receivers provide significantly increased performance when networked together to provide triangulation information. Linear estimates of range and bearing from radar sensors with errors measured in hundreds of meters can be triangulated to provide data accurate enough to track moving targets, and similar principles using TDOA and FDOA for ESM receivers can achieve similar degrees of accuracy when tracking hostile emitter signals (Figure 2.129).

As these concepts are developed, network techniques are also improving that allow the degradation of the sensor network without significant reduction in network performance, where if one sensor fails or is degraded, others will switch their functionality to maintain, or at least maximize, the network performance in accordance with intelligence gathering priorities.

In order to achieve these improvements through networked performance, it is clearly essential if one is relying on these levels of accuracy to ensure that the network itself is reliable, can provide the required bandwidth, and is immune or resistant to degradation and jamming.

Implications for C2/asset control/ownership

Ideally, networked sensors are employed to address intelligence gathering on the highest priority targets across the network with control of sensors, weapons systems, and target prioritization being the domain of the network itself rather that of single platforms.

The spatial coverage provided by a sensor network is very much dependent upon the number of connected sensors, and this requirement opens interesting possibilities in the use of increasing numbers of cheap networked sensors to provide significant improvements in geospatial coverage. Wide sensor coverage may be particularly beneficial during periods of low tension

Figure 2.129 ESM sensor effectiveness is significantly enhanced through networked capability. Here a Thales passive ESM sensor forms a node on a sensor network to triangulate and integrate sensor readings across a large geographical area. (Photo courtesy of Thales.)

where monitoring rather than precise target acquisition is required. During hostilities, more expensive sensors and sensor networks with enhanced sensitivity, jamming resistance, and improved overall performance characteristics can supplement the network to provide the detailed intelligence picture, especially in areas of particular interest where targets have been detected but not identified by the cheaper, less capable sensors.

The concept of networked sensors will in itself open up new uses for a range of multispectral sensors and the platforms on which they are deployed. The increasing proliferation of unmanned air, land, and sea platforms will provide significantly increased options for the deployment of relatively inexpensive sensors, which can be rapidly deployed in a relatively covert manner to monitor an area of interest. The inevitable demise of manned platforms from the battlespace will also significantly increase the persistence and range available to the commander in the field when making his choice about which platforms to deploy. As with sensor integration, however, the utility of unmanned platforms in hostile environments is very much dependent on their ability to operate on an autonomous basis with a high degree of immunity from jamming and degradation.

Traditionally, sensors have been constrained to the platforms on which they are mounted such as combat aircraft or ships. Given the advantages enabled by networks, the optimum solution for sensors could be in locations mounted away from the platforms in positions that could provide maximum spatial coverage and awareness of the area of interest, while being relatively unconstrained by the platform's spatial position or mission. Where the platform has been tasked with a specific mission, sensor information may be called upon from across the network to support critical mission-related information requirements during key phases. There are, however, plenty of opportunities for the sensors to support the requirements of other missions when they are not required for critical aspects of the platform's primary mission.

References

[1]   "Network Enabled Capability," U.K. MoD pamphlet V1, April 2004.

[2]   Alberts, D., J. Garstka, and F. Stein, "Network Centric Warfare—Developing and Leveraging Information Superiority," DoD C4ISR Cooperative Research Program, CCRP Publication Series, February 2000.

[3]   Boyd, Cameron, Warren Williams, Daniel Skinner, and Shaun Wilson, "A Comparison of Approaches to Assessing Network-Centric Warfare (NCW) Concept Implementation," *SETE Conference*, 2005.

[4]   U.S. Naval Network Warfare Command, "FORCEnet: A Functional Concept for the 21st Century," February 2005 (CNO and CMC).

[5]   Alberts, David S., and Richard E. Hayes, *Power to the Edge: Command and Control in the Information Age*, CCRP Publications Series, 2005.

[6]   Vane, R., et al., *Urban Sunrise*, AFRL-IF-RS-TR-2004-22 Final Technical Report February 2004.

[7]   Llinas, James, "New Challenges for Defining Fusion Requirements," *Center for Multisource Information Fusion, University at Buffalo, and CUBRC, MSU/Army Research Laboratory Knowledge Management Centre of Excellence Workshop*, October 2006.

[8]   Haffa, Jr., Robert P., and Jasper Welch, "Command and Control Arrangements for the Attack of Time-Sensitive Targets," Northrop Grumman Analysis Center Papers, November 2005.

[9]   Cares, Jeff, *Distributed Networked Operations: The Foundations of Network Centric Warfare*, Alidale Press, 2005.

[10] Burke, Martin, "Information Superiority Is Insufficient to Win in Network Centric Warfare," Joint Systems Branch, Defence Science and Technology Organisation, Salisbury, Australia, 2000.

[11] Leavitt, Harold J., and Homa Bahrami, *Managerial Psychology: Managing Behavior in Organizations*, Chicago, IL: University of Chicago Press, 1988, pp. 208–216.

[12] Leland, Joe, and Isaac Porche III, "Future Army Bandwidth Needs and Capabilities," prepared for the United States Army, RAND Corporation, Arroyo Center, 2004.

[13] http://en.wikipedia.org/wiki/Ad_hoc_protocols_implementations.

[14] Lorincz, Konrad, "HyperCast: A Super-Scalable Many-to-Many Multicast Protocol for Distributed Internet Applications," Faculty of the School of Engineering and Applied Science, University of Virginia, October 1, 2001.

[15] Watts, Duncan J., *Small Worlds: The Dynamics of Networks Between Order and Randomness*, Princeton, NJ: Princeton University Press, 2003.

[16] Ignetik, Rainer, and Paul Burn, "Application of Metcalfe's Internet Value Law to NCW Battlespace Communications Networks," IDL Society, Australia & New Zealand Chapter.

[17] Kopp, Carlo, "Understanding Network Centric Warfare," *Australian Aviation*, January/February 2005.

[18] Liepman, Skip, "Integrated Network of Combat Capability Will Create a New, Network-Centric Approach to Air Warfare and Achieving Battlespace Effects," *Military Information Technology*, Online Edition, Doc. 574, 2008.

[19] Alberts, David S., and Richard E. Hayes, *Understanding Command & Control*, CCRP Publication Series, 2006.

[20] Hypercast 2.0 Design Document, Department of Computer Science, University of Virginia, January 2002.

[21] Milgram, Stanley, "The Small World Problem," *Psychology Today*, Vol. 2, 1967, pp. 60–67.

[22] United States Government Accountability Office, *GAO Report to Congressional Committees, Defense Acquisitions, Significant Challenges Ahead in Developing and Demonstrating Future Combat System's Network and Software*, GAO-08-409, March 2008.

[23] Hayes-Roth, Rick, "Event Processing in the Global Information Grid (GIG): Orders of Magnitude Advantage in Information Supply Chains Through Context-Sensitive Smart Push ('VIRT')," professor, information sciences, Naval Postgraduate School, Monterey, CA. November 2006.

[24] "Data Sharing in a Net-Centric Department of Defense," U.S. Department of Defense Directive (DoDD), Number, 8320.2, December 2, 2004.

[25] U.S. DoD Joint Command & Control Doctrine Study, February 1, 1999.

[26] Leedom, Dennis K., and Robert G. Eggleston, "Modeling the Construction of Actionable Knowledge within an Effects-Based Targeting Process," Conference paper, April 1, 2005.

[27] Network Enabled Capability, JSP777, UK MoD, January 2005.

[28] U.S. Army War College, *Information Operations Primer, Department of Military Strategy, Planning, and Operations*, November 2006, AY07 Edition.

[29] "Irregular Warfare," USAF Doctrine Document 2-3, August 1, 2007.

[30] U.S. DoD, JP 2-01 Joint and National Intelligence Support to Military Operations, October 2004.

[31] Rose, John, Major General, "Defence Intelligence: Meeting the Challenge," *RUSI Conference*, October 2007.

[32] U.S. DoD, Interagency OPSEC Support Staff (IOSS), "Section 2, Intelligence Collection Activities and Disciplines," Operations Security Intelligence Threat Handbook, May 1996, http://www.fas.org/irp/nsa/ioss/threat96/part02.htm.

[33] U.S. Army Field Manual 2-0, Intelligence, Chapter 9: Measurement and Signals Intelligence, Department of the Army, May 2004.

[34] Preiss, M., and N. J. S. Stacy, *Coherent Change Detection: Theoretical Description and Experimental Results*, Report number: Australian DoD DSTO-TR-1851, AR-013-634, August 2006, http://hdl.handle.net/1947/4410.

[35] U.K. MoD source, http://www.army.mod.uk/intelligence/role/4217.aspx.

[36] JP 2-01 Joint and National Intelligence Support to Military Operations, October 2004.

[37] Rau, Russell A., Assistant Auditor General, Defense Intelligence Agency, *Evaluation Report on Measurement and Signature Intelligence*, June 30, 1997.

[38] Kiczuk, Bill, Raytheon Publication, *Technology Today*, Issue 1, 2009.

[39] U.S. DoD Joint Publication (JP) 1-02.

[40] Garstka, John J., Joint Staff Directorate for C4 Systems, "Network Centric Warfare: An Overview of Emerging Theory," U.S. DoD, December 2000.

[41] Joint IO Center, "History and Evolution of IO," January 2000.

[42] "Entering the Dragon's Lair," Rand study, 2007.

[43] U.S. DoD Information Operations Roadmap, October 2003, Declassified.

[44] Joint Publication 3-13, Information Operations, February 2006.

[45] Alberts, David S., et al., *Understanding Information Age Warfare*, CCRP Publications Series, 2001, p. 106.

[46] *U.S. Army Information Operations: Doctrine, Tactics, Techniques, and Procedures Field Manual*, No. 3-13, November 2003.

[47] U.S. DoD Army Doctrine, FM 3-90.

[48] www.af.mil.

[49] "USAF Information Operations," Air Force Doctrine Document 2-5, January 11, 2005.

[50] Apparhurai, James, North Atlantic Council Press Briefing, November 25, 2009, as quoted in *Janes Defence Weekly*, December 2, 2009.

[51] Paraphrased from the writings by Prussian Field Marshal Helmuth von Moltke (1800–1891) in Tsouras, Peter G., *The Greenhill Dictionary of Military Quotations*, London, U.K.: Greenhill Books, 2004, p. 363.

[52] Pirnie, Bruce R., Alan Vick, Adam Grissom, Karl P. Mueller, and David T. Orletsky, "Beyond Close Air Support Forging a New Air-Ground Partnership," RAND Corporation, 2005.

[53] Department of Defense, Global Information Grid Architectural Vision, Vision for a Net-Centric, Service-Oriented DoD Enterprise, Version 1.0, June 2007, Prepared by the DoD CIO

[54] Lail, Bryan, *Raytheon Technology Today*, Issue 2, Internal publication, 2008.

[55] Kohut, John N., "Providing Transformational C4I Capabilities to the Joint Warfighter," Program Executive Officer, PEO C4I and Space, May 18, 2004.

[56] Gentleman, W. Morven, "Planning for Interoperability," *Global Information Networking Institute, Halifax, Canada, RTO Information Systems Technology Panel (IST) Symposium*, Quebec, Canada, May 28–30, 2001.

[57] "Information Management Challenges in Achieving Coalition Interoperability," *NATO Research and Technology Organisation, RTO Information Systems Technology Panel (IST) Symposium*, Quebec, Canada, May 28–30, 2001.

[58] McDonald, Andrew, and Eleanor Hepworth, "Information Superiority: Enabling the Need to Share," Roke Manor Research Ltd., 2009.

[59] Haffa, Jr., Robert P., and Jasper Welch, "Command and Control Arrangements for the Atack of Time-Sensitive Targets," Northrup Grummon Analysis Center Papers, November 2005.

[60] Vane, R., et al., "Urbane Sunrise," AFREIF-RS-TR-2004-22, Final Technical Report, February 2004.

[61] "U.K. MoD Network Enabled Capability," JSP777, January 2006.

Endnotes

1. A packet is the unit of data that is routed between an origin and a destination on the Internet or any other packet-switched network.

2. A datagram is a self-contained, independent entity of data carrying sufficient information to be routed from the source to the destination computer without reliance on earlier exchanges between source or destination and the transporting network.

3. Actionable intelligence is intelligence upon which meaningful action can be taken.

4. Incoherent change detection identifies changes in the mean backscatter power of a scene. Typically the average image intensity ratio of the image pair is computed to detect such changes. Coherent change detection, on the other hand, identifies changes in both the amplitude and phase of the imagery that arise in the interval between collections. The sample coherence of the images is commonly used to quantify changes or detect movement [34].

5. Also known as urban military operations in urban terrain (MOUT) Advanced Sensor System (UMASS).

6. "Semantic labeling" refers to data being labeled in such a way that it can be processed by a computer, which understand the structure and meaning of the data.

7. 5-eyes refers to the grouping of Australia, Canada, New Zealand, the United Kingdom, and the United States.

8. Authorization to engage a target will usually require positive identification of the target from at least two sources, typically including visual identification, along with accurate geolocation information. The engagement authorization decision becomes more complex at longer ranges, for example in air-to-air engagements where no positive visual identification can be provided prior to the decision to engage.

The Virginia-class attack submarine Missouri (SSN 780) conducts sea trials July , 2010 in the Atlantic Ocean. (Photo courtesy of General Dynamics Electric Boat.)

NEC Concepts

Core NEC themes

Data, information, and knowledge

The challenge of NEC is to create a knowledge-based C2 environment where the relevant information is presented to the appropriate commander at the appropriate time within the OODA loop to enable optimum and rapid decision making, and subsequent decision superiority.

Such an approach has also been referred to as *knowledge-enabled capability* (KEC) [1]. KEC focuses on the challenge of turning masses of gigabytes of data and information into useable knowledge. It places more emphasis on how the information is stored, accessed, linked, and presented to enable better informed and more timely decision making. It also challenges traditional concepts of knowledge management where the strength of the knowledge- based network relies on the concept of *need to know* rather than *need to share*, and decisions are made across the network in line with command intent rather than on a centralized basis.

The strength of the knowledge-based network relies on the concept of need to share rather than need to know.

The challenge in achieving an effective network-centric environment lies in two critical areas: first, the creation of knowledge from data, and its management throughout the network and C2 process, and second, the technical connectivity and interoperability required across the network [2].

From the effective creation of knowledge comes the challenge of knowledge management. Knowledge management is the systematic process of discovering, selecting, organizing, distilling, sharing, developing and using information in a social-domain[1] context to improve warfighter effectiveness [3].

Data is the representation of facts expressed through numbers, characters, measurements, images, or instructions in a manner suitable for communication, interpretation, or processing by humans or computers. Data by itself has no meaning because it is devoid of context against which it can be measured. Data on its own may or may not be sufficient to make a particular decision, but will need to be interpreted or processed within a context or framework that gives it meaning. In an NEC context, data is capable of being sent in digital form from sensors and communications systems across networks for further processing.

Information is data placed in a contextual framework through processing and exploitation that gives it meaning from relative values or patterns and reduces uncertainty. Information is formed from the interpretation of data to form an instruction, pattern, image, or message.

Knowledge is information that when placed in a framework of observation, perception, reasoning, relationships, and experience has intrinsic value resulting in a reasonable degree of certainty and understanding about the situation and its implications.

Actionable knowledge is information that enables the decision maker to understand the situation in context through the integration of information from multiple sources and to make use of the resulting situational awareness to enable proper framing of decisions and actions within the multidimensional battlespace. Creating actionable knowledge involves the assimilation of information relating to objectives, constraints, resources, options, uncertainties, and cultural influences. Actionable knowledge shared across the network becomes the basis for joint planning and prioritization of sensors, effects systems, and assets to achieve the command intent. Shared actionable knowledge results in knowledge and decision superiority [4].

Across networks, information can be transmitted to serve various purposes, but all transmission will take place in a digital format standardized to the requisite transmission and interface standards. Different forms of information can be transmitted at differing speeds; video consumes significantly more bandwidth than text, for example, and at the recipient's end, the speed with which information can be assimilated is heavily dependent on the nature of the information, the processing, and the automation associated with turning it into something meaningful.

Information[2] may be exchanged in many forms:

- Numerical data (digital machine code);
- Text (words and numbers);
- Voice (human language);
- Static images (symbols and pictures);
- Dynamic images (video).

Effective NEC requires the consistent and reliable integration of several concepts such as *knowledge management* (KM), *network management* (NM), and *information assurance* (IA) across tactical and strategic networks (Figure 3.1). The themes outlined below address the challenges of both connectivity and knowledge management—the essential elements in creating network-enabled capability.

Information quality and information richness

The utility and effectiveness of a networked approach very much depends on the quality of the information gathered, processed and transmitted to users across the network, and its interpretation within the context with which it is gathered and used [5].

Information quality can be considered as an absolute measure of the worth or value of the information. It considers the quality of the information in its own right, and does not take into account the broader contextual setting in which the information is used.

Information adequacy, however, considers the value and completeness of the information in the context that it is required to be used for decision making or for translation into knowledge, in effect a measure of real-world

Information—The Essential Ingredient [6]

The first instruction in a famous old recipe for chicken soup is to "steal a chicken." Information is like a chicken. It's the essential ingredient in the recipe for lifting the fog of war. And like catching a chicken, getting your hands on information isn't easy. Information doesn't just lie around waiting to be exploited. It has to be coaxed out of its hiding places—databases, sensor outputs, intelligence reports, open source literature, and observant people.

Sensors, databases, and reports are sources of data, but data isn't information. Data is a carrier of information. To get at the information lurking deep within the data, we have to process that data. After we distil the information content, we have to understand the significance of that information as it relates to our mission, geospatially, historically, and militarily.

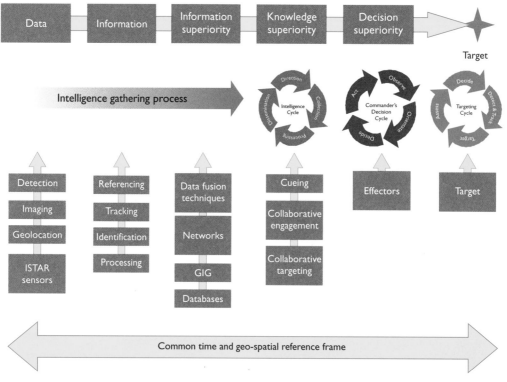

Figure 3.1 The transformation of data to information and knowledge is a critical part of all military decision-making processes.

knowledge versus ideal knowledge. The degree of adequacy will also depend on factors such as the extent to which the original data has been processed and assumptions drawn from it—especially when combined with multiple information sources.

Information in a networked environment needs by definition to be able to be fused with other information—there is no point in a nugget of rich information being produced in a form that isolates it from integration with other data. Information adequacy recognizes this in its qualitative definition.

It must also be recognized that the provision of high quality information does not necessarily result in a high-quality decision-making input if the adequacy and context of the information, and the use to which is it put, is not considered in parallel.

Data, information, and knowledge fusion

Individual sensors produce individual observations or measurements in the form of raw data (Figure 3.2). This data that must be processed and organized within a time and space framework to create an organized and

Information Quality

Currency: the time lag between the situation and the information being available.

Accuracy: the precision (granularity), completeness, and relationships between the collected information, which reduces the margin of error in the decision-making process.

Reliability: the degree to which the user can be sure that the information presented is an accurate portrayal of the situation.

Utility: the degree to which the information can be made available in a form (where further processing may be required to achieve that state) that will assist in decision making within the time required for use in the OODA loop.

interpretable standardized data set to provide information, which may be combined with further information inputs for evaluation to facilitate a higher-level interpretation of the overall context in which the information is set in order to create knowledge for the decision-making process. Data or information may be fused by combining outputs from sensors with information from other sensors, information data sets, databases, or knowledge bases, into one standardized format. The fused product can improve the accuracy, completeness, reliability, and latency compared to the prefused data source.

The fusion of outputs from sensors can take place at a number of different levels in the battlespace, typically:

Information Adequacy

Relevance: the latency of collected data and the time criticality of information gathered for decision making and its degree of relevance to the information superiority process.

Completeness: the degree to which the information gathered enables essential knowledge to be generated to achieve effective decision superiority.

Fusibility: the ability to be able to integrate or fuse information from multiple sources in a meaningful and consistent way through standard formats to enhance the utility, meaning, or accuracy of the information.

Degree of synthesis: the degree to which information has been processed from multiple sources, assumptions made about its relevance or importance, and associations and correlations made with other information sources.

Accessibility: the ability to be able to utilize and process the required information at the required time to enhance the information superiority picture.

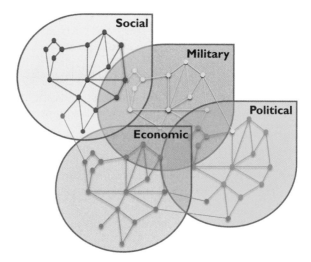

Figure 3.2   Where multiple nodes and networks are collaborating in a system of systems approach, the challenge of coordinating accurate geospatial and time reference data can be significant.

- **Data fusion:** translation of data into a common (usually digital) format and fusion of that data either at the sensor (to enhance the output of the sensor data) or at a central point for onward dissemination across a network. Data is often integrated from multiple sensor sources to produce a significantly enhanced situational awareness picture. A typical example would be the fusion of multiple passive ESM sensors to triangulate and accurately geolocate a signal source that would not be possible from a single ESM source, which can only provide data concerning the bearing of the signal.

- **Information fusion:** data sources are processed and translated into useful information from which situational awareness knowledge can be used to make decisions within the OODA loop. Information fusion often enhances the battlespace intelligence picture by combining inputs from multiple information sources. Typically these sources would be accessed through an intelligence picture from the *Global Information Grid* (GIG) and displayed in a joint operations center.

There are several frameworks in existence that define the fusion of different types of data. They cover from the "simple" fusion of data for single point targets up to complex multisensor 3-D data fusion.

Space and time information grids

Accuracy of data is one of the essential prerequisites of a highly performing synchronized network. This is particularly the case when it comes to the need for accurate ISR data to be shared between sensors and C2 nodes. The accurate reporting of target information, particularly that related to dynamically changing information about moving or transient objects, is essential if multiple sensors and effects systems from across the network are to be engaged in crosscueing targets or synchronizing assets across the *kill chain* or *defend chain* (see Appendices E and F) sequence.

Many concepts such as collaborative engagement (see Chapter 1) rely heavily on the ability to share accurate geospatial and time-referenced data in a dynamic real-time environment.

In a networked environment, establishing target tracks using multisensor data fusion relies heavily on accurate target position and timing information. Where coordinated sensors are mounted on the same platform, a common clock can be used to synchronize event information. Across a network, one node may serve as the master reference clock for collaborating nodes, but across a system of systems network, with many different types of data-handling systems, the challenges of coordinating timing data can be significant. Increasingly, multiple network synchronization can be achieved through common access to satellite timing signals.

These challenges are amplified when one takes into account the data latency or delay in receipt of data from sensors that are a function of the type of sensor, its position in the network, and the data processing and transmission capabilities available to it. Further timing errors may also be introduced into the data from errors in the initialization of equipment or sensor data against the reference grid, delays in transmitting the data, or in the use the recipient system makes of the data it receives.

In addition to synchronized timing, it is essential that all assets that provide geospatial references utilize the same framework so that accurate coordinates of both the platform and targets can be accurately passed between nodes.

Given the complexities of networked sensor and data management, the ability to achieve what is termed *gridlock*, where all participants in the data/information/sensor/C2 chain are working to the same geospatial and time reference grid, remains a significant technical challenge.

Geolocation

Geolocation, or precise position information, is key to an accurate understanding of the situational awareness picture in the network-enabled battlespace. Dynamic and accurate reporting of friendly, neutral, and hostile force locations is crucial to the C2 and decision-making process. In addition to the need to understand the location of one's own forces (blue force tracking) to avoid fratricide, the accurate location of enemy positions is particularly important for the delivery of effects onto a precise target, particularly in a cluttered battlespace where it is essential to ensure sufficient distinction between targets to guarantee that the correct target is crosscued (see Chapter 2 for a description of crosscueing) between sensors and subsequently engaged(Figure 3.3).

The most accurate location positions are achieved with military GPS position reporting. Target locations will often be calculated as a three-dimensional vector from a *Global Positioning System* (GPS)–equipped platform obtained through laser ranging or synthetic radar information.

Backup or complimentary navigation systems often include *inertial navigation systems* (INS), which measure three-dimensional acceleration and can be used to compute the path of the vehicle to which the system is fitted.

Both target reporting and the guidance of precision munitions are becoming increasingly accurate; they depend on knowing precisely the location of the

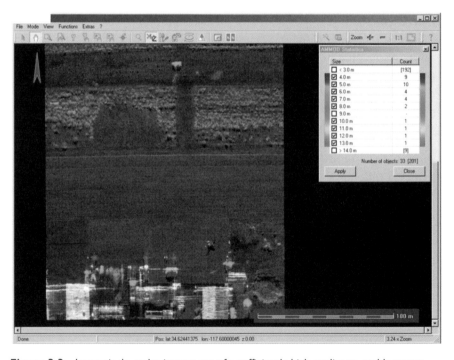

Figure 3.3   Increasingly, radar images are of a sufficiently high quality to enable target designation. Here a Lynx radar mounted on a Predator UAV is used to classify potential targets based on the automated measurement of radar cross-section. (Image courtesy of General Atomics).

target and the location of the weapon. Even weapons that traditionally have been unguided, such as artillery and mortar shells, are increasingly fitted with various forms of guidance such as GPS or semiactive laser guidance. GPS guidance is generally used for stationary targets while dynamically updated GPS or semiactive guidance can be used for engaging moving targets (Figures 3.4 and 3.5).

In addition, sensors and computing power are already at a level where each pixel on a radar image or on a video image is capable of being defined by its own grid reference.

Shared network environment

One of the key challenges in NEC is the ability to establish nodes that are capable of undertaking network transactions with diverse sources of data, yet are able to operate across multiple networks and network layers in a standardized manner.

In order for nodes to exchange information effectively, they must transact with each other using a common currency [2]. All data and interchange formats must conform to defined standards allowing data to be visible,

Figure 3.4 The GPS-guided Excalibur 155-mm artillery shell incorporates an inductive fuse setter to transfer target and fuse data in the form of a digital message containing the coordinate for the round to find just as it is launched. (Figure courtesy of Raytheon.)

Figure 3.5   An XM982 Excalibur precision-guided artillery round falls onto a suspected insurgent safe house during combat operations in the northern region of Baghdad on May 5, 2007. Soldiers of the 1st Brigade Combat Team, 1st Cavalry Division fired the round from their M109A6 Paladin howitzer on Camp Taji, Iraq. The event marked the first-ever operational firing of the XM982 Excalibur projectile.

accessible, understandable, trusted, interoperable, and responsive as early as possible in the life cycle to support mission objectives [8, 9]. When planning network interchanges therefore, data will need to be understandable through common data structures and interchange frameworks and transmittable through a common network data interchange standard (Figure 3.6).

Data visibility is achieved by "tagging" data with a descriptor called a *metadata tag*. Metadata is used to organize, locate, retrieve, and address data using descriptors and keywords to describe content. Metadata typically includes structured descriptors about information about various aspects of the data, including its content, context, origin, timing, source, geospatial data, and security and information assurance tags. Metadata associated with video images, for example, would describe what has been posted, who might benefit, and how it can be retrieved. *Discovery metadata* is a type of metadata that allows relevant data to be found using enterprise search capabilities.

Data accessibility is achieved by making data available through user groups with a shared interest called *communities of interest* (COIs) (See Chapter 3 for an explanation of communities of interest) and by publishing metadata in a structured metadata registry.

The strategic technical plan [10] (STP) developed by the USAF Electronic Systems Centre (ESC), provides a common technical vision and technical guidance to the implementation of C4ISR networks working together as one enterprise to achieve the goals of network-centric capabilities. While this is by no means the only framework available, it does provide a number of key features that are standard across most NEC approaches. The STP focuses on 6 areas to achieve network commonality:

1. Construction of a common layered architecture consisting of four discrete layers supported by a common information assurance approach:

   • Applications;
   • Information infrastructure;
   • Networks or transport layer (See the OSI model in Chapter 4 for a description of the transport layer);
   • Sensors.

2. Populate data to the network by sharing unprocessed data from physical sensors (or semiprocessed data for bandwidth-limited environments).

3. Build the network by using IP as the common protocol (use a gateway to convert data to IP format if IP is not native).

4. Share information on the network by using XML programming language to capture key data.

Figure 3.6   U.S. JTAC operators use the L3 Rover terminal, which is capable of processing data and video feeds from multiple networks and displaying them on a single screen.

5. Design for flexibility by having applications implement a service-oriented architecture to publish, subscribe, and access data between loosely coupled applications.

6. Protect the network through the use of common enterprise information assurance (IA) solutions consistent with the Global Information Grid (GIG) architecture such as the *public key* concept.

The U.S. DoD [11] specifies the use of the Internet Protocol as the standard data interchange format for network implementation, requiring that any applications or sensors that interface with the network or transmit over the network do so via an IP format (Figure 3.7). This is particularly important over extended networks such as the GIG, allowing easy interface with nonmilitary organizations, government departments, industry, academia, and public service capabilities to support worldwide operations.

Integrated networked systems

In addition to network interface standards enabling communication between nodes, individual systems also need to work as part of a larger entity where they contribute to enhancing the overall effectiveness of an extended networked set of systems sharing common data sets and processing standards.

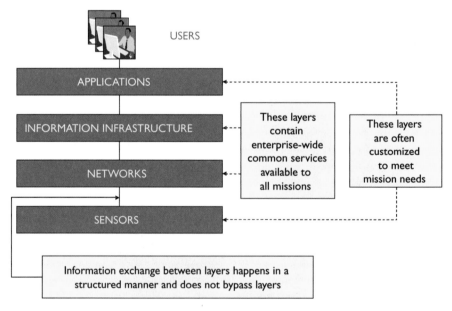

Figure 3.7 The U.S. DoD Defense Technology Plan provides a common IP-based network structure and data interchange standards.

A *system of systems* can be defined as an arrangement of interdependent systems connected to provide a capability greater than the sum of the member systems [12].

At a lower level of benefit, a *family of systems* can be defined as a capability, which is the summation of a grouped set of systems with common characteristics. Unlike a system of systems, a family of systems does not acquire new properties or capabilities as a result of such a grouping [13].

Real-time shared situational awareness

Networked sensors share near real-time information across the C2 grid enabling an increased tempo of decision making within the OODA loop and access to the joint operational picture (JOP) for all assets on the network. By sharing the JOP and matching the threat to effectors, assets across the network will benefit from both improved survivability and lethality as the most appropriate effects systems can be brought to bear on priority threats in a tempo that would be impossible with stand-alone systems.

Across the network, participants provide data and information to support a shared understanding of the disposition of enemy, friendly, and neutral forces, the interpretation of a situation, the intensions of friendly forces, and the options open to achieve the command intent that can be supported by the networked ISTAR and effects systems.

When constructing a real-time situational awareness picture (Figure 3.8), the challenge of data latency can be significant—whether that is related to data link transmission delays, delays in processing data or simply the limitations of "old" ISR data; all intelligence sources need to be woven together to identify short periods of meaningful activities from hours of battlespace intelligence and to share this intelligence across the network with relevant users.

Clarity of the battlespace picture depends on the ability to transform sensor and intelligence data into information and then into time-critical, actionable knowledge.

Network and architectural modeling

Bringing the various sources of information together in to a structured architectural framework can be a complex and challenging task. Several military conceptual frameworks have been developed to assist with the structured analysis necessary to create an efficient optimized structure; in particular to model, understand, analyze, and specify capabilities, systems, systems of systems, business processes, and associated data and information interactions.

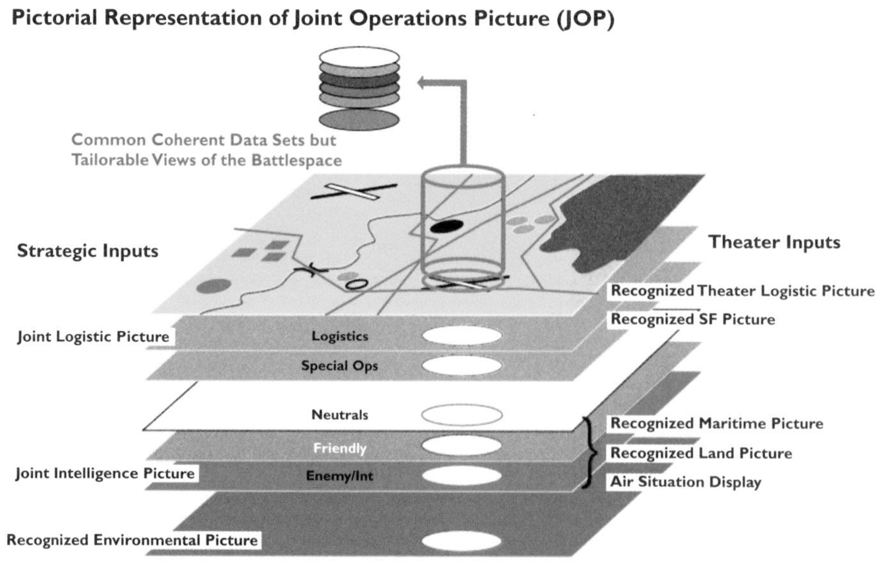

Pictorial Representation of Joint Operations Picture (JOP)

Figure 3.8 Situational awareness pictures integrate complex layers of common coherent datasets. The U.K. MoD approach is shown in this illustration [14]. (*From:* [38].)

Of particular note are the following frameworks:

- **DoDAF:** a U.S. DoD architectural framework [15];
- **MoDAF:** A U.K. MoD architectural framework developed from DoDAF [16];
- **NATO architectural framework:** drawing on elements of MoDAF and DoDAF [17].

Architectural frameworks (Figure 3.9) aim to capture the concept and purpose of the network, capture the logical data flows necessary between each functional element, and specify the data interchanges and standards required across collaborating elements of the network.

These structured models are particularly useful in that they enable the user to visualize different requirements of the architecture viewed from different perspectives, thereby ensuring a high degree of integrity once the final network is brought together.

Early models such as the U.S. DoDAF V1.5 model (Figure 3.10) defined four views around which the analysis is centered:

- **All views:** summarizes the scope and context of the architecture;
- **Operational view:** identifies what needs to be accomplished by whom;

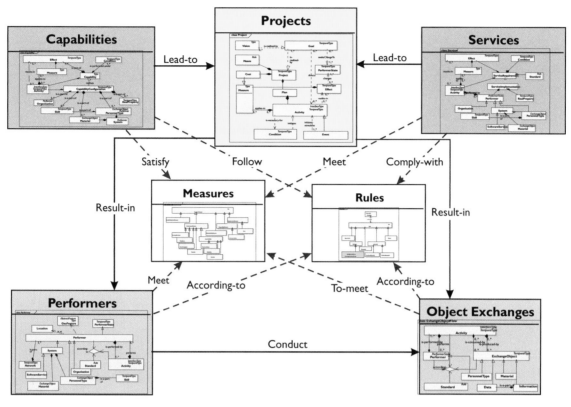

Figure 3.9 Interactions between different elements of a network are clearly captured and specified in architectural models [18].

- **Systems and services view:** relates subsystems, services, and characteristics to operational needs;
- **Technical standards view:** specifies standards and conventions across the solution.

More recent work has developed this approach further and the latest DoDAF, MoDAF, and NATO models now all use 7 views to describe the architecture and its implementation. The DoDAF 7-view model is shown in Figures 3.11 and 3.12 and is typical of the latest thinking in this concept.

Interoperability

Interoperability is critical to the success of NCW and stems from the need for different systems to work together as part of the same network at appropriate levels of security and interaction across collaborating military forces and government or civilian networks. Interoperability needs to be considered in the context of operational, technological, and procedural compatibility.

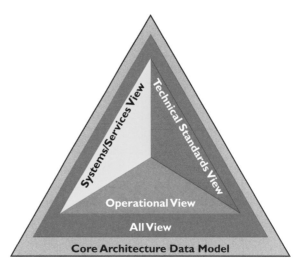

Figure 3.10 The U.S. DoD architecture framework (DoDAF) defines four different views to enable the specifications to be developed of different aspects of the network [19].

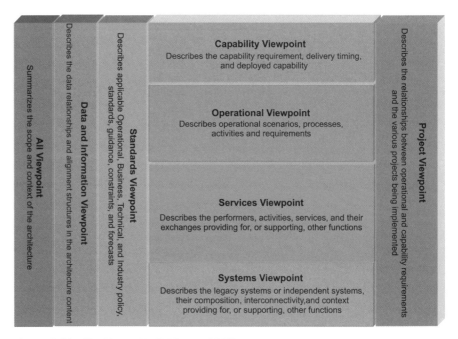

Figure 3.11 The 7-View DoDAF model [18].

The overriding goal is the need for all participating coalition systems to provide and receive information to enable the compilation of a single integrated picture of the battlespace displaying enemy, friendly, and neutral assets, potentially leading to shared and integrated C2 structures. Given the

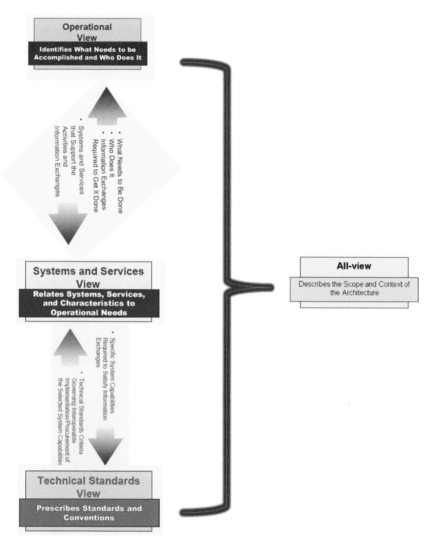

Figure 3.12 Linkages between the views are defined to enable precise analysis, definition, and specification of different aspects of network operation [19].This diagram illustrates the connections between the 4 views in the V1.5 DoDAF model, but the principles are equally applicable to the 7-view models.

Communities of Interest—Key Attributes [23]

- Composed of stakeholders cooperating on behalf of various organizations, with emphasis on crosscomponent activities;
- Members committed to actively sharing information in relation to their mission and/or task objectives;
- Recognize potential for authorized but unanticipated users, and, therefore, strive to make their data visible, accessible, and understandable to those inside and outside their community.

diversity of coalition networks, and in particular legacy systems, it is essential that all systems, or at least network interfaces, must be capable of interfacing with modern IP-based networks in order to participate in and across multiple networks and systems.

Interoperability standards and protocols are often established across *communities of interest* (COIs). COIs are most likely to be functional or joint entities that cross organizational boundaries and are usually formed to meet a specific data-sharing mission or fulfill a specific task.

An example of a COI might be a meteorology COI or a joint task force. COIs should include producers and consumers of data, as well as developers of systems and applications (Figure 3.13) [22].

Interoperability requires a seamless sharing of data, information, and knowledge through an assured, protected network (referred to as information assurance or IA) linking C2, ISTAR, and effects systems across coalition forces. This can only be achieved across diverse systems by standardizing the interoperability instructions and making them available to collaborating systems at the design stage to ensure that interoperability is

Interoperability: The ability of systems, units, or forces to provide services to and accept services from other systems, units, or forces and to use the services so exchanged to enable them to operate effectively together [27].

Figure 3.13   Communities of interest can include a wide range of joint-force platforms created for a specific mission or role. Here *USS Mobile Bay* (CG 53), *USS Russell* (DDG 59), and *USS Shoup* (DDG 86) perform a pass-in-review with the Nimitz-class aircraft carrier *USS Abraham Lincoln* (CVN 72). *Lincoln* and embarked Carrier Air Wing Two (CVW-2) in the Western Pacific during a scheduled deployment in 2009.

built-in from the outset using standards-based communication architectures and repeatable patterns[3] of data interchange.

Data interchange itself (Figure 3.14)is a complex task involving the standardization of key data parameters and frameworks. All data destined for interoperability purposes needs to be properly characterized and referenced against a common time and geolocation framework to ensure that it can be used across different systems with consistent reference metadata parameters. Source data needs to be time stamped to ensure that any data latency issues experienced across the network are accounted for. Multiple systems operating across the same network, especially in a system of systems environment, will need to ensure that they cannot only process disparate types of information, but that the information can also be processed, adapted, and integrated for use in the network structure into which it is introduced (for example, structured around a service-oriented architecture environment and stored in a structured manner for access by a wide variety of networked users).

In addition, particularly for critical tasks, data needs to comply with information assurance (IA) protocols to ensure that the quality of information shared with other nodes is to the requisite standards.

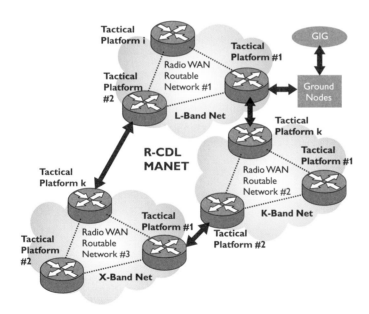

Figure 3.14 Interoperability between networks operating on different frequencies is essential for an integrated warfighting network. Here, data links are established between different networks through the use of a radar data link, which transmits information between collaborating platforms acting as nodes on a network. (Illustration courtesy of Raytheon.)

Technical connectivity through standardized security protocols and data transfer and synchronization protocols, preferably using self-synchronization routines, however, is only one challenge. Once connectivity has been established, data transfer needs to be structured in keeping with the concept of operations (ConOps) employed by the network. Information on the network needs to be accessible by other nodes, which also need to find and utilize information available on the net using information push and pull techniques as appropriate (see Chapter 3).

The appropriate management and accessibility of information according to security classifications is key to effective interoperability. Operations involving multiple participants, and in particular multiple coalition partners, need to employ processes to handle multiple security classifications, and to restrict information accessibility to authorized participants. In order to handle such complexities, data handling processes need to keep track of the classification of all information and the security access levels for each node. This needs to be undertaken in a dynamic environment in which information is continuously moving across communications networks in numerous directions at once, and may be processed or fused by various nodes, thereby changing the classification level.

Future networked information management concepts envisage that information will routinely be cleansed or downgraded to lower security classifications using information management services resident on the network, thereby enabling access by as many participants as possible.

Processing of data at each node aims to add more value to the information on its journey across the net through association with additional information from various sources. Processing functions need to able to process the information to defined standards and retransmit the processed data as standardized information on the network with associated metadata. Where information is normally sent in the form of unprocessed or nonstandard data, it will need to be reformatted before being published to the network.

Interoperability requires that all nodes are able to contribute to C2 decisions across the net, respond accordingly, and have other nodes respond accordingly in keeping with the tasking assigned to each node.

It should be noted that without interoperability (at all levels) the ability to achieve information superiority, shared situation awareness, and decision superiority is significantly degraded. Additionally, coordination of coalition assets, in particular through the shared understanding of command intent, can be corrupted to an extent that NEC operations become a drain on resources as manual information-sharing processes substitute for self-synchronized automated information inputs within the OODA loop.

In addition to interoperable technology standards, interoperability also demands that operation policies, tactics, techniques, procedures, and a culture of shared information must also be harmonized.

Reconfigurability of network assets

The concept of reconfigurability (Figure 3.15) supports the need for agility and speed of response within the OODA loop enabling networks and the functionality of nodes to rapidly reconfigure to meet changing mission needs. Such agility permits networks to work together with minimum disruption and confusion, enabling the dynamic creation and configuration of mission groups that share awareness and coordinate and employ a wide range of systems for a specific mission.

Reconfigurability requires an ability to synchronize assets across the battlespace based on prioritization of sensor or effects tasks. The current approach is for systems to be able to self-synchronize, forming ephemeral networks that form and disband as the need presents itself. Self-synchronization permits a high degree of reconfigurability to suit dynamic C2 requirements and is particularly important as the command structure becomes more decentralized.

Persistence

The acquisition and dissemination of battlespace information is critical in enabling swift and decisive military decision making and the delivery of the

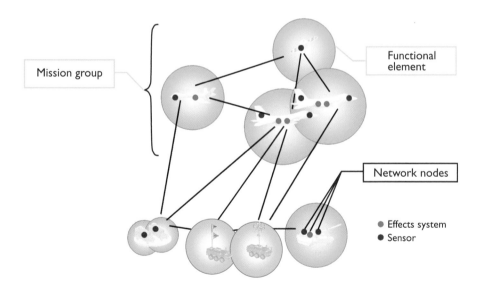

Figure 3.15   Reconfigurability needs to occur between nodes, functional elements, and mission groups to enable geographical centers of ISTAR and effects systems to be created to meet the command intent.

desired effect. Modern warfare requires dynamic situational awareness and this also extends to the concept of time, where information about adversary activities needs to be available 24/7 in all weather to enable rapid decision making within the OODA loop. Such requirement demands a high level of persistence and availability across the sensors and networks within the battlespace to detect any changes to the situational picture as they occur (Figure 3.16).

The requirement today is to understand the situational disposition of all enemy forces, at sea, on land, or in the air and to detect any change to their movements or intentions in near-real time over long periods, denying the enemy a place to hide, particularly in complex terrain where persistence is particularly important for area dominance.

Information dominance includes the need to acquire and disseminate information about the position, intentions, and movement of all forces within the battlespace and about the environment in which they are deployed. This information often needs to be gathered at long ranges by sensors that must stay undetected, or at least out of range of the enemy forces' lethal envelopes, if they are to provide a persistent surveillance capability.

Persistent surveillance can be described as "the systematic observation of the battlespace with sufficient precision and frequency that targets will not be able to move or change significantly without being noticed." This has also been described in simpler terms as "[g]iving the enemy nowhere to hide."

Reconnaissance	Persistence
• Periodic, "snapshots" in time.	• Continuous, enduring contact and "dwell."
• Stovepiped, Hierarchical collection.	• Multimode collection with broad access.
• A few sensors support a few missions.	• Sensors support entire enterprise.
• Analysts see data first and pass.	• Data available across network to all.
• Target-centric collection and analytic focus.	• Deep systemic and relationship focus.
• Analytic templates and assessment.	• Patterns, inference, case-based models.
• Data sets remain within stovepipes.	• Data integration—horizontal and vertical.
• Driven by predetermined requirements.	• Data and analysis on demand.

Figure 3.16 A paradigm shift: from reconnaisance to persistence.

In the Cold War era, the intelligence picture was gathered from snapshots of military capability whose smallest entity was a missile system or a ship, enabling the formation of a reasonably accurate picture of the adversary's capabilities. In fourth generation warfare, however, that is not the case. Targets tend to be small, well hidden, fleeting, and dispersed across an asymmetric environment, making them difficult to even recognize as targets. Rather than identifying targets at a platform level, we now need to find and follow targets that are small and hard to distinguish, such as individual people if we are to determine their purpose and threat. A central concept to generating this knowledge is that of persistence and this is usually applied to the surveillance of the battlespace through use of ISTAR technologies.

Network-enabled capabilities have greatly enhanced the reach and persistence of battlespace assets that can remain connected to the intelligence network for long periods of time, sensing and sharing ISTAR information and feeding it into the overall intelligence picture. Tasking and retasking assets can also be performed as part of an integrated sensor and effects network where assets can remain on station and ready to be tasked as appropriate for flexible missions as the need unfolds.

Persistent surveillance (Figure 3.17) can be a complex challenge, requiring a broad range of sensors covering a range of spectral frequencies, often deployed at long distances for long periods of time. The persistent nature of

Figure 3.17   Unmanned platforms such as the Northrop Grumman Global Hawk can remain on station for over 24 hours, providing persistent surveillance of wide areas of interest.

the surveillance requirement means that in order to contend with the evolving nature of the threat, a wide range of sensors and platforms will inevitably be required. These range from space-based sensors to airborne and land-operated sensors, with increasingly advanced technologies enabling resolutions capable of observing individuals, their movement, and their behavior within crowds.

In today's ISTAR environment, persistent surveillance can only be achieved by linking these sensor technologies to real-time, high-bandwidth ephemeral networks, which form to provide the battlespace commander with the required granularity of information to complete the mission and dissolve when the persistence is no longer required, or when information is required with a differing degree of resolution.

No single platform can deliver total persistent surveillance. Persistent surveillance demands an enterprise-wide ISTAR architecture that provides the right collector in the right place at the right time for the duration of the surveillance period demanded (Figure 3.18).

Another enabling requirement is the exchange between legacy and new ISR systems. This is not just a technical challenge, but will also require a capability development framework where IP can be shared without contributors having to reduce their input to the net the lowest common denominator of data utility.

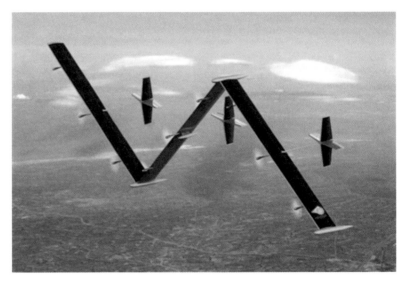

Figure 3.18   DARPA's Vulture program is developing an unmanned aircraft capable of remaining on station uninterrupted for over 5 years, a capability ideally suited to persistent surveillance.

Collaborative sensing, tracking, targeting, and engagement

Collaborative engagement principles recognize that a network of sensors, effects, and C2 nodes, by acting together as a single entity, can prioritize time-critical engagement decisions for the network as a whole and achieve rapid and simultaneous engagement of multiple targets.

A collaborative network-based approach also provides an increased offensive capability through the ability to engage high-value fleeting targets where a narrow engagement window exists, and an improved defensive capability where the combined strength of the network's ISTAR sensors and effects systems is more capable of defeating threats compared to standalone platform-centric defense systems.

The degree to which true collaborative networking can be achieved is driven by several factors that determine the effectiveness of contributions made by each network node.

- **Coordination:** Shared time and geospatial grids, shared awareness, collaborative sensing and effects.
- **Coherence:** The integration of numerous entities, protocols, relationships, and values, particularly timing and geospatial awareness, in order to achieve a common objective.
- **Collaboration:** Better prioritization of ISTAR and effects missions across the battlespace, improved survivability when operating in hostile environments through improved threat information and avoidance.
- **Speed:** Collaborative engagement relies on near real-time data exchange where speed of sensing, transmission and data fusion is of the essence.
- **Resilience:** The ability for parts of the network to be degraded without a significant adverse impact on the outcome of the engagement.

Within a collaborative engagement framework, these advantages provide a *force multiplication* effect through the ability to prioritize and rapidly engage targets across a network.

Close collaboration enables the network to adapt the participants, their sensors, and effects systems in patterns that reflect a holistic approach to targeting and intelligence gathering to maximize the effectiveness of the force across the battlespace.

Rather than a single platform having to rely on its own ISR sensors and weapons systems, the concept of collaborative engagement allows networked sensors and effects systems to present themselves as nodes in a near real-time network, and share and receive sensor information from other nodes across the network. Building on this principle, the integration of weapons systems across the network permits a platform under attack to engage the threat by cueing a weapons system from another networked platform taking into account the weapons status and readiness of participating weapons systems (Figure 3.19). This integrated network approach provides a significant step forward in engagement capability permitting, for example, over the horizon engagement of targets or initiation of defensive action against sea-skimming missiles well beyond the range of the sensors on the platform on which the weapons are housed, with one platform launching a missile and another guiding it to its target.

As collaborative engagement processes (Figure 3.20) mature, multisensor and multispectral sensor data will be passed to dynamic data fusion engines, which will match target signatures to a multispectral target reference database. A set of operating rules will govern automated sensor tasking, bandwidth allocation, and even engagement decisions to match the threat priorities.

While modern networks are increasingly designed with collaboration in mind, putting the principles of collaboration to work across multiple sensors, networks, and effects systems for the greater good of the network priorities network remains a significant technical and doctrinal challenge.

Figure 3.19   The AGM-88 High Speed Anti-Radiation Missile (HARM) is capable of loitering in an area until a radar emitter is located for attack. Here an F-16 launches a missile.

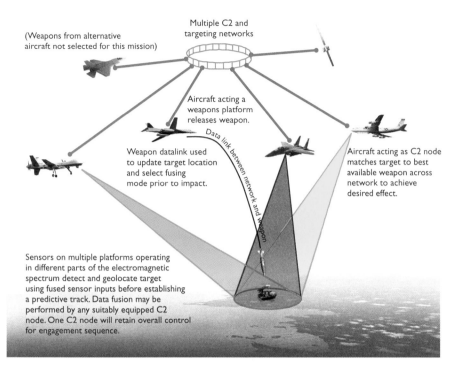

Figure 3.20 Collaborative sensing, track formation, and updates are key building blocks for the complete engagement cycle.

Although theoretically the allocation of sensor and engagement assets across the network should reflect the priorities of the network as a whole, those decisions are very much subject to the perspective of the networked C2 node involved in the prioritization rather than necessarily the perspective of the commander in the field. Such centralized decision making tends to restrict rather than encourage the contribution of assets to the network, and even when assets are allocated to priority missions, the allocation can be late on in the planning cycle, resulting in late decisions about whether the mission will take place, or even missions being planned without reference to the availability of wider network assets.

Collaborative sensing

The ability to fuse sensor data from multiple air, land, and sea sources and integrate it into a common networked picture has driven one of the most important changes in military capability since the advent of the computer (Figure 3.21).

The use of collaborative principles enables data from different sensors to increase the probability of detection of targets with low signatures such as

Figure 3.21 Collaborative sensing enables sensors from different domains to share situational awareness and to coordinate the crosscueing of priority targets across the kill chain [24]. Here the first of class of the new Lockheed Martin designed, littoral combat ship *USS Freedom* LCS 1 is shown undergoing sea trials. The LCS class of ships integrate and fuse data from multiple on-board and off-board sensors to form a composite situational awareness picture that can be shared across the network with collaborating platforms.

stealthy platforms, or those at long ranges, by forming a virtual sensor network that can exploit the detection characteristics of different spectral sensors and detect and track targets that fall below the detection thresholds of individual sensors through passive, active, or multistatic means.

Collaborative engagement

Collaborative engagement is at the heart of network-enabled doctrine. It recognizes that integrated sensor and effects networks generate a level of military capability that far exceeds that which could be generated by stand-alone assets.

Sharing sensor data across the network produces significantly improved targeting quality information, which can be shared directly across the network for use in collaborative engagements. The improvement in speed and accuracy of this approach over stand-alone sensors enables a significant improvement in the rapidity with which the identification of target locations can be achieved, typically between 1 and 10 seconds [25] at targetable levels of location accuracy along with a significant increase in the number of targets identified.

For offensive roles, collaborative engagement enables a mission planner to deal with specific targets while at the same time being able to dynamically replan targeting priorities to take into account targets of opportunity, fleeting targets, which need a rapid engagement sequence, or to deal with emerging threats as the battle unfolds. Theoretically, collaborative engagement techniques can also intelligently predict track and target dynamic motion and can provide the user with automated decision making to ensure that threats are effectively prioritized.

For defensive roles, a collaborative engagement approach provides a greatly enhanced defensive capability giving much longer sensor ranges, improving awareness to threat situations, and providing multiple opportunities to engage threats well beyond the range at which they can effectively engage target-friendly forces. This will provide platforms within the network information that they are under attack well before they are even able to see the threat, and importantly enable them to cue sensors in a particular direction and use the weapons systems of other platforms to assist with their defense. Collaborative engagement data can also be fed into the Global Information Grid (GIG) (For an outline of the GIG see Chapter 4) to compliment the larger battlespace picture, where assets outside the immediate control of the collaborative network may need to be tasked to address emerging or time-critical threats.

Collaborative engagement principles have also highlighted the emerging importance of flexibility in on-board and off-board planning as part of the OODA loop. Within collaborative engagement systems (Figure 3.22), although the platform may be on a preplanned mission, the platform, its ISR sensors, or weapons may be diverted to deal with higher-priority targets.

"The sum of all wisdom is a cursor over the target."

General John. P. Jumper, USAF Chief of Staff

Figure 3.22 F-22 Raptors are able to share information with each other to the extent that each knows the others fuel and weapons status, thereby permitting the most effective engagement options to be allocated to each aircraft to best deal with the situational awareness priorities.

Time-sensitive targeting

Arguably, all targeting is time sensitive to varying degrees. Even static targets will have a window of opportunity where an enemy team is located in a building, or where the weapons system will have an enhanced probability of delivering the desired effect in an environment where the enemy's defensive systems can be appropriately dealt with at a specific time by supporting strike assets. At the other end of the scale are targets, often mobile or moving, with very short opportunities to detect, track, and engage and emitters, for example, which may switch on and off for short periods of time in an erratic manner. Increasingly, time-sensitive targets may even be specific individuals or groups such as one or two combatants who may have exposed themselves and their activities for a brief period of time, thereby requiring immediate engagement within a short time window.

...the war on terrorism is predominantly a war on time-sensitive and time-critical targets [26].

Attacking time-sensitive targets is inherently difficult and complex. It involves technically advanced and geographically distributed equipment. It requires personnel trained to perform technically demanding tasks and the development of a harmonious organization of personnel and equipment to execute the commander's intent. The time-critical C2 process needs to balance the need to coordinate priorities and actions with the need to respond quickly to new priorities, which often conflicts with earlier priorities. The coordination of assets and intelligence inputs adds time to the decision-making process. The more inputs that are included, and the more they are dispersed, the more time is added. An essential command function

Figure 3.23 Collaborative engagement of moving targets represents a complex technical challenge. Here a Boeing small diameter bomb destroys a Russian-made rocket launcher during a live fire test at White Sands Missile Range, New Mexico, United States. (Photos courtesy of Boeing.)

associated with time-critical targeting is therefore the adjudication of the conflict between timeliness and coordination. In addition, the timely prosecution of time-sensitive targets demands the allocation of sensor platforms and shooter platforms to operate in loitering modes in the vicinity of where such targets are expected to appear (Figure 3.23). This allocation of scarce resources competes with the efficient and effective prosecution of the main combat action [27].

Time-sensitive targeting (Figures 3.24 and 3.25) is a challenge because it requires rapid information sharing across all stages of the kill chain sequence, using multiple sensor inputs to enable decision-making information to be shared across the OODA loop and correspondingly rapid C2 decisions made about the engagement of time-sensitive targets.

By efficiently netting and synchronizing sensors in real time and combining detections, it is possible to increase the probability of detecting and identifying a target, while also reducing the time it takes to accurately locate targets by more than 90%. This refined information updates a common, shared network database so that the problem of multiple reports on a single target is dramatically reduced [29].

(a)   (b)

Figure 3.24   Even static targets can be considered as time sensitive. Here a Boeing small diameter bomb destroys an A-7 Corsair target after penetrating a simulated bunker during a live bomb test flight at the White Sands Missile Range in New Mexico. (Photos courtesy of Boeing.)

Figure 3.25   While TSTs are often considered a challenge for air power, nonair assets, particularly special operations forces, can certainly accomplish target discovery and destruction, but their greater value, as demonstrated in combat operations in Afghanistan and Iraq, is likely to be cueing air assets to track and attack TSTs [28]. Here a JTAC operates with an A-10 Warthog CAS aircraft.

Various frameworks exist, which map the role of network-enabled systems within a time-sensitive targeting environment. Typically they comprise of the acquisition of the target, its identification and tracking, decisions concerning prioritization, and effects allocation and the assessment of poststrike damage assessment.

Time-critical targets (TCTs) (Table 3.1 and Figure 3.26) are a subset of time-sensitive targets, which are defined as those targets that pose or will soon pose a clear and present danger to friendly forces or are highly lucrative, fleeting targets of opportunity [31]. TCTs may be either mobile, moving, or static and typically include targets such as vehicle-mounted theater ballistic missiles (TBMs) that can be prepared for launch in less than 30 minutes, insurgents setting up indirect fire attacks in less than 10 minutes, buildings and facilities containing high-value personnel, or C2 assets whose disruption is key to securing the command intent or protecting coalition forces.

Several projects have been developed in recent years to exploit collaborative engagement technologies, of which network-centric collaborative targeting (NCCT) and collaborative engagement capability (CEC) are the most widely used.

Detect	Decide	Engage	Assess
Receive TCT cue	Prioritize in target list	Retask platforms to mission	Collect
Assess	Commit to kill TCT	Position platform	Exploit
Task network sensors	Pair weapon/platform/ sensor to target	Hand off to weapon	Decide TCT negation
Collect	Coordinate/deconflict mission	Deliver weapon	Remove from target list
Exploit		Weapon effects	
Nominate	Update platform		
Identify	Mission plan		
Geolocate			

From: [30].

Table 3.1 Time-Critical Targeting Principles and Engagement Sequence

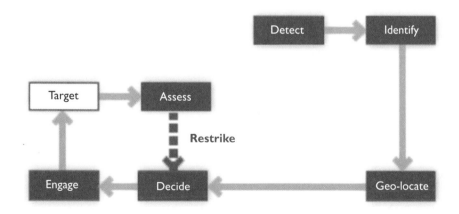

Figure 3.26 Time-critical targeting principles and engagement sequence.

Network-centric collaborative targeting (NCCT) is principally a USAF program to integrate sensors in support of time-sensitive targeting of ground targets. The aim is to obtain very accurate and very fast targeting quality information by fusing data from multiple sensor sources such as radar, ELINT, and SAR/GMTI data, and to share the information back across the network in order to provide timely detection, identification, and geolocation of high-priority targets to combatant commanders (Figure 3.27).

The NCCT system operates around a hub and spoke wideband network to enable the sharing of data, coordination of sensor activity, and the provision of rapidly correlated results between dissimilar collection and decision nodes. Coordination and orchestration of interactions and allocating engagement priorities is achieved through a common set of multisensor, multidiscipline, and network management rules benefiting from standardized network messages and formats, correlation software, and data rules of interaction.

Recognizing that many participating sensors and C2 assets may utilize different communications protocol standards, a platform-interface module (PIM) converts the disparate data standards and operational architectures to a common IP-based standard with a NCCT network controller providing common control across participating nodes.

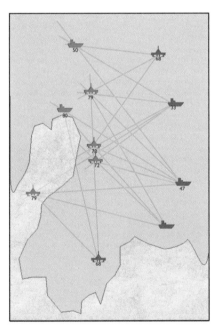

Figure 3.27 Typical collaborative engagement display showing networked triangulation of contacts (red) by multiple sensor platforms (green).

The collaborative engagement sequence focuses on the find, fix, target, and assess portions of the kill chain and is controlled by a platform-based ISR sensor manager (ISM), which synchronizes NCCT operations using automated updates of network tasking priorities based on command priorities and air tasking orders (ATOs) (Figures 3.28 and 3.29).

Another key program in this domain is the U.S. Navy's Collaborative Engagement Capability (CEC) program. The original purpose of CEC was to integrate the sensors and naval surface to air missile systems among Aegis cruisers equipped with SPY-1 radar to provide an extended engagement envelope across a deployed U.S. fleet enabling incoming targets to be engaged through layered defenses from multiple ship and aircraft platforms with an increased probability of kill (Figure 3.30). Its use has, however, been extended to enable suitably equipped collaborating platforms to participate in the rapid engagement of high-value targets. In essence, CEC allows one platform to track an object and another to engage it in a maritime domain, even if the engaging platform cannot "see" the target using its own ship-based sensors.

The CEC network (Figure 3.31) cues and orchestrates multiple intelligence, surveillance, and reconnaissance sources across all platforms within the network, effectively taking control of sensors and weapons systems to

Figure 3.28 NCCT integrates dissimilar ISR assets including rivet joint, joint surveillance and target attack radar system (J-STARS), airborne warning and control system (AWACS), deployable common ground station (DCGS), U-2, and army guardrail aircraft. The Guardrail aircraft shown here integrates Comint and Elint sensors to provide enhanced signal classification and recognition, fast direction finding, precision emitter location, and targeting data to collaborating network assets. (Photo courtesy of David Loera at Burner Aviation Photography.)

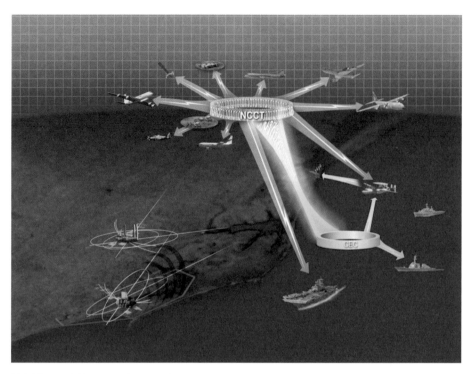

Figure 3.29 NCCT and CEC networks can link to extend coverage and deal more effectively with time-critical targets. (Image courtesy of L3.)

Figure 3.30 Aegis-class guided missile cruiser *USS Normandy* (CG-60) on patrol in the Persian Gulf in June 2005. The SPY-1 antennas are shown on the forward and aft superstructure.

prioritize data gathering and engagement sequences. The networked data from active sensors is independently processed at each collaborating unit by common software processes to form a composite track picture.

Figure 3.31 The CEC system has been fitted to carrier-based E-2C Hawkeye airborne surveillance and control aircraft to assist in the naval air defense mission by extending the engagement envelope with extended sensor coverage outside the range of the SPY-1 radars and filling in gaps in radar coverage between SPY-1 equipped ships.

Each sensor and weapons system is effectively assigned its own IP address and can receive commands and provide data across the network, distributing the data across existing communication systems such as JTIDS, GCCS, and other CEC networks with very low data latency rates to distribute the engagement quality data to network assets. Mission-specific, geographic, coalition, or other subordinate networks can be formed and linked together to form a transparent and scalable network, which can itself be linked into a Global Information Grid.

The CEC system uses the Joint Composite Tracking Network (JCTN) to perform crosscueing of mixed sensor types such as GMTI, radar, and Elint sensors, producing composite sensor tracks that are compatible with the C2 systems on the platforms on which the track information is shared.

Use of the point-to-point JCTN between collaborating platforms permits the exchange fused plot data to produce a shared air picture, which is real time and of sufficient accuracy and quality to allows the weapon system of one platform to engage objects without itself necessarily having generated the sensor data (Figure 3.32). Sharing of data in this manner also enables greatly improved geolocation accuracy, and the ability for one platform to request image or more detailed situational awareness from zones of particular interest.

Figure 3.32 Collaborative engagement exploits existing networks to maximize speed of reaction and joint targeting processes. (Image courtesy of L3.)

CEC units are equipped with omnidirectional beacon antennas, capable of exchanging data at speeds of up to 10 Mbps on the JCTN network, which allow for directed LOS data exchange at ranges of approximately 50–60 km.

Synchronization

Synchronization can be described as the orchestration of networked assets to ensure they are aligned to their military objectives and supported by an appropriate force structure to enable them to achieve the command intent.

Effective network synchronization is very much dependent upon the implementation of an effective C2 process, which in turn requires rapid and dynamic assimilation of network inputs and coordination of C2 tasks to continually adapt force structures and adapt tasks across time and space dimensions to achieve the command intent.

Given a common command intent, synchronization can be viewed as the coordination of:

- Command and control (C2);
- Intelligence gathering and engagement priorities;
- Tempo (the timing of such actions);
- Situational awareness (a common picture shared across the network);

- Information (collected, processed, stored, and disseminated);
- Effects (matching of effects to command intent).

Full synchronization across the extended network involves both ISR and effects assets. Synchronized ISR assets are required to ensure that optimization is achieved between available networked sensors to reflect the specific data collection requirements in accordance with network priorities, while synchronized effects are required to ensure that targets are prioritized and dealt with using the best placed and most effective effect systems available to the commander.

Important considerations in planning synchronization include the following:

- **Data latency and time grids:** These grids ensure that data is synchronized to take into account the moment when the measurement was taken. Even when sensors are measuring the characteristics of the same target, they do not all produce data outputs that are synchronized to the exact moment in time. Additionally, if data has to travel through several network nodes or has been processed since collection, additional time delays will be added, which need to be taken into account. All data, therefore, needs to be synchronized around the same time grid.
- **Space reference grids:** Targets need to be accurately geolocated so that the coordinates can be used for targeting purposes or for future reference. Sensors may record the location of targets as being in slightly differing locations, with errors becoming exaggerated particularly at long distances. An accurate and definitive target location is required for targeting purposes.
- **Data standards:** To enable ISR and effects synchronization, the key ingredient common to both is that of data interchange and handling standards—both across the network where nodal interface standards need to be harmonized and within databases and other intelligence sources benefiting from standardized approaches using xml programming, IP data sharing, and common postprocessing data formats.
- **Security and accessibility:** Data required for synchronization across a network needs to be accessible to all units that need to either handle it or process it. Security classifications may change as data is processed, and this needs to be taken into account in terms of accessibility by critical nodes. The network also needs to be arranged to enable suitably structured data access, typically involving a mixture of push/pull information access typically using a service-oriented architecture approach.
- **Bandwidth requirements:** Suitable bandwidth is required to link network nodes, and capacity requirements will be reduced if a structured information sharing approach is used. The exact bandwidth requirements will be determined by type of data needing to be transmitted, with the

volume of data varying over time according to the mission requirements and collection priorities. Additional bandwidth requirements may also be imposed by the degree of processing of data (postprocessing) that is required once it has been transmitted from the collector. (Many collectors now process data to transform it into information, but other assets may collate, fuse, and interpret it, for example, to create accurate situational awareness pictures for further distribution across a network.) Information processing and reintroducing that data to the network will take place at different nodes across the network. If data collection and accuracy (or granularity) requirements are matched in time and location to the specific mission requirements to ensure that detailed information is collected only when and where it needs to be, the actual information process, if structured appropriately, can be quite efficient, and bandwidth requirements can be reduced as a result.

- **Integration of data collection process with synchronized effects:** This is required to maximize the probability of the desired outcome on the target by ensuring coordination within and between assets to allocate the most appropriate effects available within the network through dynamic distributed planning, coordination, and execution. Related to this is the concept of *effects-based planning* (EBP), which focuses on the use of military and nonmilitary effects including appropriate use of information operations across the battlespace.

In addition to ISR and effects synchronization, the need to provide synchronized support services and deployment of sustaining resources, such as logistics, medical, and intelligence assets, is an essential part of the synchronization process. The ability to automatically schedule resupply units based on actual or anticipated usage of weapons and supplies from data gathered from across the network is an essential part of battle winning network-enabled technologies.

Effects-based operations

Effects-based operations (EBO)[4] (Figure 3.33) describes a process for achieving a defined strategic outcome (effect) on an adversary, through the synergistic, multiplicative, and cumulative application of the full range of military and nonmilitary capabilities at the tactical, operational, and strategic levels [32] by targeting more precisely an adversary's will, capability, and cohesion.

Effects may range from psychological warfare, to nonlethal effects, and to the use of varying degrees of lethal force. EBO, in particular, recognizes the importance of delivering a specific measured effect on a precise target and recognizes the value of precision weapons in the need to avoid collateral damage.

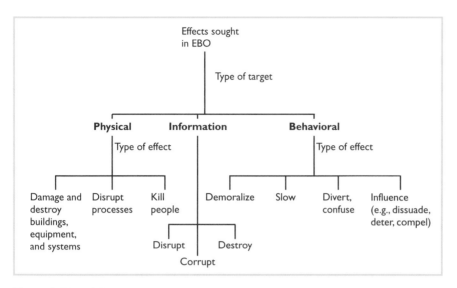

Figure 3.33 EBO can be used as an approach to understand and reduce the adversary's critical capabilities and coherence allowing for decisions in intensely dynamic and complex situations. In this context the effect is the physical functional or psychological outcome, event, or consequences that results from specific military or nonmilitary action [33, 34].

NEC will unlock the full potential of EBO by allowing decision superiority from enhanced situational awareness, increased cooperation and coordination between HQs at all levels, and coordination with nonmilitary actors.

Engagement in EBO requires the identification of the effects that will lead to campaign success (the ends), the optimal combination of capabilities to achieve these effects (the ways), and the capabilities themselves (the means) (Figure 3.34) [35].

The concept of EBO involves the use of capabilities and propagation of effects across space, time, and different domains of the DIME (diplomatic, information, military, economic) construct (Figures 3.35–3.37) [37].

Shared intelligence

Intelligence is derived from the compilation and fusion of many sources of information from across the battlespace (Figure 3.38). The challenge of an intelligence provision is to ensure that the information required for situational awareness and decision making is captured in a timely manner and presented in a format that is relevant to the decision-making process represented by the OODA loop and other longer-term strategic planning processes.

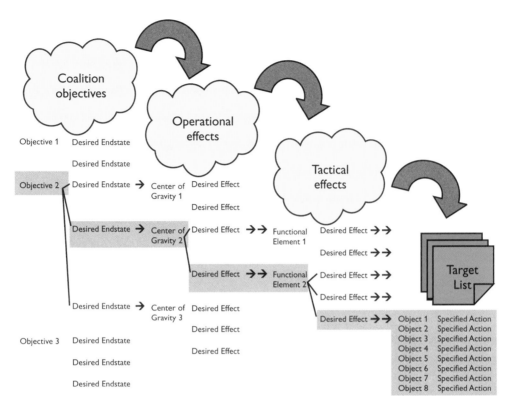

Figure 3.34   Abstract decomposition of an effects-based operation [36].

Figure 3.35   A precision air strike takes out an identified insurgent stronghold as 3rd Battalion 1st Marines, 1st Marine Division moves forward through the city of Fallujah, Iraq, during Operation Al Fajr (New Dawn) in November 2004.

Figure 3.36 A member (bottom right) of the Combined Weapons Effectiveness Assessment Team assesses the impact point of a precision-guided 5,000-pound bomb through the dome of one of Saddam Hussein's key regime buildings in Baghdad.

Figure 3.37 Most platforms are designed to deliver a range of precision effects appropriate to the specific mission. Here a RAAF F-111 from No. 6 Squadron at Air Force Base Amberley displays the impressive range of explosive ordnance it is capable of delivering to a target.

BATTLESPACE TECHNOLOGIES: NETWORK-ENABLED INFORMATION DOMINANCE

Figure 3.38 Intelligence information is typically the product of a data gathering and fusion process involving the collection of information from multiple sources of varying degrees of latency accuracy and relevance. (Image courtesy of Thales.)

Typically the intelligence-gathering process involves the collection, processing, fusion, transmission, analysis, and interpretation of information from multiple sources to assist with the valuation and assessment of actual and emerging threats in the battlespace, the evaluation of the threat and the threat's course of action, and the optimized tactical and strategic response and scenario planning.

Accurate intelligence collection and analysis enables what is termed the *intelligence preparation of the battlefield* (IPB) to understand the challenges ahead in their broadest context and to assist with the planning of the optimum course of action (Figure 3.39). This requires a broad range of intelligence inputs from the environment or physical domain within which operations will take place through to a clearer understanding of the threat and the threat's options for action, and insight into how the planning scenario may evolve.

 Being able to collect, assimilate, and interpret information more quickly than the opponent is crucial for information and decision superiority. In an NEC context this means being able to collect information from multiple sensors and intelligence sources, process information to varying degrees, and make it available to everyone who needs it, in a form that they can use, in a secure, relevant, and timely manner (Figure 3.40).

A shared intelligence picture enables significant additional benefits to be gained through processing and interpretation of information patterns often requiring multiple sensor and intelligence inputs to make connections and

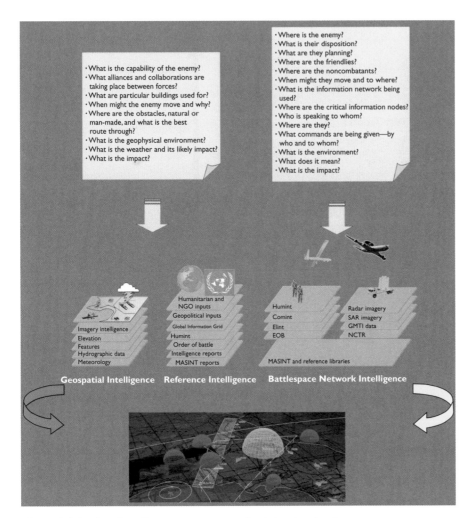

Figure 3.39 Fused intelligence picture providing joint situational awareness.

establish patterns or changes. These techniques are particularly useful in countering unconventional threats where threat detection, especially in an urban environment, can be difficult to achieve.

Amongst these techniques are:

- **Pattern of life:** mapping normal activity and movement patterns to establish air, land, and sea traffic patterns and highlighting deviations from those patterns, which may merit further investigation. Where sensors record patterns over long periods, movements may be tracked back to establish where vehicles or ships have originated from, and where they went to, following particular incidents. Sensor types required: electro-optical, GMTI.

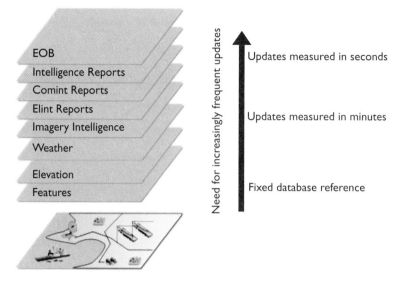

Figure 3.40 Accuracy, frequency of updates, and granularity of intelligence information also vary from layer to layer.

- **Pattern matching:** looking for suspicious patterns of activity near roads, bridges, infrastructure assets. Early detection enables activity to be interdicted and address either with direct action or through following those involved. Sensor types required: electro-optical, GMTI.

- **Change detection:** commonly applied to optical systems where you are looking for a change in magnitude of particular parameters to provide intelligence on what has occurred in a particular area of interest. As sensors evolve, change detection will increasingly be applied to measurements involving three-dimensional shape, texture, temperature, and color. Sensor types required: electro-optical, ground-penetrating radar, SAR.

- **Coherent change detection (CCD):** refers to a radar system that looks for the change in phase between two images. For instance, if a surface moves by a millimeter or so because of footprints in the ground, then there may be no change in amplitude of the reflected signal, but CCD should pick it up as a change in phase. CCD is normally seen as a radar or optical technique but theoretically possible for laser applications.

Reach back and reach forward

Reach back involves passing C2 information to the headquarters well away from the front line and enables battlespace assets to be connected to operational headquarters and to intelligence and analysis agencies that may be located anywhere on the globe. For many operations involving expeditionary warfare, the operating headquarters, particularly in the opening stages of an engagement, may be geographically removed from the theatre of operation itself. NEC enables near real-time coordination and C2 functions to be performed from any location on the globe.

Reach forward enables remote operational or brigade headquarters to take direct involvement in controlling remotely deployed assets such as UAVs, or issuing C2 instructions directly to deployed units from a remotely located headquarters.

Both reach forward and reach back concepts involve the extensive use of satellite communications to relay data and information around the globe (Figure 3.41).

Resource management

In truly homogeneous sensor and communications networks, the efficient allocation of resources across a network such as bandwidth, processing capability, transmission speed, sensor modes, and coverage area is a key determinant not only of network effectiveness, but also of the ability to support information superiority requirements.

Several issues influence network resource management. Sensor networks containing a diversity of sensor types, operating modes, and coverage patterns need to prioritize power, processing, and sensing capabilities,

Figure 3.41 Long-range UAVs such as the Global Hawk are fitted with a high-bandwidth satellite communications antenna to enable access to near real-time sensor data and control to be maintained virtually anywhere on the globe. Here a Global Hawk Block 30 arrives for testing at Edwards Air Force Base.

together with routing the distribution of raw or processed sensor data to the target node or nodes in the most efficient manner. Increasingly, nodes process data at the source to extract critical information and it is this that is then transmitted across the network rather than large volumes of raw data to be processed at another node.

A key goal in networked sensor fusion is therefore to understand the factors that determine efficient and timely information creation and distribution. This includes the need to minimize the burden on the network, defined by multiple factors such as the *power burden* (associated with sensor tasking), the computational burden (associated with the stage at which data is fused), and the *communications burden* (associated with the stage at which the data is sent to the next stage of the processing chain and the routing across the network topology). For example, depending on the available bandwidth, data can be fused at various processing stages. For example, radars can send data for onwards processing at the plot level (10–100 kb/s), or can process the plots, create tracks, and transmit track data, which requires very low transmission throughput (1–20 kb/s).

Network data fusion requires careful assessment to limit the burden and to determine the most efficient locations where the data will be fused, where decisions will be made, how node sampling and communications protocols will be constructed, and the numbers of fusion and decision nodes to optimize information gathering and dissemination. The patterns and methodologies used will also be largely determined by both the physical characteristics of the network such as bandwidth and sample rates and the use to which the fused data is put—that is to say whether it is used to support time critical targeting, or whether it is used for situational intelligence gathering.

Extended battlespace networks will typically comprise of several subnetwork systems collaborating together to achieve a common purpose. Such an arrangement is known as a *system of systems* architecture, and comprises of multiple sensor networks which need to be managed as if they were a single entity. Each system will need to take account of multiple inputs and prioritize its behaviors accordingly, even requiring several functions to be performed by the sensor at any one time. For example, an AESA radar may prioritize a SAR processing function in support of a high-value mission while also undertaking air-to-air surveillance in another sector at the same time (Figure 3.42). In systems of systems approaches, networks will typically contain several disparate sensor systems, which need to be managed as a functional group, even though the group may be performing several functions at any one time, theoretically for several networks.

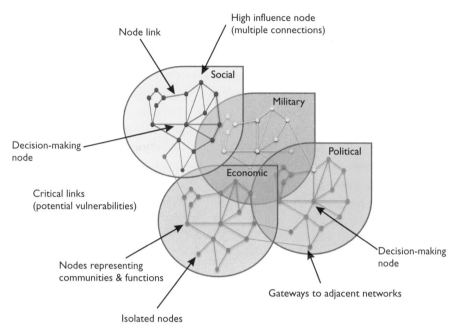

Figure 3.42 Resource allocation across a network needs to take into account multiple internal factors and external situational inputs to determine overall network priorities.

Information management and accessibility

Network-enabled warfare provides many challenges as well as operational advantages. The need to share information across networks in increasing volumes requires a structured approach to collection, storage, and dissemination of information in multiple formats, particularly in an environment where interoperability between one's own armed forces and allies needs to be addressed. The scale, size, and complexity of this task are compounded by the need to collate information from multiple sources and multiple formats, while preserving the security and integrity of the data as it is transformed into common formats for onwards dissemination (Figure 3.43).

Information management (IM) activities seek to provide the right information to the right individual at the right time in a usable form to facilitate situational understanding and decision making. The focus is on friendly information and information systems [31].

As networked approaches to modern warfare become more widespread, more systems need to be able to communicate with each other to share common data interchange formats and to provide agility as networks form

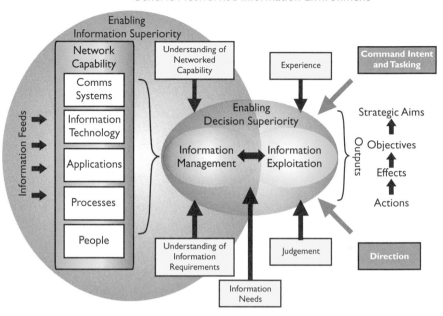

Figure 3.43   Information management and exploitation is at the heart of achieving decision superiority [38].

and disband to meet mission requirements. Similarly, information exchange processes need to be adaptable in terms of who participates as well as the roles each participant plays. Increasing use of IP-based protocols and search engines are increasingly being used in information management roles to access timely and relevant data (Figure 3.44).

Information management and accessibility concepts focus around key areas of information handling and storage, typically aiming to standardize the functions to [40]:

- *Store*, *catalog*, and *retrieve* all information produced by any node on the network in a comprehensive, standard repository so that the information is readily accessible to all nodes and compatible with the forms required by any nodes, within security restrictions.
- *Process*, *sort*, *analyze*, *evaluate*, and *synthesize* large amounts of disparate information while still providing direct access to raw data as required.
- Provide each decision maker the ability to depict situational information in a tailorable, user-defined, shareable, primarily *visual representation*.
- Provide distributed groups of decision makers the ability to *cooperate* in the performance of common command and control activities (Figure 3.45) by means of a *collaborative work environment*.

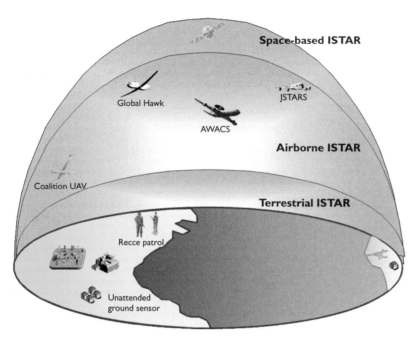

Figure 3.44   The concept of layered surveillance is important in managing the collection of intelligence to the desired degree of granularity and timeliness.

Network information management inevitably involves interfacing with legacy systems, or systems that operate using different data exchange protocols. This typically involves an interface gateway that converts the required data into whatever format is appropriate before storage or onwards transmission. Such an approach enables multiple legacy and current NCW systems to interface effectively in order to share sensor information and collaborate in network systems such as NCCT and CEC.

For time-critical applications, information needs to be shared in near-real time, and the speed at which information can be converted into the required format, stored, and retrieved is a key factor in creating an effective networked system. This, in turn, represents another challenge where time-critical targeting and C2 information from multiple sensors must be fused against a common time reference grid to ensure accuracy of data input.

Timeliness of information in a network-enabled environment is about getting the right information to the right node at the right time, and in this context, for mission critical information requirements, consideration also needs to be given to information resilience and redundancy. Issues such as whether the information can be accessed from multiple sources, whether the database need to be duplicated, whether the collection and dissemination routes need to be duplicated, and so on, need to be addressed in the protocol design.

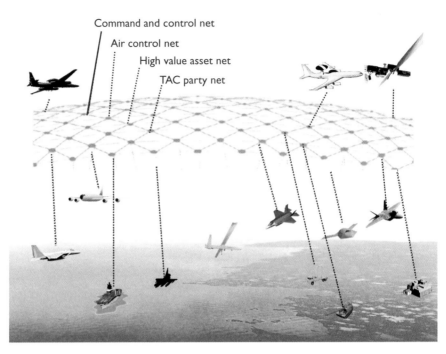

Command and control net

Air control net

High value asset net

TAC party net

Figure 3.45   The principles behind effective information management recognize that an organization's information superiority advantage depends on how well, and how widely, the network creates, accesses, and augments its own collective knowledge.

The protocols by which information is exchanged across the network may also vary across command and control approaches. While *information push* still has a place in networked warfare, the sheer volume of information that is now shared across a network has necessitated a different approach to information management, cataloging, and access. Improvements in search engines, data storage, and information association have all helped to increase the breadth, depth, and variety of information sources that can usefully be accessed as input to the OODA loop process. As a result, information can now be tagged (using metadata principles) and organized to make it easy to find, and stored in standard IP-based network structures such as that provided by the GIG and Integrated Broadcast System (see Chapter 4). Such a structured approach to information provision has enabled participating forces to specify what they need and receive efficient updates in accordance with their needs, while posting information on the network in a similarly structured manner for access by other users.

Data and information for situational awareness and decision-making processes can be accessed by two different approaches to information access [41, 42]:

• **Information delivery:** Using an information push approach, analytical processes actively assess how certain types of information improves

Information Push and Information Pull

Conventional C2 approaches have historically been structured around the concept of *information push* where the system that created the information was responsible for deciding what to share with whom, when to do so, how it should be presented, and if and how often to update and the information. Because the information provider has limited information about the value of the information to network participants (it cannot know the precise real-time requirements of all participants in the battlespace), information push requires complex and time-consuming planning and execution, relying on common data interchange formats and standard operating procedures. The standardized message protocols severely limited the flexibility of message transmission or query, especially involving nonstandard information or data requests, which may potentially add significant delays into the OODA loop.

Information pull is a more efficient means of distributing the data as it conceptually relies on the asset requiring the information to ask for information relevant to its situational needs. This ensures that only information that is relevant to the mission is transmitted, rather than hoping that within the broad range of information transmitted (as with information push) a small percentage will be of value. Information pull processes require network participants to post information on the network in a structured manner for others to access and analyze to support specific mission requirements. Information quality and information assurance of posted information become particularly important in such arrangements.

mission outcomes and form associations so that algorithms operating across extended networks can watch for the relevant information and push it to operators to assist with decision making. Global services applications require discovery and delivery services to support information push principles. A subset of information delivery is *information subscribe*. Subscription broadcasts are the next step in information delivery systems as they allow users to specify the type of information, geographical coverage, or other categories to be delivered directly to them as soon as it becomes available.

- **Information discovery:** Using an information pull approach, source information is tagged with semantic metadata so operators can pull what they deem relevant. Global services applications are required to discover and deliver information to users to support information pull principles. A subset of information discovery is *information post*. Information posting enables participant to place information in a structured manner for access by other network participants and also permits users who are authorized to know what has been made available and how to access certain classes of information using standard data sets, for example, related to certain missions, geographical areas, or intelligence requirements.

Information accessibility involves the storage, processing, access, and exchange of information for all users on the network taking into account timeliness, relevance, and security classification of shared information. It requires common data interchange formats and careful consideration of the security levels of information shared and processed across the network.

Information assurance

Information assurance (IA) (Figure 3.46) is closely related to accessibility and consists of measures that protect and defend information and information systems by ensuring their availability, integrity, authentication, confidentiality, and authenticity. This includes providing for restoration of information systems by incorporating protection, detection, and reaction capabilities [31]. IA is important because users need to know that the information they are using can be trusted to contribute to critical decision-making processes. Users who do not trust the quality of information

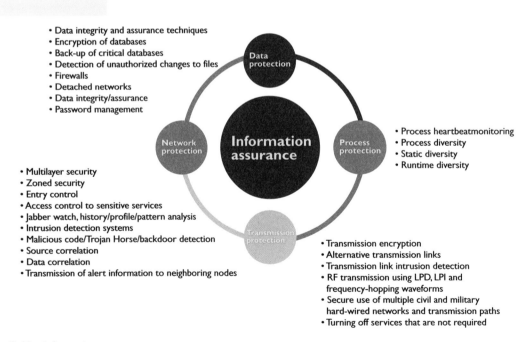

Figure 3.46   Information assurance.

available or who do not have access to reliable information sources are more likely to take time checking the information, be more cautious about making decisions, and hold back in passing the information or its implications on to others across the network.

Information assurance can itself be seen as covering a range of information-related aspects:

- **Security:** integrity of data stored or transmitted. Identification, authorization, and accessibility for appropriate users with read/write access to the data in an appropriate secure manner.
- **Availability:** provision of adequate network paths and bandwidths in the face of network degradation and attack to provide timely and reliable access to data and services by authorized users.
- **Coherency:** uncorrupted data from a storage, transmission, and processing perspective.
- **Traceability:** the assurance of being able to track data back to its source to ensure that it has not been corrupted intentionally or unintentionally.
- **Relevance:** timeliness or latency of correct and relevant information.
- **Reliability:** the degree of assumptions, measurement inaccuracies, or time delays that can introduce errors to the processing of that information and in the way the information is then utilized and relied upon. The cumulative reliability of information sources must retain sufficient accuracy to effectively contribute to the decision-making process.
- **Integrity:** the protection of data and information from unauthorized intentional or unintentional change.
- **Authentication:** the certainty of positive user and receiver identification and the conformation of security levels that may be employed between them. Authentication includes the concept of nonrepudiation—the ability to prove a sender's identity and prove delivery to the recipient.

IA protects information from attack and exploitation by integrating information assurance processes, operational protocols, and technical capabilities to deliver uninterrupted and secure information to authorized users across the extended network. Integrated IA processes protect information throughout its lifecycle from its collection at network nodes, during processing and transmission, through to storage, cataloging, and distribution to network users.

At the same time IA provides the means to efficiently reconstitute vital information distribution and information-dependent services following disruption from a deliberate attack, data corruption, or equipment failure.

Information Assurance

Defined as measures that protect and defend information and information systems by ensuring their availability, integrity, authentication, confidentiality, and nonrepudiation. This includes providing for restoration of information systems by incorporating protection, detection, and reaction capabilities [3].

NEC vulnerability

While the advent of interconnected networked systems has transformed military operational concepts, it has not come without its vulnerabilities, both in terms of implementation and also to exploitation from external threats.

Operationally, NEC presents significant challenges in the management of huge volumes of data, the risk of information overload, the attendant demands for data processing, and the injection of a constant flow of new information into the OODA loop decision-making process. Exponential demands for bandwidth for data-hungry applications often leads to insufficient bandwidth and data latency issues. Additionally the connectivity between networks and subnetworks, and the authority and decision-making processes, especially in a coalition environment, represent real challenges to effective network operations.

For antagonists of networked forces, their reliance on electronic architectures and data networks, often easily accessed via the Internet, is certainly seen as a weakness for exploitation by many opponents. The heavy reliance on civil and military systems and IT networks, if poorly protected, can leave entire national infrastructures open to complete paralysis and exploitation by cyber attack. The U.S. NIPRNet (Nonsecure Internet Protocol Router Network), for example, provides an Interface between U.S. defense networks and the Internet; in recognition of the vulnerabilities associated with such gateways, the number of access points in now being significantly reduced [44].

Exploiting vulnerabilities in computer networks, often referred to as *cyber warfare*, is increasingly being recognized as a weapon in its own right. Cyber warfare involves disrupting information flows, data storage, and communications systems on which an opponent relies, in order to undermine the operation of critical military and civilian systems. Through such techniques it is technically possible to disrupt an entire country's infrastructure and bring the country to a halt without the use of any overt military force.

While forces that heavily depend on network infrastructures will often use more advanced weapon and sensor systems than an opponent, they will also have a correspondingly greater requirement and reliance on information integrity and accessibility. These requirements extend not only to the accuracy, quality, and latency of the information, but also to the need to protect the wider information network and its data integrity (Figure 3.47).

As our reliance on network-enabled technologies and mission critical data increases ...so too does our vulnerability to exploitation.

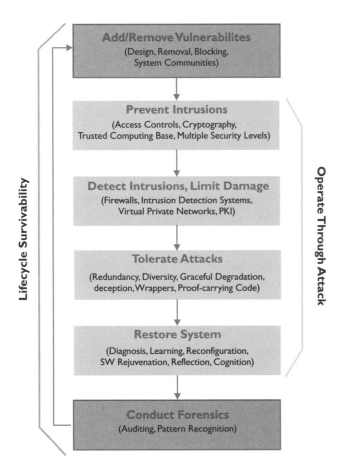

Figure 3.47 Computer networks need to prevent intrusions and deal with them to minimize damage should they occur. Effective network protection requires a structured multilayered approach to provide maximum protection and to enable operation during an attack [45].

The transmission and integration of information across networks is particularly dependent upon secure and uninterruptible high-bandwidth C4 networks linking coalition forces. This is an essential building block if a common integrated situational awareness picture is to be achieved. The importance of this task and associated vulnerabilities can also be recognized by the adversary who will make the task significantly more challenging by intentional interference such as jamming of transmissions or sensors and destruction of key nodes across the network.

Even without interference from adversary forces, the sheer density of communications traffic across the electromagnetic spectrum can present a significant challenge for the establishment of a high-integrity network even in peacetime.

Vulnerabilities of networked systems include [46]:

- Overdependency on information;
- Difficulty in recognizing when information has been corrupted;
- Data latency in decision making;
- Cyber attack (potential access to the whole network);
- Increased systems, procedures, and operational complexity;
- Poor interoperability with nonnetworked allies.

Clearly, given the potential for the data within networked systems to be corrupted, it is important to understand the failure modes and vulnerabilities across the network. Preserving network integrity involves planning for multiple redundancies, preparing alternative sources of targeting and ISTAR information, and considering the adequacy of reversionary modes in order to ensure that the network-enabled forces and associated C2 does not become incapacitated if elements of the network are degraded or destroyed.

The U.S. DoD's defense in-depth strategy acknowledges that a robust multilayered defense is needed to protect from network attack. Recognizing that the network is as much as an asset that needs protecting as any physical network, the DoD approaches the problem through recognizing that the network will need to be fought much like any other weapons system available to the military commander [47].

The U.S. DoD *Information Operations Roadmap* recommends that the defense in depth strategy should include [47]:

- Robust network defensive infrastructure—classified data protected in separate zones, firewalls, and security layers;
- Networks structures configured to act as barriers to attackers;
- Situational awareness across the network to ensure attacks can be rapidly identified;
- Network areas that can be isolated to prevent and limit damage across the network;
- The ability to recover from network attacks consistent with the provision of minimum operational service.

"Networked C4ISR is dependent upon automated decision-making and support, broadband networks, and electromagnetic capabilities, with a corresponding increase in associated vulnerabilites that should be planned for and managed."

—U.S. DoD, Information Operations Roadmap, October 30, 2003

Counter-information warfare and computer network operations

Computer network operations (CNO) are defined as actions taken through the use of computer networks to disrupt, deny, degrade, or destroy information resident in computers or computer networks. CNO comprises of computer network attack, computer network defense, and related computer network

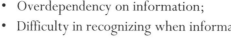

exploitation enabling operations [43]. There are three related areas of CNO, which support counterinformation warfare operations [31, 39]:

- **Computer network defense (CND):** Actions taken through the use of computer networks to protect, monitor, analyze, detect, and respond to unauthorized activity within information systems and computer networks.

- **Computer network attack (CNA):** Actions taken through the use of computer networks to disrupt, deny, degrade, or destroy information resident in computers and computer networks, or the computers and networks themselves.

- **Computer network exploitation (CNE):** Enabling operations and intelligence collection capabilities conducted through the use of computer networks to gather data from target or adversary automated information systems or networks. CNE includes the assessment of both adversary and friendly sources of access and vulnerability for the purpose of securing one's own networks, predicting and recognizing likely threats, understanding and mapping adversary networks and critical nodes of the adversary's networks, network access techniques, planning, and execution of techniques in support of CNA and CND.

Computer network defense

CND employs various hardware and software-based technologies to defend friendly information from adversary efforts to destroy, disrupt, corrupt, or usurp it. Typically it will include both preattack and postattack activities to plan, direct, and execute actions to prevent unauthorized activity and to plan, direct, and execute responses to recover from unauthorized activity should it occur.

Typical approaches to providing robust network defense include the following [28]:

- The *architecture* should first maximize its intrusion prevention and detection capabilities using mature security technologies and techniques.
- The architecture must tolerate Byzantine failures. This is because malicious faults can asynchronously occur in any replica and yield Byzantine failures[5].
- *Static diversity*, or implementing a function in multiple ways, should be used to avoid common vulnerabilities. For example, research has made it practical to automatically generate diverse executables from the same source code.
- *Runtime diversity*, which implements a function differently at different times, will make it harder for attacks to succeed. For example, a system could be designed to automatically change its configuration from time to time to confuse the attacker.

- *Attack isolation and containment* will prevent damage from spreading and bind the set of elements that a system must reconstitute after an attack.
- Correlating *alerts* from multiple intrusion sensors will allow a system to better diagnose, isolate, and adaptively respond to each attack.
- *Adaptive response* will enable a system to respond appropriately to different types of attacks.
- *Graceful degradation* will prevent an abrupt or catastrophic loss of service during an attack.
- *Self-regeneration* after an attack will automatically restore full functionality and level of service. Automation will speed the process and make it reliable.

A well-designed architecture provides concentric *layers of protection* to mission applications, system operations and associated security management, placing management functions in the most highly protected zone. These zones are replicated in a Byzantine fault-tolerant manner.

Middleware[6] should also be designed from the outset for survivability. Security mechanisms should be integrated around a common multicast protocol to enhance integrity, access control, resiliency, and ensure graceful degradation[7]. The middleware design should ensure redundant protocols so that transport protocols can be changed dynamically to suit the messaging requirements. The middleware provides redundant channels that connect each mission application to the core zones of the architecture. If all channels to the core fail, the middleware can attempt to attach mission applications directly to one another.

Security design also needs to be approached in a structured manner, with session keys and cryptographic credentials being used to manage access control. Messages, which are checked for valid size, frequency, and signature, can be briefly held in escrow so that if the publisher appears corrupt, a message is not forwarded. Heartbeat signals are generated by the middleware to indicate that each mission application is alive and beating at the correct frequency

Policy-driven *protection domains* can help protect system, process, and network components from attack. Domains are used to isolate network functions, limit access privileges, prevent corrupted or infected processes from accessing critical resources, isolate application-specific resources from the interference, and disallow actions that exceed privileges. Any breaches of these rules then results in the generation of an alert, and appropriate further protective responses are initiated.

Responses to the detection of an attack on the network can be reactive or proactive. For example, if sensors detect a process' attempt to transition to the directory root, the response might be to kill the offending process (a reactive response). However, if sensors detect file corruption, the system may decide to check and restore files (a more proactive response). If the system determines that a host is compromised it may disconnect the host and reconfigure the system to an earlier configuration.

Information assurance technologies used in support of computer network defense provide communications security through vulnerability assessment tools, intrusion detection software, and a multilayered security infrastructure.

Given the wide range of sources of attack, and multiple threats to networked information, CND must provide the ability to rapidly identify, process, and respond to diverse and often unanticipated threats initiated from anywhere in cyberspace from multiple hostile sources, potentially in a coordinated manner.

It is not enough just to detect intrusions—a system needs to decide on a course of action that will effectively respond to an attack. Once an attack is identified, the priority is one of reaction response times to ensure that the area of the network under attack is isolated, that access routes used by a hacker are disabled, and that through dynamic reconfiguration of the affected network an alternative network or data source can be made temporarily available to authorized users.

As networks move away from fixed hierarchical and centrally managed structures towards mobile self-forming ephemeral networks, computer network defense is becoming particularly challenging. Such networks form a dynamic structure, unbounded by central administrative control, where the numbers and type of participants and the associated network topology frequently change. Such networks operate without central authority, and members seek permission to join the network in an increasingly autonomous manner without reference to a central authorization process—often through wireless communication protocols. Such diversity in network structures, characteristics, and protocols significantly challenges the effectiveness of contemporary security systems beyond their immediate network (Figure 3.48).

Network defense capabilities are evolving rapidly in both military and civilian environments as new open source software and improved network analysis tools are developed to counter increasingly sophisticated attacks. However, the starting point for CND relies on the implementation of best practice in the design of information systems. A structured and rigorous approach to

Figure 3.48   BAE Systems IT monitors network operations across the U.S. from its Network Operations Center in Herndon, Virginia.

adherence of sound principles will significantly reduce vulnerabilities to attack and exploitation across the network (Figure 3.49).

Looking further into the future, we can expect systems to reason about attacks, develop more effective responses to new attacks, and improve their survivability over time by identifying and removing vulnerabilities [28].

The lack of a clear definition as to what constitutes a cyber attack along with limited international legal standards provides no clear consensus on when a cyber attack constitutes an act of war, or guidance as to an appropriate response.

Figure 3.49   A secure Internet network is secured for transportation on board a C-17 Globemaster III aircraft.

Computer network attack

CNA is the employment of network-based capabilities to destroy, disrupt, corrupt, or usurp information resident in or transiting through networks. Networks include telephony and data services networks. Additionally, computer network attack can be used to deny, delay, or degrade information resident in networks, processes dependent on those networks, or the networks themselves [49]. A primary effect is to influence the adversary's decisions in favor of the desired outcome sought by the force employing CNA methods.

In future wars, information highways, C4I systems, and associated networks will be priority targets for attack. Of particular note is that future adversaries will certainly not constrain their attention to obvious military targets; increasingly, the disruption of economic targets such as civilian infrastructure systems, stock markets, logistics, transportation, and energy distribution systems can be undertaken relatively easily and can cause massive and widespread disruption.

Even during periods of nonovert conflict, computer networks daily come under constant attack from global military and civilian sources to examine areas of weakness and exploit penetration opportunities discovered as a result. China and Russia, for example, are known to scan the Global Information Grid on a regular basis to look for weaknesses and test network defenses.

A well-planned network attack generally starts with *network reconnaissance* to probe the network for weaknesses and to find nodes that are most important to attack or exploit in the event of conflict, or for general intelligence gathering on the information contained in the network.

Typically network reconnaissance includes techniques such as port scanning, network probing and exploration, and network service mapping. Such information-gathering operations will often exploit automated network communication protocols to disguise probing operations with a view to establishing network and port maps and seeking information such as the type of operating systems and firewall in use. Such information can then be analyzed to check for vulnerabilities or to identify targets for further exploration, ultimately with a view to gaining access to secure networks and computers.

Computer networks can come under attack from a variety of different sources:

- Physical attacks;
- Cleared insiders;
- Attacks on critical infrastructure;
- Electromagnetic pulse attack;
- Cyberattacks;
- Crypto attacks.

Within the cyber domain, there are a number of different hostile activities that can present a risk to the integrity of the network and the data within it. The depth and persistence of each activity will determine the severity and implications of the attack. Key hostile activities can include:

- Monitoring and recording;
- Data and network mapping;
- Corruption/manipulation of data;
- Error insertion;
- Malicious code insertion;
- Input data saturation;
- Bots, worms, viruses, and Trojan horses;
- Spoofing/phishing.

Attackers often hide the identity of computers used to carry out an attack by falsifying the source IP address from where the activity originated, often routing attacks through several different computers often in different countries. This makes it more difficult to identify the sources of the attack and sometimes shifts attention onto innocent third parties.

As intrusion attempts become more and more sophisticated, the discovery that an intrusion has taken place often comes well after the event and may leave little evidence describing the true nature of the intrusions or the source from where they originated. It is relatively easy to access any network of choosing that is connected to the internet (virtually all networks and computers have a connection to the internet). The great advantage of these types of attack for a foreign power (or hacker) is that they can be conducted well away from the sovereign territory on which they are conducted with relative anonymity. Attacks can even be staged from a location that is geographically far removed from where the attacker is based so that it appears as if it the attack is being undertaken from a different location. Tracking down the source of the attack is often a complex, time-consuming, and costly exercise.

The challenge in identifying that an attack has occurred is to identify patterns of network activity and other indicators that alert the network manager

Worm: A piece of self-replicating malicious mobile code that spreads through a network without human interaction. Because they are self-propagating, worms can spread extremely quickly. Typically, worms do not alter or delete files, rather, they reside in memory, eat up system resources, and slow down computers.

Trojan horse: A hidden piece of malicious code added to a seemingly useful and benign program. When this program runs, the hidden code may be performing malicious activities like allowing back door access to your computer by hackers or destroying files on your hard disk. Trojans are commonly used to introduce spyware or worms into a system. The main difference between a Trojan and a virus is that Trojans are unable to replicate.

Malicious code attacks: Introduction of new code in to existing system—more than just a virus (includes Spyware).

Insertion of false data: Introduced either through sensors connected to a network or through hacking in to the network itself, false data can be introduced that may only become evident after the event has occurred. Typically this will involve the creation of false information about targets, capabilities, and authorizations or will interfere with data that would otherwise alter the network operator or associated C2 systems to the presence of the enemy.

Spoofing/phishing: Directing users to lookalike Web sites where they are encouraged to access a familiar-looking Web page through inserting their user names and passwords, which in due course will then be used to enter the real Web site or portal.

Botnet: Describes a compromised network where software robots, or *bots*, have been installed via worms, Trojan horses, or backdoors, usually under a coordinated control infrastructure. A botnet's originator (termed a *bot herder*) can control the group of computers remotely and nodes of over 1-million affected computers have been discovered. Advanced bots can automatically scan their environment and replicated themselves across the infected network, exploiting security vulnerabilities to penetrate further in to a system, making it more valuable to the bot herder.

before a damaging attack takes place, rather than trying to trace an attacker once the damage has been done.

Information attacks may be undertaken against carrier networks, specific communications networks, C2 systems, databases and targets of a political, social, economic, R&D, or military nature without the need for sophisticated or costly resources. Most attackers initially aim to exploit weaknesses in the access points, firewalls, and backdoors, which in complex networked systems are often difficult to totally secure and protect from intrusion.

Use of commercial off the shelf software and hardware is increasingly being used in information-critical software in systems. The pressures of commercial software production often leaves loopholes in access points, not just at a software application level, but also into the heart of the networks, often providing relatively easy access through firewalls at the entry point and often across multiple layers [21].

Entry to communications and sensor networks may also be gained through wireless access at different frequencies depending on the type of node being attacked (Figure 3.50). In order to achieve the goal of information superiority, it is necessary to dominate and control the entire electromagnetic spectrum. Specifically, as outlined in the *U.S. Roadmap for Computer Network Attack*, the ability to "disrupt or destroy the full spectrum of

Cyber Security

Cyber security policy includes strategy, policy, and standards regarding the security of and operations in cyberspace, and encompasses the full range of threat reduction, vulnerability reduction, deterrence, international engagement, incident response, resiliency, and recovery policies and activities, including computer network operations, information assurance, law enforcement, diplomacy, military, and intelligence missions as they relate to the security and stability of the global information and communications

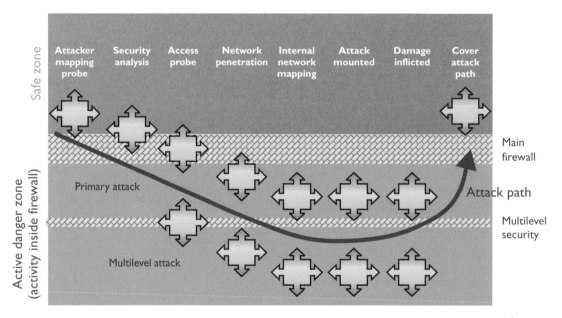

Figure 3.50 Network attacks typically follow a predictable pattern of probing entry points before penetrating successive layers, often waiting to see whether their activity has been detected at each stage of the intrusion [7].

globally merging communications systems, sensors and weapons systems dependent upon the electromagnetic spectrum" [20] should be a key objective.

Of course countries wishing to achieve military advantage through CNA also need to protect against such activities being undertaken by opposing forces.

Computer network exploitation

CNE involves maximizing the information gained from attacks on an opponent's computer networks, or alternatively trying to penetrate one's own networks before someone else finds a way in so that appropriate software and network modifications can be made.

The challenge facing an attacker in penetrating the target network will vary depending on the route taken by the attacker and the defenses employed across the firewall and access points.

Effective exploitation of computer networks allows structured exploitation of information-gathering opportunities, while denying adversaries the information needed to make timely and relevant decisions or leading them to make decisions favorable to friendly forces.

References

[1]   Hobbins, T., "Cultural Shift," *C4ISR Journal*, August 1, 2007, http://www.isrjournal.com.

[2]   Joint Staff, "US Net-Centric Operational Environment Joint Integrating Concept," Version 1.0, Washington, D.C., October 31, 2005.

[3]   U.S. Department of Defense Directive 8500.1, "Information Assurance," October 24, 2002, section E2.1.17.

[4]   Leedom, D.K., and R.G. Eggleston, "Modeling the Construction of Actionable Knowledge Within an Effects-Based Targeting Process," U.S. Air Force Research Laboratory, Wright-Patterson AFB, OH, 2005.

[5]   Alberts, D.S., and R.E. Hayes, *Understanding Command & Control*, CCRP Publication Series, 2006.

[6]   Bially, T., "Lifting the Fog of War," DARPATech conference presentation, 2005.

[7]   InfowarCon '99, Workshop 1, "Information Operations, IW for the Warfighter," 1999. (Paper VIII: Stategies for Achieving Information Resiliency" by Paul Zavidniak; Paper V: "Commonalities of Military and Commercial IW" by Anita D'Amico.)

[8]   Department of Defence, "Data Sharing in a Net-Centric Department of Defense," directive number 8320.02, December 2, 2004.

[9]   Defence Information Systems Agency, "Net-Centric Enterprise Services (NCES)," June 15, 2007.

[10] "USAF Electronic Systems Centre Strategic Technical Plan v 2.0," March 10, 2005.

[11] "U.S. DoD Defence Technology Area Plan," Office of the Director of Defense Research and Engineering, Washington DC, January 1997.

[12] "U.S. DoD."

[13] http://akss.dau.mil/dag/Guidebook/IG_c4.2.6.asp

[14] U.K. MoD, "Network Enabled Capability," JSP777, Jan. 2005.

[15] "DoD Architecture Framework Version 2.0," May 28, 2009.

[16] "UK MoD Architectural Framework MoDAF V1.2," June 2008.

[17] "NATO Architectural Framework V3," Feb. 11, 2010.

[18] Wilczynski, B., "Delivering an Information Advantage," *Integrated EA Conference*, 2009.

[19] "DoD Architecture Framework Version 1.5," April 23, 2007.

[20] *US Roadmap for Computer Network Attack*, October 30, 2003.

[21] Partnership for Critical Infrastructure Security, "Consensus Roadmap for Defeating Distributed Denial of Service Attacks, Version 1.10," February 23, 2000.

[22] DoD 8320.02-G, "Guidance for Implementing Net-Centric Data Sharing," April 12, 2006.

[23] David S. Alberts, John J. Garstka, and Frederick P. Stein, *Network Centric Warfare—Developing and Leveraging Information Superiority*, CCRP Publication Series, 2000.

[24] "Cyberspace Policy Review: Assuring a Trusted and Resilient Information and Communications, Infrastructure," Washington, D.C.: White House, 2009, p. iii.

[25] Boothby, A. RUSI C4ISTAR Conference presentation, 2006.

[26] Haffa, Jr., R.P., and J. Welch, "Command and Control Arrangements for the Attack of Time-Sensitive Targets," Northrop Grumman Analysis Center Papers, Nov. 2005.

[27] "Joint Pub 1-02," U.S. DoD Dictionary of Military and Associated Terms, April 2010.

[28] Bracewell, T., "Intrusion-Tolerant Systems," *Raytheon Technology* Today, Issue 2, 2007.

[29] "Network Centric Collaborative Targeting Program Achieves In-Flight Milestones," *Business Wire*, July 7, 2004.

[30] Department of Defense, "Network Centric Warfare," Report to Congress, July 27, 2001.

[31] U.S. Army War College, "Information Operations Primer," AY07 Edition, Dept. of Military Strategy, Planning, and Operations, November 2006.

[32] "USJFCOM definition," U.S. DoD Dictionary of Military and Associated Terms, April 2010.

[33] Llinas, J., *New Challenges for Defining Fusion Requirements*, Workshop, October 2006.

[34] Jobaggy, M.Z., "Literature Survey on Effects-Based Operations," TNO Physics and Electronics Laboratory, Netherlands Organisation for Applied Scientific Research, 2003.

[35] "UK Future Maritime Operational Concept (FMOC)," UK DWP 03, 2005.

[36] Leedom, D.K., and R.G. Eggleston, "Modeling the Construction of Actionable Knowledge within an Effects-Based Targeting Process," conference paper, April 1, 2005.

[37] Ling, M.F., "Nonlocality, Nonlinearity, and Complexity: On the Mathematics of Medelling NCW and EBO," 22nd International Symposium on Military Operational Research, 29 August – 2 September 2005.

[38] "U.K. MoD Network Enabled Capability," JSP777, Jan. 2006.

[39] Air Force, "Information Operations," Doctrine Document 2-5, August 5, 1998.

[40] US DoD, "FORCEnet—A Functional Concept for the 21st Century," 2005.

[41] Hayes-Routh, R., "Event Processing in the Global Information Grid (GIG): Orders of Magnitude Advantage in Information Supply Chains through Context-Sensitive Smart Push (VIRT)," Naval Postgraduate School, Monterey, CA, November, 2006.

[42] Defence Information Systems Agency, "Net-Centric Enterprise Services (NCES)," June 15, 2007.

[43] "Joint Publication 3-13, Information Operations," February 2006.

[44] "2008 Report to Congress by the US-China Economic and Security Review Commission," 2008.

[45] Bracewell, T., "Intrusion-Tolerant Systems," *Raytheon Technology Today,* Issue 2, 2007.

[46] "Network Enabled Capability," MoD pamphlet V1, April 2004.

[47] U.S. DoD, *Information Operations Roadmap*, October 2003.

[48] *Jane's International Defence Review*, August 2009, p. 5.

Endnotes

1. In this context, the term social domain refers to the unstructured nature of information generation, storage, and processing, where users may receive broadcast information (information push) or may request specific information from multiple databases and sensor sources (information pull).

2. In this context, the term information is used loosely. For a definition of data, information, and knowledge, see Chapter 3.

3. A pattern can be defined as a repeatable way to solve a class of problems that exist in an operational domain such as C2, logistics, or ISR. (NCOIC definition from [48].)

4. Sometimes also referred to as effects-based approach to operations (EBAO).

5. In fault-tolerant distributed computing, a Byzantine failure is an arbitrary fault that occurs during the execution of an algorithm by a distributed system such as crash failures and send, receive, and omission failures. Following a Byzantine failure, the system may respond in an unpredictable and inconsistent manner.

6. Middleware is a computer software that connects software components or applications by a set of services, allowing multiple processes running on one or more machines to interact.

7. Graceful degradation or fault tolerance is the property that enables a system to continue operating effectively in the event of the failure of one or more of its functions or components. If its operating quality decreases at all, the decrease is proportional to the severity of the failure, rather than resulting in a total breakdown.

Staff Sgt. Jeremy Emond inspects the satellite dish of the virtual secure Internet protocol router, nonsecure Internet protocol router access point system installed October, 2009, at Combat Outpost McClain in Afghanistan. (U.S. Army photo/Pfc. Melissa Stewart.)

NEC Techniques and Technologies

4

Data fusion

Data fusion is a multilevel, multifaceted process dealing with the registration, detection, association, correlation, and combination of data and information from multiple sources to achieve refined state and identity estimation and complete and timely assessments of situation (including threats and opportunities) [1].

Decision processes incur significant delays when humans manually manipulate data to provide direct or cognitive integration of multiple sources of data. Moreover, information is lost or modified in the process so that the original meaning or value is damaged [2]. The challenge of automated data fusion is therefore one of speed and accuracy, while information fusion must ensure that relevant features and facts are preserved during processing for human-in-the-loop decision making.

Generally, data fusion (Figure 4.1) can be divided into several categories [3]: *low-level fusion*, where raw data from a common signal type is combined; *intermediate-level fusion*, also known as feature fusion, where the vectors from processed data sources such as radars and motion and IR detectors are combined and at a video level where static features such as edges, corners, lines, and texture parameters are fused into a unified target dataset; and *high-level fusion*, or *decision fusion,* which combines decisions or confidence

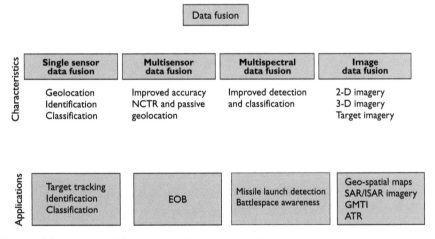

Figure 4.1 For both information and knowledge fusion, at some point in the process, data will have been fused to form a more comprehensive and accurate picture of the environment that is being measured or reported.

scores coming from several sources such as voting methods, statistical methods, or fuzzy logic–based methods.

Typically data fusion engines take data sets from multiple sensors (and sensor types) and fuse data in near real time to form target tracks and integrate these tracks with other ISR sensor data sets or other processed situational awareness data to form a more complete situational awareness picture and improve the inputs into the decision-making process.

When effectively integrated, data fusion across multiple sensors exploits the strengths of individual sensor characteristics to provide comparative target data sets and, when employed across a network, also benefit from the different sensor-target geometries to greatly improve the target resolution, position accuracy, and track quality, enabling earlier identification and engagement of threats.

Virtually all targets can be identified by their specific characteristics—data fusion using inputs from multiple sensors operating across different electromagnetic frequencies can significantly enhance the chances of target identification and tracking.

The task of data fusion has become much easier in recent years with the rapid advancement in processing speeds and transmission bandwidths. Early data fusion principles relied on a client-server approach to sharing the data, but more recent efforts have moved towards the concept of *mobile agents* [4] to achieve rapid data fusion across distributed networks and are typical of fusion principles found in the NCW environment.

While various data fusion models are available, the approaches taken to fuse data into a cohesive data set generally follow similar steps. Typical of the methods available and representative of the approach adopted by many is the model developed by the U.S. Joint Directors of Laboratories (JDL) [5], which defines the functional fusion elements and the stages of the fusion process. It has been revised several times in recent years to reflect advances in sensor and network performance.

Data fusion involves discrete steps to combine data into a useable picture, and typically models such as the JDL model [6] and subsequent revisions [7] define several levels of processing (Table 4.1 and Figure 4.2):

- **Level 1, Data fusion/object refinement:** Processing performs the sensor-to-sensor data registration, correlation-association, and state/identity estimation.
- **Level 2, Situation assessment:** Performs situation assessment based on all data.

- **Level 3, Impact and threat assessment:** Assesses the threat content of the data.
- **Level 4, Process refinement:** Includes management of the sensors and the internal fusion processes to optimize the process (objective functions for optimization include sensor emissions, target update rates, estimation accuracies, network bandwidth requirements, and load).
- **Level 5, Cognitive representation:** Includes specific situational adaptation to take account of the context in which the data is being processed.

Within each level, the following processing is typically required [9].

Level 1: Data fusion

- **Data preparation:** Performed to remove background noise and apply sensor-specific corrections (e.g., bias, gain, and nonlinearity corrections) to the data. Where different sensors are used, this stage will also harmonize the data formats ready for integration in a common format. This level encompasses various uses of multiple measurements in signal and feature processing including feature extraction in image fusion and signal detection and parameter estimation in ELINT and MASINT observations.
- **Data registration:** Places all sensor data in a common time and space coordinate system to correct for different time sampling, geometrical perspectives, and image planes of different sensors.

Data Fusion Level	Data Association Process	State Estimation Process	Product
Level 0: Subobject data association/ estimation	Assignment (observation to feature)	Detection	Estimated signal state
Level 1: Object refinement	Assignment (observation to entity)	Kinematic/attributive state	Estimated entity state (ID and track)
Level 2: Situation assessment	Pattern recognition/relationship (entity to entity)	Relation/identified pattern	Estimated situation state
Level 3: Impact assessment	Evaluation (situation to actor's goals)	Prediction/estimate intent/COA analysis	Estimated situation utility
Level 4: Process refinement	Resource management/ library refinement (task to resource)	Optimization/control/a priori state estimation (patterns)	Action/library update
Level 5: Cognitive representation	Information discovery: multiexpert collaboration (entity/situation/action-to-information representation)	Shared understanding/collaborative decision making	

Source: [8].

Table 4.1 Data Fusion Techniques Employ a Continuous Process of Assessment and Refinement to Produce Information Suitable for Decision-Making Processes

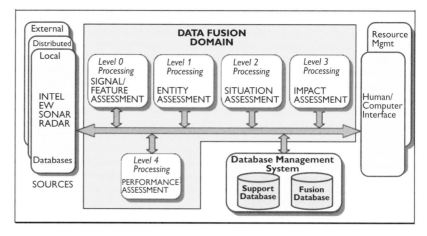

Figure 4.2   The data fusion process uses structured and discrete steps to fuse data from multiple sensor types.

- **Data integration:** The process of fusing processed data from all sensors to derive a refined data set including quantified error and uncertainty measurements in order to provide data estimates that can be processed to determine target detection events.

- **Detection event processing:** The determination of whether processed data represents a valid target detection event. Such a decision may be based upon the evaluation and "voting" of multiple individual sensor decisions or it may be based upon the combination of raw (preprocessed) data from multiple sensors (often referred to as predetection fusion).

- **Target data association:** The process that labels data by categories through comparing sensor measurements using temporal, spectral, or spatial properties to find the closest category match. The sensor data and category are then associated with the corresponding data from other sensors and are assigned to a predefined classification such as track files, targets, entities, and events.

- **Track or entity correlation:** New measurements are correlated with predicted states of all known targets, entities, and events to determine if each measurement can be assigned to update an existing track, used to initiate a new track, or eliminated as a false alarm.

Level 2: Situation assessment

- **Target identification:** All of the associated data for a given target is used to perform automatic target recognition and to classify the target by threat type. A target may subsequently be established as a friendly track or entity but will still need to be tracked to maintain situational awareness.

- **Situational context processing:** Tracks, entities, and events are placed in the context of the environmental scene to take into account terrain, target mix, and spatial arrangement in the assessment of their characteristics and behavior.

Level 3: Threat assessment

- **Threat identity, state, and behavioral assessment:** The output product from the situational context processing, together with the target classification data, is used to determine as precisely as possible the specific platform type (and therefore its offensive and defensive capabilities), its current state (whether, for example, an aircraft is locked onto a target or is simply in a surveillance mode), and its dynamic movement (in order to predict whether the characteristics represent a threatening behavioral pattern). Such an assessment can then be used to determine the threat to the sensor platform or to the network as a whole.
- **Situational threat assessment:** Once each target has been assessed, the targets are placed in the environmental context and a relative assessment is made to determine current and potential threats (defensive) and opportunities (offensive). This will enable engagement priorities to be determined relative to the survival of the platform and the network and against mission priorities.

Level 4: Sensor tasking

- **Data enhancement:** To reflect the priorities across determined in earlier stages, sensors may be tasked to refine the clarity of the situational picture, for example, by tasking observations to be made from a different perspective or sensor type or by increasing the track update rate. In a networked environment where sensors on one platform may be tasked by commands from another, the challenge is to ensure that the sensors are available, that the new tasking is appropriate to the intentions of the platform that is being tasked, and that it can be undertaken and shared across the network for the duration required.

Level 5: Cognitive and user refinement

Data fusion methods (Figure 4.3) may also be categorized by the extent to which the fusion algorithms benefit from intelligent learning, where knowledge is developed from situational measurements and is fed back into the fusion algorithm to improve computed results. Such input is referred to as *expert knowledge* and is typically expressed in terms of *if, or, then,* and logic for data fusion processes, often supported by fuzzy logic theory.[1] *Neural networks*[2] may also provide useful inputs where their knowledge has been

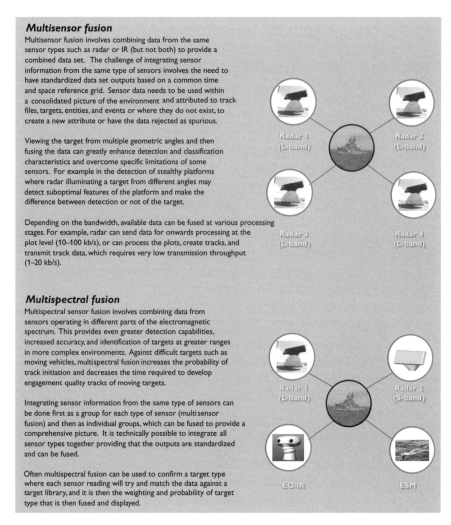

Multisensor fusion

Multisensor fusion involves combining data from the same sensor types such as radar or IR (but not both) to provide a combined data set. The challenge of integrating sensor information from the same type of sensors involves the need to have standardized data set outputs based on a common time and space reference grid. Sensor data needs to be used within a consolidated picture of the environment and attributed to track files, targets, entities, and events or where they do not exist, to create a new attribute or have the data rejected as spurious.

Viewing the target from multiple geometric angles and then fusing the data can greatly enhance detection and classification characteristics and overcome specific limitations of some sensors. For example in the detection of stealthy platforms where radar illuminating a target from different angles may detect suboptimal features of the platform and make the difference between detection or not of the target.

Depending on the bandwidth, available data can be fused at various processing stages. For example, radar can send data for onwards processing at the plot level (10–100 kb/s), or can process the plots, create tracks, and transmit track data, which requires very low transmission throughput (1–20 kb/s).

Multispectral fusion

Multispectral sensor fusion involves combining data from sensors operating in different parts of the electromagnetic spectrum. This provides even greater detection capabilities, increased accuracy, and identification of targets at greater ranges in more complex environments. Against difficult targets such as moving vehicles, multispectral fusion increases the probability of track initiation and decreases the time required to develop engagement quality tracks of moving targets.

Integrating sensor information from the same type of sensors can be done first as a group for each type of sensor (multisensor fusion) and then as individual groups, which can be fused to provide a comprehensive picture. It is technically possible to integrate all sensor types together providing that the outputs are standardized and can be fused.

Often multispectral fusion can be used to confirm a target type where each sensor reading will try and match the data against a target library, and it is then the weighting and probability of target type that is then fused and displayed.

Figure 4.3 Data fusion requires the fusion of data from multiple sensors operating at different wavelengths as well as the fusion of data from the same type of sensors on multiple platforms.

gained from the analysis of real-life situations from which logic and processing patterns can be derived.

Information fusion

Where information fusion is required, it will already have been derived from data that has been processed and placed in a contextual framework. Typically, information fusion will therefore involve the fusion of postprocessed sensor and intelligence data from disparate sources with differing conceptual, contextual, typographical, and space-time frameworks. Information fusion (Figure 4.4) is also called *high-level data fusion* or *decision*

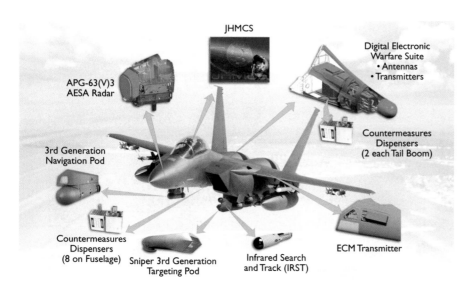

Figure 4.4 Information fusion from multiple active and passive sensors is essential to create a comprehensive situational awareness picture around the platform. In addition to on-board information fusion, additional information obtained via data links to other real-time and nonreal-time sources will also help to supplement the accuracy of the picture and target priorities. The on-board sensor suite for the Boeing Advanced F-15 is illustrated in this picture. (Image courtesy of Boeing.)

fusion, where decisions coming from several sources are correlated by a variety of statistical means to provide a new decision or an indication of a degree of confidence in a fused decision.

Image fusion

The fusion of multispectral images from sensors located on different platforms (Figure 4.5) can be technically difficult as each sensor will have its own spectral and geometric view of a target and, most likely, differences in resolution of the image. Fusing imagery of a target to produce a coherent picture from different sensors requires a four-dimensional space/time grid and complex processing.

Typically, image fusion can incorporate inputs from multiple sensors operating in different parts of the electromagnetic spectrum. The task complexity can be reduced if sensors are colocated on the same platform to ease the challenge of geospatial integration across differing time references. The subsequent single channel output can then be either displayed to an operator or further incorporated into a networked fusion processes. This is particularly important for time-sensitive targets when the number of available spectral bands providing target information becomes so large that it is impossible to view the images separately. This kind of fusion requires a

Figure 4.5   Platforms have numerous physical, spectral, and RF features that can be used to classify them against a library of known characteristics.

Automatic Target Recognition

Automatic Target Recognition (ATR) typically uses multispectral sensor fusion to provide early detection, track initiation and accurate determination of the target class and identity. Typically fusion processes will process inputs from multiple sensors types such as EO/IR, ESM, and radar and correlate sensor measurements against spectral signatures in target libraries to improve identification accuracy.

ATR makes use of current environmental information (terrain, weather, urban activity,etc.) to improve the accuracy of targeting decisions.

In a networked environment consideration is also given to where the data fusion is carried out. This is particularly important in reducing the transmission of high-volume, low-content, raw (unprocessed) data across the network and to speed up the kill chain process thereby maximizing the kill probability, especially for fast moving or fleeting targets.

Typically three methods of ATR can be used:

Spectral emission matching: Platform targets in particular will offer numerous multispectral characteristics that can be measured from different angles by sensors positioned across networks and fused to form a composite target information picture and matched against an emissions characteristics library which typically compares waveforms and frequencies with known target RF emission characteristics.

Non-Cooperative Target Recognition (NCTR): using detailed analysis from radar returns, multispectral signatures, and RF signals to analyze the features of targets to enable it to be identified by type and sometimes by individual platform.

Image Feature Matching: Images may also need to be fused from multiple sensors during automatic target recognition processes so that features can be extracted from the image and compared against an image library to accurately identify the target (Figure 4.6). Platform targets in particular will offer numerous multi-spectral characteristics that can be measured from different angels by sensors positioned across networks and fused to form a composite target information picture. In such cases, a logical processing sequence is followed to construct a geometrically accurate model of the target that can be referenced against the geometric, spectral and feature models. Given that the target may be viewed from any angle, the first step is to warp the image data to a common coordinate system before extracting target features for matching. Once the target type has been identified further processing may also consider the situational context including terrain, target disposition and vectors to classify the nature of the threat and to prioritize the engagement sequence [82].

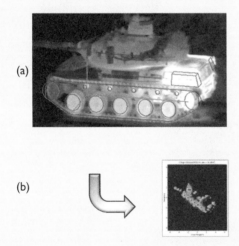

Figure 4.6   (a) Target feature extraction forms the principle of image-based automatic target recognition [82]. (b) Matched against sensor image such as a SAR radar image of a tank shown here. (Figure courtesy of QinetiQ.)

precise (pixel-level) registration of the available images [3] (Figures 4.7 and 4.8).

While image fusion of a target using sensor data taken from the same geospatial point is challenging, more complex scenarios add to the processing requirements and can introduce a further degree of latency in to the production of the image. A target such as a building or a vehicle can be viewed by a platform on the move such as a UAV to build up a

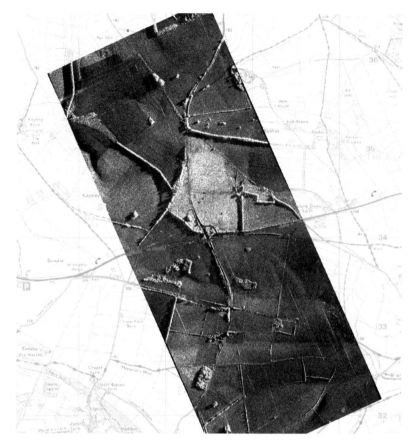

Figure 4.7 Image fusion is most effective when overlaying sensor information in a multidimensional grid, such as the overlay of GMTI and SAR data on a topographical map. In this picture, every pixel and GMTI track from the SAR radar image has its own geolocated map reference. (Image courtesy of QinetiQ.)

comprehensive image reference of target from different reference angles (Figure 4.9). Image fusion from multiple angles is called *multiaspect image fusion*. Multiaspect image fusion introduces the possibility of building an image library for automatic target recognition using feature or image recognition or for using the data to undertake 3D modeling of the scene being surveyed.

3D modeling is most effectively undertaken using active sensors where the range from the sensor to different parts of the target can be detected. In dense or built-up areas, target reconstruction is often impossible from a single SAR measurement alone. However, the reconstruction quality can be significantly improved by a combined analysis of multiaspect data. Further improvements are also possible using active sensors with very narrow beams such as Lidars (laser radars) for fine discrimination of target features and range (Figure 4.10).

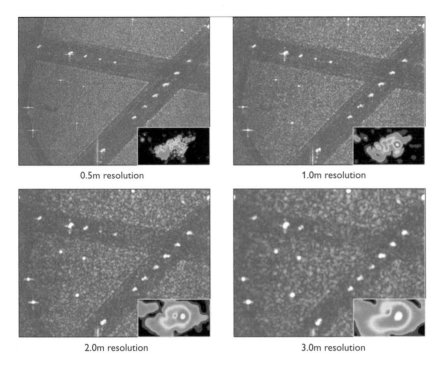

0.5m resolution	1.0m resolution
2.0m resolution	3.0m resolution

Figure 4.8   Vehicles imaged at various resolutions (inset: a single tank).

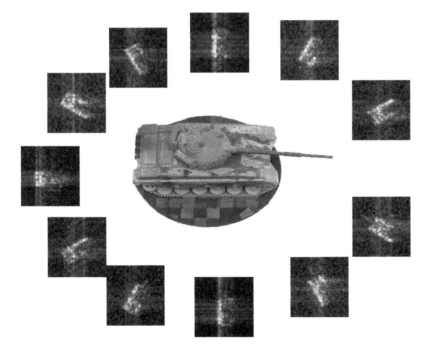

Figure 4.9   Variation of SAR signature with look angle for a Russian T-72 tank.

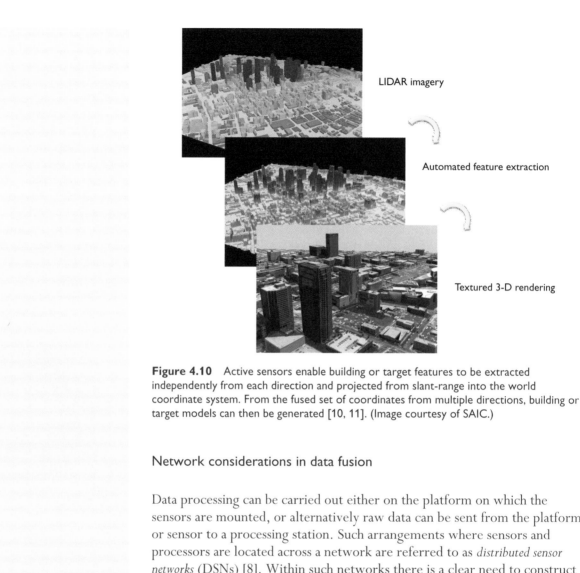

LIDAR imagery

Automated feature extraction

Textured 3-D rendering

Figure 4.10   Active sensors enable building or target features to be extracted independently from each direction and projected from slant-range into the world coordinate system. From the fused set of coordinates from multiple directions, building or target models can then be generated [10, 11]. (Image courtesy of SAIC.)

Network considerations in data fusion

Data processing can be carried out either on the platform on which the sensors are mounted, or alternatively raw data can be sent from the platform or sensor to a processing station. Such arrangements where sensors and processors are located across a network are referred to as *distributed sensor networks* (DSNs) [8]. Within such networks there is a clear need to construct information flow and decision-making protocols that maximize the efficiency of the network and the various elements within it for optimum fusion performance.

Transferring raw data, particularly for sensors such as radars, EO, and EW systems requires a huge amount of bandwidth and can be more effectively achieved if partial processing is carried out on-board the platform prior to transmission.

Data fusion in a distributed sensor network (DSN) is slightly different to that found in hierarchical sensor fusion in that each node, depending on its function, may need to be capable of data fusion in its own right. No node is central to the fusion process and a standardized modular approach is typically used for specific fusion processes between families of nodes. Measurement errors and background noise also need to be taken into account in the fusion process, and the processing of outputs from multiple sensor nodes will improve the overall network accuracy.

Given the huge amount of information available to the networked decision maker, there is an increasing challenge of how to use the sometimes overwhelming knowledge that is created to best effect. Much emphasis in networked systems is placed on the task of ensuring the distribution of relevant, timely, and appropriate information throughout a mission, in particular in a format that enables its rapid assimilation and use in decision making. For example, a single seat fighter aircraft equipped with multiple sensor types and access to networked situational information needs to present information to the pilot in a manner that ensures a manageable workload in a high-density operational environment (Figure 4.11).

Data fusion has also driven the standardization of processing methods, symbology, and text handling to harmonize and facilitate interaction and data sharing between different networks and C2 systems. Typically, networks that need to openly share and fuse data are constructed around a *Service Oriented Architecture* (SOA) approach enabling users and nodes to publish, subscribe, and access data between loosely coupled open-standard network applications. Communications protocols are increasingly constructed around IP standards, sometimes necessitating preprocessing to format nonstandard data into IP formats. Information is then shared on the network by using XML to capture key data (typically using both a community of interest rich schema and an enterprise-loose coupling schema).

Figure 4.11 Single seat fighters such as the F-15E Eagle integrate a vast amount of sensor and network data into a condensed picture for the pilot. Here an F-15E Strike Eagle soars over the mountains of Afghanistan in support of Operation Mountain Lion in April 2006.

Since the introduction of DSNs (Figure 4.12), multiple architecture arrangements have been developed to improve network efficiency, including the hierarchical and committee organization [13], the flat tree network [14, 15], the deBruijn-based network [16], and the multiagent fusion network [17].

While improving different aspects the performance of DSNs (Figure 4.13), all these approaches use a common network computing model that supports many standard distributed systems processes and architectures, such as Remote Procedure Calling (RPC), Common Object Request Broker Architecture (CORBA), and Service Oriented Architecture (SOA). Such an arrangement is referred to as the client/server model; the client (individual sensor) sends data to the server (processing element) where data processing functions are carried out.

IP Markup Languages and Protocols

Markup Language

A markup language is an artificial language using a set of annotations to text that give instructions regarding the structure of text or how it is to be displayed. The term markup is derived from the traditional publishing practice of "marking up" a manuscript, which involves adding symbolic printer's instructions in the margins of a paper manuscript.

Extensible Language

An extensible language allows the user to define the mark-up elements.

Hyper Text Markup Language (HTML)

HTML, or HyperText Markup Language, is the predominant markup language for internet web pages. It provides a means to describe the structure of text-based information in a document—by denoting certain text as links, headings, paragraphs, lists, and so on—and to supplement that text with interactive forms, embedded images, and other objects. HTML is written in the form of tags, surrounded by angle brackets

eXtensible Markup Language (XML)

Is an open standard programming language that enables the sharing of structured data across different networked applications especially via the Internet. Its self-documenting format describes structure and field names as well as specific values.

Transducer Markup Language (TML)

Is a programming language developed to describe any transducer (sensor or transmitter), its inputs and outputs in terms of a common model, including characterizing and time-tagging not only the data but also providing XML formed metadata describing the system producing that data, and the data inputs, physical and logical relationships used to produce the output. TML can process data from simple stationary in-situ transducers to high bandwidth dynamic remote devices such as a synthetic aperture radar system.

Figure 4.12   An unattended ground sensor on trials with the U.S. Army. Target detection, classification, localization, and tracking are all typical applications in distributed sensor networks. By integrating observations from different sensors at different locations, the networked system will automatically trigger network alarms whenever a certain target is detected.

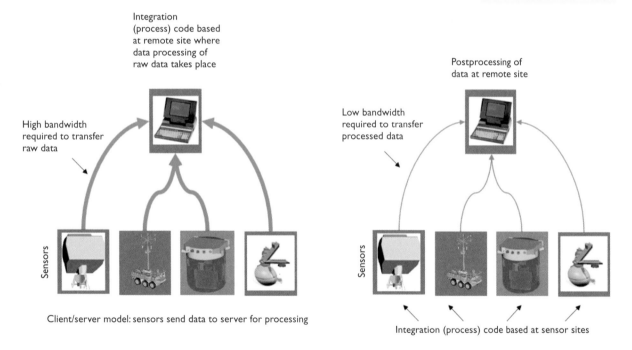

Integration (process) code based at remote site where data processing of raw data takes place

Postprocessing of data at remote site

High bandwidth required to transfer raw data

Low bandwidth required to transfer processed data

Sensors

Sensors

Client/server model: sensors send data to server for processing

Integration (process) code based at sensor sites

Figure 4.13   (a) Traditional and (b) mobile agent distributed sensor networks.

The principle of *mobile agents,* however, takes a different approach. Rather than moving data to processing elements themselves for data integration, as is typical of a client/server model, the concept of mobile agent distributed sensor network (MADSN) moves the processing function to the data locations. This reduces network bandwidth requirements and provides an effective means for overcoming network constraints, since the transfer of large volumes of data are avoided and the output data from the computations can be sent instead, reducing the network bandwidth requirements to a minimum [18].

The mobile agent concept allows processing algorithms to operate on an autonomous basis and provides for the software to migrate from node to node, or sensor to sensor, performing data processing and also reporting on an autonomous basis. Mobile agents provide the right user with the right information in a timely manner from across the Global Information Grid (GIG), taking advantage of multiple intelligence databases, user communities, ISTAR networks, and Web-based government and NGO information sources.

Traditional software by comparison can typically only process data when it is called to do so by other software routines. An alternative approach to truly autonomous software operation is to ensure that sensors perform preprocessing to a defined standard so that the output from all sensors is compatible with common data standards that can use the information, irrespective of the type of sensor from which it originates.

Network-level data fusion

At a conceptual level, networks need to fuse and analyze data in order to automatically organize services and applications, position them in the network, and advertise their existence, adapting them as needs change [19].

In order to accomplish this at a network level, data fusion is generally structured around a Service Oriented Architecture (SOA) approach, an approach to defining integration-architectures based on the concept of service, where services are provided in the form of a collection of applications, data, and tools that interact via message exchanges [2]. The challenge in a global information environment is to ensure that information can be integrated across different operating systems and different transmission standards, and a SOA approach addresses that challenge in a standardized, structured manner.

SOA is an approach to structuring software architecture and interfaces with other systems that enable each interface or node to provide and receive data in the form of a service provision from other entities or nodes. SOA separates service-based functions into distinct data provision functions, which are then made accessible over a network in order that users can combine and reuse them to produce user-specific applications [20]. Such services communicate with each other through a mesh of software services by passing data between themselves and by coordinating activity between services according to the functional requirements for data provision requested by users on the network.

Such an approach enables multiple users to interact with and use data resources and processing capabilities from across the network (Figure 4.14). In an SOA environment (Figure 4.15), nodes on a network make resources and data available to other participants in the network as independent services that participants can access using standard information exchange protocols. Most SOA approaches therefore use an open systems, Web-based protocol approach to provide industry-standard architectures and languages; however, other nonstandard approaches can be used if the gateway or interface to the network presents the information in a predetermined repeatable format.

Unlike traditional node-to-node architectures, SOAs comprise flexible, loosely structured but highly interoperable services with the ability to communicate across the network either directly to another node or via other nodes to retrieve or send data. These services exchange information based on a structured protocol, which is independent from the underlying platform, network, or programming language. Once the SOA software code is written, it can easily be duplicated to perform the same functionality for similar nodes in the network as the nodal interfaces are structured around defined standards and are independent from the underlying implementation of the service logic.

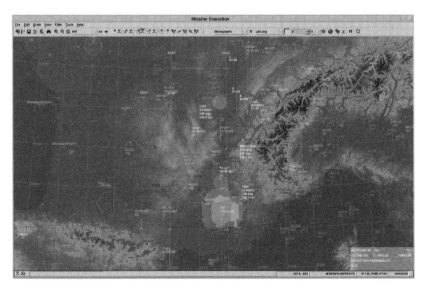

Figure 4.14 The NATO Air Command & Control System fuses data from multiple national networks to form a comprehensive Single Integrated Air Picture (SIAP) for participating nations. (Image courtesy of Thales Raytheon Systems.)

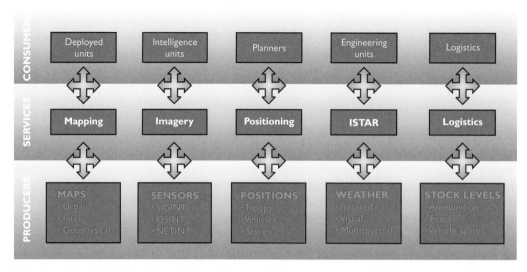

Figure 4.15   Examples of C4ISR services, their producers, and their consumers in a SOA environment. The situation picture is an aggregation of the map, sensor data, and positioning services. All services can be accessed by authorized consumers [21].

Service Oriented Architecture Definitions

A service-oriented architecture (SOA) is an application framework that takes everyday business applications and breaks them down into individual business functions and processes, called services. An SOA lets you build, deploy and integrate these services independent of applications and the computing platforms on which they run.
—IBM Corporation

Service-Oriented Architecture is an approach to organizing information technology in which data, logic, and infrastructure resources are accessed by routing messages between network interfaces.

—Microsoft Corporation

SOA relies on the ability to identify across the accessible network, services, functions, and their capabilities. In order to achieve this, SOA structures this information in a directory that describes the services or functions available in its domain. This enables a function in one domain to perform a service for another domain in response to a suitably structured request. As SOA is designed to ease connectivity and work with structured data sharing, it also has the capability to provide for detailed and granular services to be rolled up into a cohesive coarser-grained service that can meet functional requirements at a higher level. Further extensions to SOA address business processing, workflow management, content management, peer-to-peer services, and portal enquiry services.

Data and information fusion networks can be designed to fulfill a number of roles, often categorized by the time sensitivity of the information or data that they carry. In the United States, these networks generally fall into one of three categories (Figure 4.16):

- **Joint Planning Network (JPN):** JPN networks carry a high volume of nonreal-time processed data and information such as operational readiness status, logistics information, order of battle, mission planning, and mission status information.
- **Joint Data Network (JDN):** JDN networks carry near-real-time cueing and weapon engagement coordination information to provide unambiguous composite track information and details of target type.

Joint Network Architecture

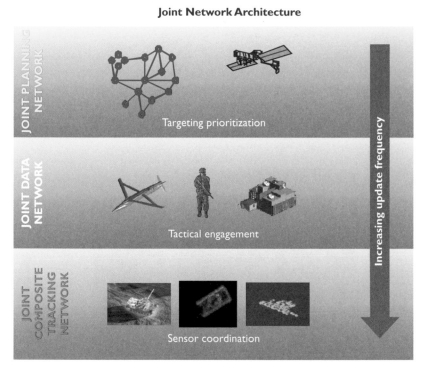

Figure 4.16 Joint Network Architecture is structured to support varying degrees of data latency and accuracy.

- **Joint Composite Tracking Network (JCTN):** JCTN networks carry real-time, accurate precision sensor data to enable rapid and timely processing and fusion of multisensor inputs across a specific network. Such real-time sensor fusion provides the mechanism to engage targets prioritized across the network, including those outside the range of the networks' own sensors and effects systems, by tasking other collaborating units who may be within range of their effects systems, but who may not have detected the target with range of their own sensors.

Data fusion displays

Typically, situational awareness information displays fuse data so that the viewer is presented with a situational awareness picture formed of the fused data from multiple sensors. The determination of the information displayed will take into account the weighting of the various sensor inputs in determining the target classification and location.

Confirmed targets are presented on a composite electronic order of battle (EOB) display on a digital map background which can be combined to (fuse)

Data Fusion in Land Combat Systems and the U.S. Land Warrior Program

In common with the principles embedded in networked information systems, land combat systems seek to fuse data from multiple sources to provide a comprehensive real-time situational awareness picture (Figure 4.17).

Figure 4.17   Sgt. 1st Class David Thomas looks through his helmet-mounted display introduced as part of the U.S. Land Warrior program.

To provide warfighters with actionable information, the data from the various distributed intelligence, reconnaissance, and surveillance (ISR) and other sensor assets are subject to complex data processing, filtering, correlation, aided target recognition, and fusion. The sensor data management (SDM) software organizes all the sensor data—including detection reports—and tracks information as received from the sensor packages. Data is then processed and fused to synthesize information about the object, situation, threat, and ongoing intelligence, reconnaissance, and surveillance (ISR) processes. In addition to receiving data from organic sensors, SDM has the capability to receive sensor data from nonorganic sources, including current forces and Joint, Interagency, and Multinational (JIM) sources. SDM will perform sensor data format conversions to output the data in standard data formats [22].

inputs from multiple sensors such as ELINT, COMINT, and radar sensors resulting in a higher-order Electronic Order of Battle (EOB) representing friendly, enemy, and unknown force deployments, together with (where this has been deduced) the relationship between these units as represented by their communications interactions. A typical EOB display layout is shown in Figures 4.18 through 4.20.

Applications of data and information fusion

Information superiority provides friendly forces with a competitive advantage only when it is translated in a timely manner into superior knowledge and decisions. A key challenge in this process is, of course, the ability to handle and process huge volumes of data and to display them in the form of information useful to create knowledge. Accordingly, the processing and display of data to achieve a near real-time common situational awareness picture of the battlespace are now at the very forefront of military research.

Achieving an integrated situational awareness picture requires the fusion of data (Table 4.2) (or information) from multiple types of sensors across multiple networks to gain a complete picture of the battlespace, including the disposition and intentions of enemy, friendly, and civilian assets. Sensors are directed to the area of interest and data is collected before being processed and disseminated both directly back to the tasking source and across the network to start the process of conversion from data to knowledge (Figure 4.21). Data fusion occurs in multiple steps and in multiple layers

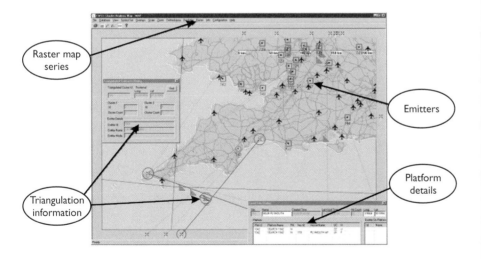

Figure 4.18   A composite EOB display.

(a)   (b)

Figure 4.19   (a) F-35 Pilot vehicle interface (PVI). This is what the F-35 pilot will see on his display panel where sensor data from the aircraft and from the network is fused to provide a complete situational awareness picture and automated target engagement sequencing. ID "Hostile" = Next to Shoot. Red triangles show targets entered into a shoot list with numbers showing automatic engagement sequence. Targets are automatically identified, classified, and assessed for their hostile intent or threat status before the engagement sequence is determined. (Image courtesy of Northrop Grumman). (b) The first two F-35A Lightning II aircraft on a systems test flight. (Photo courtesy of Lockheed Martin.)

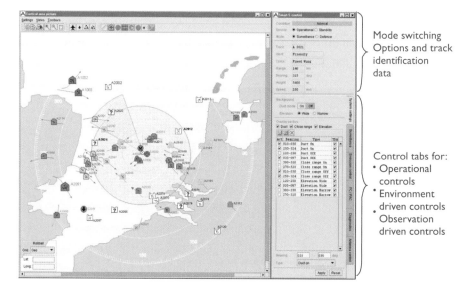

Figure 4.20   A modern naval radar display using multisensor data fusion to provide known, unknown, and hostile target information. (Image courtesy of Thales.)

Fusion examples	Single sensor-type	Diverse sensor types
Single platform	UAV with EO/IR payload (data available for fusion across a network)	F-22, Eurofighter with on-board multisensor data fusion
Multiple platforms	Networked Sigint system, SAR radar change detection mapping	Networked air C2 system providing Single Integrated Air Picture (SIAP)

Table 4.2 Varying Degrees of Fusion Occur on Different Platform Types

Figure 4.21 Sensor fusion on platforms and between platforms is increasingly important in achieving a real-time situational awareness picture across the network. Here, a Lynx HAS Mk8 helicopter of 815 Naval Air Squadron attached to Type 23 frigate HMS *St. Albans* is secured on the flight deck. The radar, EW, and EO sensors are clearly visible on the nose of the helicopter. (Image courtesy of U.K. MoD.)

across numerous sensor and network nodes to enable it to be fused in a timely manner and used in the compilation of a shared situational awareness picture or to perform a specific role within the OODA loop.

Multisensor data fusion, where ISR data is produced from multiple platforms, is a key concept of network-centric warfare (NCW). It is aimed at coordinating ISTAR assets, databases, and effects systems across the battlespace to achieve a common objective. It permits dramatic improvements in the accuracy of ISR data when compared with reliance on the data from a single platform or sensor and enables all platforms within a network to access near-real-time data produced as a composite picture from multiple multispectral sensors.

Developing shared battlespace awareness requires that sensors (or rather, the information they generate) be linked in a time-referenced network with careful planning of how best to translate the data they collect into information and then into knowledge for the decision-making process. The challenge is to determine how, where, and when data will be fused (Figure 4.22) and to understand the implications of the reliability and integrity that is then placed on that information as part of the decision-making process.

An efficient fusion engine will deliver many advantages such as improved detection range due to the ability to match weak signals from multiple sensors to improve the detection probability (Figure 4.22); earlier track establishment and engagement targets before the launch platform (Figure 4.23) can be detected; improved performance against jamming and countermeasures (it is very hard to camouflage an object across all parts of the spectrum) and improved performance in adverse environmental conditions where smoke, rain, or fog can cause poor contrast in the visible spectrum and rain can cause low thermal contrast. A combination of thermal and infrared imaging will often give a significantly enhanced picture [3].

Figure 4.22 Sensors on-board a particular platform often operate in a coordinated manner and will often fuse data before placing it on the network as a confirmed bearing or track. Here an EP-3E "Aries II" from Air Reconnaissance Squadron One (VQ-1) equipped with multiple electronic sensors is deployed to Afghanistan.

In addition to the advantages of shared situational awareness (Figure 4.24), data fusion across multiple platforms enables earlier and more accurate target identification and increases the accuracy of detection, in particular, the position accuracy and identification of specific target type and the reduction of identification errors—something that is increasingly important where engagement decisions are made well beyond visual ranges. Additionally, the availability of fused information from multiple sources provides options for platforms that do not wish to alert others to their presence by accessing a situational awareness picture without having to use their own active sensors, thereby reducing their chance of detection (Figures 4.25 and 4.26).

At an effects level (Figure 4.27), weapons systems lethality increases disproportionately with data fusion speed, accuracy of inputs help to dramatically shorten the kill-chain process, and the sensor-to-shooter timeline decreases through faster and more ubiquitous sharing of information, improved prediction of target characteristics, and weapons engagement decisions and guidance.

Figure 4.23   The need to detect platforms hiding in one part of the electromagnetic spectrum underlines the importance of multispectral fusion to detect targets. Here an artist's impression shows the effectiveness of optical camouflage using a system of cameras to detect the surrounding background patterns and a skin of LEDs to display the correct image when viewed from different angles. Such a system provides constantly variable camouflage at optical frequencies. (Image courtesy of U.K. MoD.)

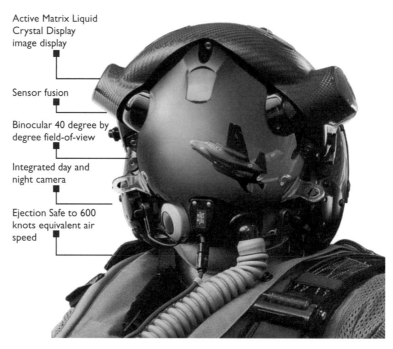

Active Matrix Liquid
Crystal Display
image display

Sensor fusion

Binocular 40 degree by
degree field-of-view

Integrated day and
night camera

Ejection Safe to 600
knots equivalent air
speed

Figure 4.24   On the F-35 Lighting II the pilot is presented with a comprehensive situational awareness picture on to the visor of his helmet from the seamless fusion of data from multiple active and passive sensors.

Figure 4.25   Single platform data fusion occurs in the F-22 Raptor (shown) where the pilot is presented with a battlespace picture derived from multiple sensors (radar, EW, IFF, data link, infrared search, and track) with data being fused to present battlespace picture of significantly enhanced accuracy and reliability with early target designation of friend or foe.

Figure 4.26 The NATO Air Command & Control System fuses over 46 different types of sensors and over 300 radar inputs into a single integrated Air C2 system. (Photo courtesy of Thales Raytheon Systems.)

Radio communications

Network-centric warfare (NCW) requires that all participating resources have access to the appropriate communications channels to facilitate information sharing and to coordinate C2 activities. Communications between networked nodes and systems are essential if dynamic control is to be maintained across the force structure. From ISTAR to the delivery of a suitable effect to the target, effective communications are the glue in the C2 processes that initiate and control these missions.

Given the diverse range of operating environments in which operations are conducted, the range of communications methods varies significantly depending on the data transmission rates required, the physical constraints of the platform and the environment in which the platform is operating.

Communications methods may exploit a wide range of frequencies across the electromagnetic spectrum (Figure 4.28). The electromagnetic spectrum covers a wide range of frequencies from extra low frequency (ELF) bands at frequencies of 3–30 Hz (1 Hz = 1 cycle per second) to extra high frequency bands (EHF) operating at frequencies between 30–300 GHz. High frequencies are capable of transmitting high volumes of data over long ranges, while low frequencies are capable of transmitting signals over long ranges but at low data rates. While in theory any part of the electromagnetic

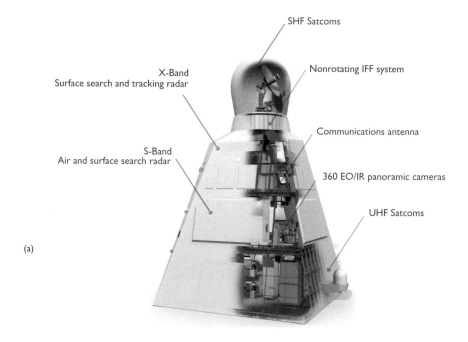

(a)

(b)

Figure 4.27 (a, b) Naval sensor fusion is well illustrated through the concept of the integrated sensor mast, where sensor data is fused to improve performance in adverse environmental and operational conditions. Typically, smoke or fog can cause poor contrast in visible light frequencies and high humidity such as rain and sea spray can cause low thermal contrast in infrared cameras. By combining both types of sensors with radar data, better overall performance is achieved in adverse conditions. (Images courtesy of Thales.)

spectrum may be used to transmit information, these frequency extremes tend to be used for specific applications, while frequency ranges between the extremes tend to find more common use in regular communication and data link systems.

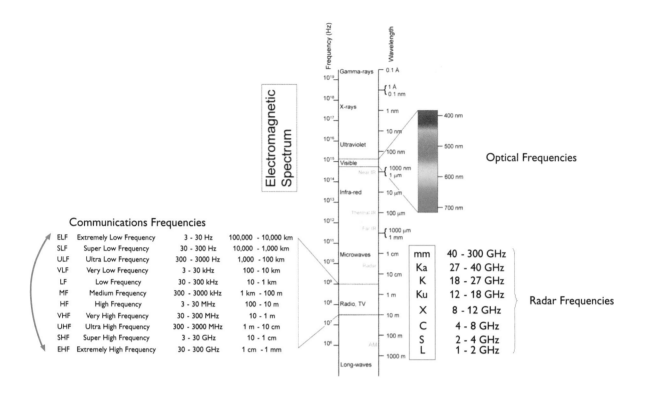

Figure 4.28 The electromagnetic spectrum.

There are various international standards for the designation radio frequencies across the electromagnetic spectrum that are used for radar and radio communications. The more common frequencies and their designations are shown in Figure 4.29.

Radio propagation [151]

The propagation (or transmission) characteristics of radio waves at different frequencies and under different atmospheric and environmental conditions determine the range and quality of reception of the signal and are therefore of great importance when planning a radio communications network.

The earth's atmosphere is divided in to several layers (Figure 4.30), each of which has their own propagation characteristics. The exact heights of these layers vary according to the time of day and are principally determined by variations in temperature across the day.

"(U) We must improve network and electro-magnetic attack capability. To prevail in an information-centric fight, it is increasingly important that our forces dominate the electromagnetic spectrum with attack capabilities."

—U.S. DoD, Information Operations Roadmap, October 30, 2003

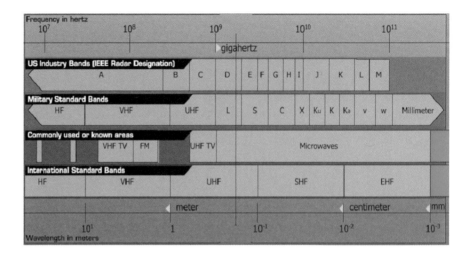

Figure 4.29 Commonly used radio and radar frequency bands. (Image courtesy of Thales.)

There are two principal layers in the atmosphere that affect radio signal propagation: the troposphere and the ionosphere, both of which act on radio signals in very different ways and enable signals to reach well beyond line-of sight distances.

The ionosphere

The layer called the ionosphere is filled with charged particles that can reflect radio waves at certain frequencies in the HF band and low VHF bands. This is called *skywave propagation* and radio waves reflected in this manner are known as *skywaves*. When radio waves at frequencies in this bandwidth reach the ionosphere, they are reflected (refracted) back towards the earth. When the radio waves reach the earth, they may again be reflected from the earth's surface back towards the ionosphere and back down to the earth several times more, providing communication distances that enable signals to be heard at certain points virtually anywhere on the globe. At the points where the radio waves reach the ground, they can be received and decoded. The distance from the transmitter to the first bounce is called the *skip distance*.

Higher HF frequencies tend to be refracted off higher ionospheric layers and hence the skip distances are longer. At a certain point the frequency becomes too high to be refracted and passes through the ionosphere in to space, with little or no reflection taking place. This is called the *critical frequency* (Figure 4.31).

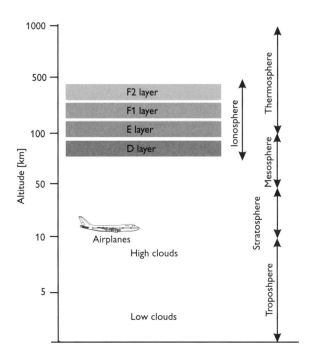

Figure 4.30 The atmosphere comprises of multiple layers, all of which exhibit their own transmission characteristics. (Image courtesy of Ian Poole.)

The angle at which the signal reaches the ionosphere (angle of radiation) will also determine the skip distance, both through the refraction angle off the ionosphere and the need to avoid the critical angle, above which the frequency will not be refracted.

The distance that HF signals travel depends on a range of factors, principally frequency (which determines the layer of the ionosphere from which the signal is reflected), transmitter power, transmission angle relative to the ground, and the number of "bounces" achieved between the ionosphere and the ground as the signal travels on its journey (Figure 4.31).

The ionosphere, in turn, is comprised of several layers:

- **D Layer:** The lowest region of the ionosphere, rather indistinct, and existing only during the day and extending from about 40 to 65 km (25 to 40 miles) above the earth. Affected by low level atmospheric conditions such as lighting.
- **E Layer:** Forms the middle layer of the ionosphere, extending from about 90 to 150 km (55 to 95 miles) above the earth and improving long-distance communications by reflecting radio waves back towards the earth in the range of 1–3 MHz. Also known as the *Heaviside layer*. Significantly weaker at night.

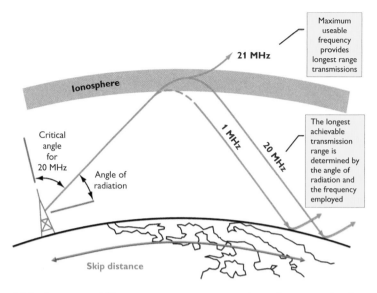

Figure 4.31   Long-range HF communications are more dependent upon the choice of frequency than the power of the transmission. For any given distance, time of day, and atmospheric conditions, there will be a specific band of frequencies that will enable optimal communications with frequencies outside that range working poorly or not at all. An increase in power may help if the frequency is too low, but using a higher, more suitable frequency is more likely to improve the range as it will be reflected from a higher level in the ionosphere and achieve a longer skip distance. The highest frequency which may be used for reliable HF skywave communications is known as the maximum usable frequency (MUF). (Adapted from: [23].)

- **F Layer:** Forms the upper layer of the ionosphere although it is not observable uniformly in the same pattern every day. The principal reflecting layer for HF transmissions during the summer is the F1 layer, extending at night from about 190 to 400 km (120 to 250 miles) and during the day from about 145 to 400 km (90 to 250 miles) above the earth. During the day it forms two layers (F1 and F2) and collapses to a single layer at night. The F layer is also known as the *Appleton layer*.

Atmospheric transmission characteristics vary between the day and night, and from the summer and winter, due to changes in the characteristics of the D, E, F layers in the ionosphere (Figure 4.32).

During the day:

- **D Layer:** Low frequencies are reflected but high frequencies tend to pass through.
- **E Layer:** Signals as high as 20 MHz can be reflected, while higher ones pass through.
- **F Layer:** During daylight hours there are two layers, F1 and F2. The F1 layer is much more predominant during summer months.

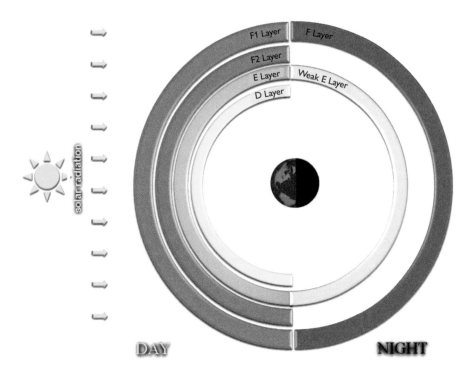

Figure 4.32 Ionosphere layers.

During the night:

- **D&E Layers:** Virtually disappear and signals that would be reflected from the ionosphere at lower atmospheric levels now are refracted in the higher F layer. This results in greater skip distances and better reception at longer ranges than in the daytime hours. The skip distance is greatest during the night when the ionosphere is the highest.
- **F Layer:** At night the F1 and F2 layers collapse to a more stable single layer starting at around 190 km (120 miles) through to 400 km (250 miles) reflecting higher frequency signals and increasing the skip distance resulting in longer ranges and more reliable communications than during daylight hours.

In addition, the characteristics of these layers vary greatly in altitude, density, and thickness with changes in the levels of solar activity. The F2 layer is most affected by solar radiation, which increases during sunspot activity and solar disturbances. As solar radiation increases, the F layer increases in density, as does the skip distance, which is the distance a radio wave travels and usually includes at least one bounce in the ionospheric layers (Figure 4.33).

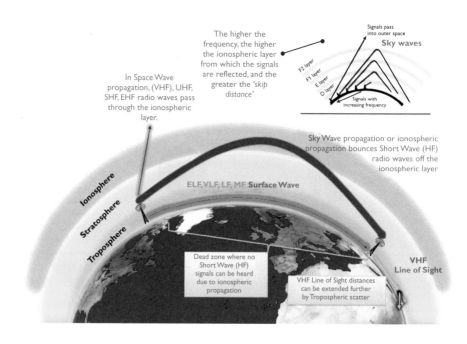

Figure 4.33 Propagation mechanism and transmission ranges vary significantly across frequency bands.

The troposphere

Frequencies below HF tend to be heavily attenuated by the D layer during the day and optimal propagation paths tend to be confined to the lower atmosphere.

Ground waves are comprised of three elements:

- The directly transmitted wave limited to line-of-sight distances;
- Wave reflections from the troposphere and from the earth;
- The surface wave which tends to follow the curvature of the earth.

Ground wave propagation is adversely affected by changes in terrain (better propagation characteristics over water) and with increasing frequency. Ground wave propagation is most effective at lower HF frequencies below 2 MHz, where it forms a very reliable communications link, little affected by seasonal weather patterns or atmospheric local disturbances.

Frequency bands in the UHF region and beyond tend to be limited to line of sight distances due to the heavy signal attenuation in the upper frequency bands.

Anomalous tropospheric propagation: ducting

In normal atmospheric conditions temperature will drop by 1 degree per 300 meters of increasing altitude. However, in certain climatic conditions such as hot and calm weather, especially over large expanses of water, evaporation can cause the upper atmosphere to become warmer than the lower layers of air below it. This temperature inversion, or ducting (Figures 4.34 and 4.35), acts as a waveguide that can significantly increase the distance over which the signal can travel, often extending ranges by several hundred kilometers.

A surface duct can be anywhere between 10 and 30 meters above the earth's surface.

Elevated ducts are formed in layers and are created when cold air is trapped between layers of warmer air. The cool air acts as a waveguide and refracts the radio wave over long distances. This occurs mainly over large inland seas where evaporation over the sea at night takes place and is moved upwards by warm air coming out from the land.

Ducting effects can be rapidly weakened by strong winds or rain.

Other influences on propagation characteristics

In addition to the effect of atmospheric layers on propagation characteristics, transmission ranges and signal quality can be affected by water vapor in the atmosphere, solar wind, physical obstructions such as mountains and buildings, the earth's conductivity, the curvature of the earth, the season, the

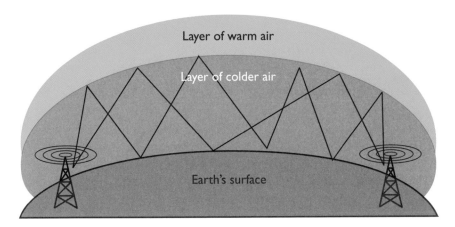

Figure 4.34   Anomalous propagation: surface duct.

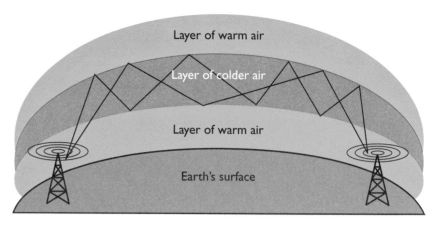

Figure 4.35 Anomalous propagation: elevated duct.

temperature, and the time of day. Such factors may all have an adverse or positive effect on the signal transmission and reception depending on the frequency in use.

All radio waves are partially absorbed by atmospheric moisture (Figure 4.36), with higher frequencies being particularly susceptible to atmospheric attenuation which reduces, or attenuates, the strength of radio signals over long distances.

Signal modulation

Radio and data link transmissions carry information by varying a combination of the amplitude, frequency, and phase of the carrier wave. In modern communications, these variations are used to create digital signal transmissions using binary ("0" or "1") signals in encrypted formats to transmit the information. This process is referred to as signal modulation (see Appendix C for details).

Signal transmission and attenuation

As electromagnetic energy in the form of radio waves spreads, the signal becomes weaker (Figure 4.37). This is due to two principal factors:

- The decrease or attenuation in the intensity of electromagnetic radiation due to absorption or scattering of the radio waves by particles in the atmosphere;
- The decrease in intensity due to the inverse-square law of geometric spreading.

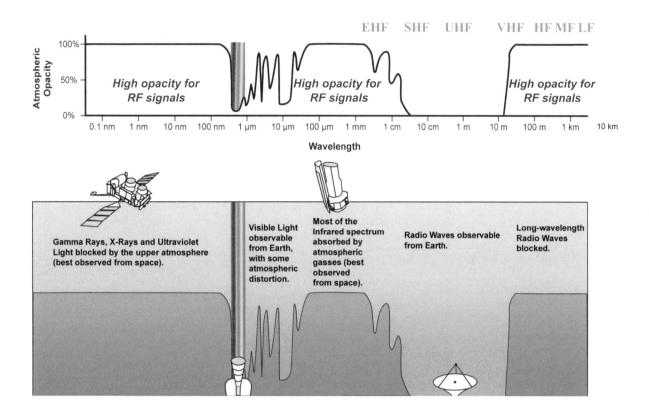

Figure 4.36 Above 30 GHz (1-cm wavelength), the absorption of electromagnetic radiation by the earth's atmosphere is so great that the atmosphere is effectively obscured to higher frequencies, until the atmosphere becomes transparent again in the infrared and optical frequency ranges. Beyond those high optical frequencies, terrestrial RF communication becomes opaque once more.

Spectrum management

Given the increasing need for radio-frequency communications for both civilian and military purposes in the optimal parts of the electromagnetic spectrum, there is a continuous need for spectrum deconfliction to ensure that discrete parts of the spectrum are allocated to specific users and specific functions. This is important if reliable communications channels are to be established and if the performance of existing communications systems are not to be degraded.

In theory, a *frequency management plan* or *spectrum management plan,* organized by the joint force commander for each operational theater, will coordinate the allocation of frequencies to ensure an integral and effective plan. In practice this is often difficult to achieve, as many forces will use unregulated communications systems that are either classified or that will remain

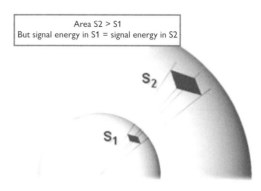

Figure 4.37 The *inverse square law* shows that signal strength in a uniform antenna reduces proportional to the surface area of an imaginary sphere. That means: the power density on the surface of a sphere is inversely proportional to the square of the distance traveled by the radio wave from the antenna.

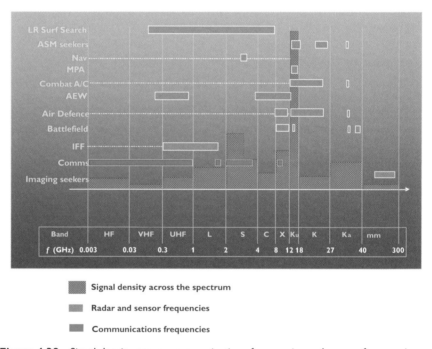

Signal density across the spectrum

Radar and sensor frequencies

Communications frequencies

Figure 4.38 Signal density across communications frequencies and sensor frequencies needs to be carefully managed through a process of deconfliction to avoid interruption and interference. (Image courtesy of Thales.)

uncoordinated in deployed environments. The picture is further complicated with the many active electromagnetic radiating sources found in the battlespace such as radars, counter IED jammers, and electronic warfare systems (Figure 4.38).

Radio Frequency Propagation[3]

Radio signals at different frequencies exhibit different propagation characteristics depending on the region of the atmosphere through which they pass. Low frequency radio signals can travel around the globe while high frequency signals are often constrained to line-of-sight distances.

Comms Band	Frequency Range	Principle Propogation Modes
EHF	30-300 GHz	Line of sight, troposcatter
SHF	3-30 GHz	Line of sight, troposcatter
UHF	300-3000 MHz	Line of sight, sometimes with tropospheric enhamcement
VHF	30-300 MHz	Line of sight, sometimes with tropospheric enhamcement, sporadic E, meteor scatter at low end of spectrum, auroral reflection towards higher latitiudes.
HF	3-30 MHz	Sky wave and ionosphere (E, F1, F2 regions), also sporadic
EMF	300-3000 kHz	Ground wave, sky wave, and ionosphere (particularly E region)
LF	30-300 kHz	Ground wave, waveguide formed with ionosphere
VLF	3-30 kHz	Ground wave, waveguide formed with ionosphere

Source: [24].

Table 4.3   Propagation Across the Radio Spectrum

Low frequency signals tend to follow ground wave propagation patterns.

Medium-frequency (MF) transmissions, including medium-wave and short-wave radio signals, exhibit different transmission characteristics during the day and night. During the day, MF signals travel by ground wave, following the curve of the earth over distances of several hundred kilometers. During sunset hours, the D layer virtually disappears and enables the signals to reach the ionosphere and benefit from skywave propagation, enabling AM radio broadcasts for example to be heard over much longer distances than is the case during the day.

Most long-distance HF radio communication (between 3 and 30 MHz) is achieved through skywave propagation.

Skywave propagation can be helpful in transmitting HF and VHF signals over long distances as the wave repeatedly bounces from the sky to the ground.

At high frequencies, UHF and EHF signals tend to adhere more to line-of-sight propagation rules. UHF signals, for example, do not benefit from the effect of ionospheric reflection, but their range and propagation may improve or deteriorate by tropospheric ducting (where the troposphere serves as a waveguide in its own right) as the atmosphere warms and cools throughout the day.

Line-of-sight transmission distances are limited by antenna height and curvature of earth. VHF and UHF transmissions, however, may extend slightly beyond line-of-sight ranges due to tropospheric scatter where radio signals are reflected further on their journey through variations in the refractive index in the lower atmosphere. Due to this characteristic, the radio horizon is about 80% greater than line of sight because of diffraction effects.

Space waves are waves at high frequencies that penetrate all the ionospheric layers without refracting.

Future solutions to the allocation of bandwidth across an increasingly busy RF spectrum include smart communications systems capable of learning and allocating transmission to a relatively quiet and interference-free part of the spectrum, a process called *dynamic spectrum allocation.* Other techniques such as interference cancellation, data compression (to reduce bandwidth requirements), advanced processing techniques, and innovative directional antenna to reduce RF clutter in areas where the signal is not required to be received will also become commonplace [25].

Note that while the ELF, SLF, ULF, and VLF bands overlap the audio frequency bands of approximately 20–20,000 Hz, sounds are transmitted by the physical compression and expansion of gases in the atmosphere, and not by electromagnetic energy; as such, they do not appear on electromagnetic spectrum or radio frequency charts.

Frequency applications

International standard definitions of communications frequencies[4] are divided in to 11 frequency zones. In practice, communications systems may span adjacent communications bands; they are not constrained to operate within the discrete bands listed in Table 4.4.

Extra-high frequency (EHF)

Operating in the 30–300-GHz frequency band, the EHF band is the highest frequency band of the communications frequencies and borders the infrared spectrum of the near-optical frequency bands. Due to its high frequencies, it is capable of transmitting high data rates; however, transmission ranges are severely restricted due to atmospheric attenuation caused by water vapor and oxygen absorption, in particular in the 57–64-GHz range where the resonance of oxygen molecules in the lower atmosphere causes significant losses in signal strength. At higher altitudes where oxygen and water concentrations are significantly lower, data transmission rates are less affected, and EHF is therefore particularly suited to high data rate satellite communications.

EHF wavelengths (Figure 4.39) are small, between 10 mm and 1 mm, and the frequency band is sometimes therefore referred to as the millimeter-wave band. The small wavelengths of the EHF frequency band allow relatively small antennas but with high bandwidths, narrow directional beamwidth, and highly directional antenna characteristics. The narrow beam characteristics make millimeter-wave transmissions an appropriate choice for relatively short-range, high data rate, point-to-point communications, and

Comms Band	Frequency Range	Wavelength Range	Communications Applications	Typical Data Rate	Typical Communications Range
EHF	30–300 GHz	10 mm–1 mm	High frequency microwave communications, satellite datalinks short-range point-to-point communications, wireless datalinks	100 Mbps	20–40 Miles
SHF	3–30 GHz	100 mm–10 mm	Missile guidance, wideband comms, satellite comms, microwave datalinkswireless LAN	250 kbps	30–40 Miles
UHF	300–3,000 MHz	1m–100 mm	GPS, satellite communications, pointto-point microwaves, datalinks, telemetry, digital broadcast, GSM, wireless LAN, Bluetooth, GPS signal broadcast	56 kbps	15–100 Miles
VHF	30–300 MHz	10m–1m	Tactical line-of-sight communications, DAB, commercial FM radio	10 kbps	25–50 Miles
HF	3–30 MHz	100m–10m	Tactical over-the horizon communications, short wave broadcasts	2400 bps	30–300 Miles
MF	300–3,000 kHz	1 km–100 m	Medium wave (AM) broadcasts	500 baud	100–1,000 Miles
LF	30–300 kHz	10 km–1 km	Long wave (AM) radio broadcasts	75 baud	1,000–5,000 Miles
VLF	3–30 kHz	100 km–10 km	Submarine communications	50 baud	>5,000 Miles
ULF	300–3,000 Hz	1,000 km–100 km	Submarine communications	20 baud	>5,000 Miles
SLF	30–300Hz	10,000 km–1000 km	Submarine communications	10 baud	>5,000 Miles
ELF	3–30 Hz	100,000 km–10,000 km	Submarine communications	1–2 baud	>5,000 Miles

Table 4.4 Communications Frequency Bands

the narrow beamwidth means that even where many antennas are operating together there is little interference between transmissions.

EHF frequencies are commonly used for mobile satellite communications systems, although the antennas are highly directional and need to be carefully aligned, especially on moving platforms.

Super-high frequency (SHF)

The 3–30-GHz frequency band is sometimes also considered as part of the millimeter-wave frequency range with wavelengths ranging from 1 to 10 cm. The high frequencies also enable high-speed data communications, and SHF

Figure 4.39   The B-2 Spirit bomber is fitted with EHF satellite communications to provide a secure channel for sending and receiving intelligence to and from the Global Information Grid.

frequencies are often used for communications with satellites and for high-speed point-to-point terrestrial data links (Figure 4.40).

The U.K. SKYNET fixed satellite down link, mobile satellite services use 7,250–7,300 MHz. The uplink is engineered at 7,900–8,400 MHz with 7,900–7,950 MHz for the mobile uplink; this system remains the primary military link to U.K. forces overseas.

Many short-range high resolution radars such as air defense radars also operate in this frequency band, along with video broadcast systems.

Ultrahigh frequency (UHF)

The 300–3,000-MHz frequency band is widely used in two-way line of sight terrestrial radio systems. It has a relatively short wavelength of between 10 cm and 1 meter and a high data transmission rate. The combination of high-speed data transfer and broader beamwidth than EHF and SHF transmissions makes it a good choice for practical use in the field.

The UHF band is the most important part of the spectrum for command and control of NATO operations (Figure 4.41).

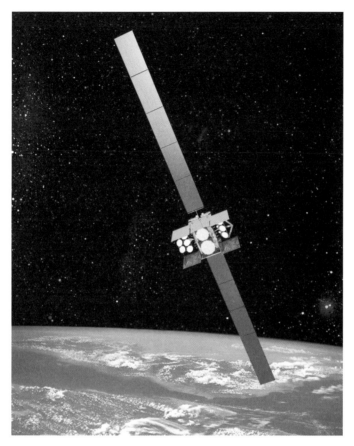

Figure 4.40 The Wideband Global EHF/SHF communications satellite is the successor to the Defence Satellite Communications System-III and provides about 12 times the bandwidth of the satellite it replaces. A constellation of five satellites are in orbit at a height of 22,300 miles and a range of frequency-specific antennas enable coverage across various EHF and SHF bands.

UHF communication frequency bands also start competing with radars at these higher frequencies (Figures 4.42 and 4.43).

Very high frequency (VHF)

The 30–300-MHz frequency band is a common choice for FM voice radio transmissions for aircraft, ships, and vehicles at ranges up to and slightly beyond line-of-sight distances as shown in the following formula.

VHF is less prone to interference by atmospheric noise, weather phenomena, and electrical static from the atmosphere or from electrical equipment than lower frequencies and is less affected by buildings and physical objects than are higher frequencies.

Figure 4.41   Tower-mounted UHF directional antennas used for UAV data links and a fixed satellite dish for high-speed communications.

An approximation to calculate the *line-of-sight horizon distance* is [26]:

- Distance in miles = $\sqrt{1.5 \times A_f}$ where A_f is the height of the antenna in feet.

- Distance in kilometers = $\sqrt{12..746 \times A_m}$ where A_m is the height of the antenna in meters.

High frequency (HF)

The 3–30-MHz frequency band is also known as the HF or short-wave band or decameter band, with a wavelength ranging from 1 to 10 decameters (10 to 100 meters).

Although technically possible to achieve long-range communications through ionospheric skywave propagation in the HF band, the reliability of the signal over long distances varies considerably as it is heavily influenced by many atmospheric and geographic factors. At best, worldwide communications can be achieved, but at worst, where skywave propagation is not possible, communications distances are limited to ground wave paths.

Medium frequency (MF)

The 300–3,000-kHz (3 MHz) frequency band contains the broadcast band from 500 kHz to 1,500 kHz and is often referred to as the medium wave band. It is used for over the horizon communications using the skywave principle.

Figure 4.42   A 50-foot telescoping mast assembly of a UHF omnidirectional antenna used for tactical ground-to-air communications is erected for a demonstration. This antenna uses a symmetrical array of dipole antennas to create a uniform transmission and reception pattern.

Figure 4.43   A tank transporter and armoured engineering vehicle in Kandahar, Afghanistan, June 2009. The multiple omnidirectional V/UHF communications whip antennas are clearly visible.

Figure 4.44   The PCS-5 VHF/UHF line of sight radio covers the 30–400-MHz frequency range and is designed to meet the U.S. DoD's requirement for a lightweight, secure, network-capable, multiband, multimission, antijam, voice/imagery/data communications capability.

Low frequency (LF)

In the 30–300-kHz frequency band, wavelengths range from 10 kilometers to 1 kilometer and the band is often referred to as the long wave band. Low frequency radio signals tend to use ground wave propagation principles and follow the curvature of the earth to achieve long ranges of several thousand kilometers. Long wave propagation by reflection (refraction) from the ionosphere is also possible, where the reflection take places in the D or E layers of the ionosphere [27]. The band is typically used as a backup for long-range maritime broadcast communications.

The disadvantage with the low frequency band is the need for large antennas to achieve suitable transmission patterns (a function of the long wavelength). The height of antennas differs by application but generally requires T-shaped antennas over 50m in height or mast antennas over 150m in height. Long-range transmissions in this band also need to be supported by high power transmitters, where the range of the ground wave will be directly proportional to the transmission power employed.

Very low frequency (VLF)

The 3–30-kHz frequency band is used for very long-range ground wave communications as lower frequencies penetrate materials such as water more effectively than higher ones. These characteristics have favored the VLF band for transmission with submarines at or near the surface. However, the bandwidth in the lower frequency bands diminishes rapidly, enabling only very basic (short) messages and low transmission speeds to be used. Due to the long transmission ranges, VLF communications are also used for long-range communications with deployed naval surface forces, although its utility is limited due to low data bandwidths (Figure 4.45).

Figure 4.45 The E-6B Mercury aircraft operates in two roles: as a national command post for C3 communications and in a TACAMO role (take charge and move out) to provide secure communications between the U.S. National Command Authority and nuclear submarine and silo missile sites. In its TACAMO role, its primary mission is to maintain communication links with U.S. strategic forces through the receipt, verification, and retransmission of emergency action messages (EAMs) to U.S. strategic nuclear forces. It does this through being equipped to communicate on virtually all frequency bands from very low frequency (VLF) to super high frequency (SHF) using a variety of modulation and encryption techniques over a wide range of networks. In the United States, this airborne communications capability has largely replaced the land-based VLF and ELF broadcast sites that are themselves vulnerable to a nuclear strike. The aircraft is equipped with a 7-km trailing wire antenna stabilized by a drogue parachute to effect the low frequency transmissions.

As with other low frequency bands, large antennas and high transmission power (often measured in megawatts) are necessary to achieve long transmission ranges.

Extremely low frequency (ELF)

ELF covers radio transmissions in the 3–3,000-Hz frequency band and is used almost exclusively for communications with submerged submarines.

Due to the electrical conductivity of salt water, submarines are effectively shielded from most electromagnetic communications; however, signals in the ELF frequency range can penetrate much deeper, and accordingly ELF communications tend to be used for more deeply submerged submarines.

ELF communications tend to be one-way communications because of the difficulties of providing the antenna size or power needed to transmit such a signal, and messages sent are often in the form of an instruction to the submarine to rise to a depth where faster data rate communications systems can be used.

Satellite communications

Communications satellites are the most common type of satellite in orbit. The height of the satellite above the earth enables them to communicate over very long ranges, thereby overcoming the limitations of the curvature of the earth's surface and, to a great extent, the limitations of atmospheric propagation.

Communications satellites broadcast and receive on a range of frequencies depending on their mission. Space communications networks using point-to-point links tend to use high frequency signals benefiting from narrow beams and high bandwidths, while others used for mobile communications or direct broadcast may use lower frequencies, which travel further and have a wider beam pattern (and hence greater geographical coverage) but at the expense of lower data transmission speeds (bandwidth).

Satellite data links and communications systems typically operate at UHF, SHF, and EHF frequencies (Figure 4.46). The choice of satellite frequencies will depend on the physical limitation of the platform requiring the communications link, the data transmission bandwidth requirements, the beamwidth and coverage, and the degree of encryption required across the link (Figure 4.47).

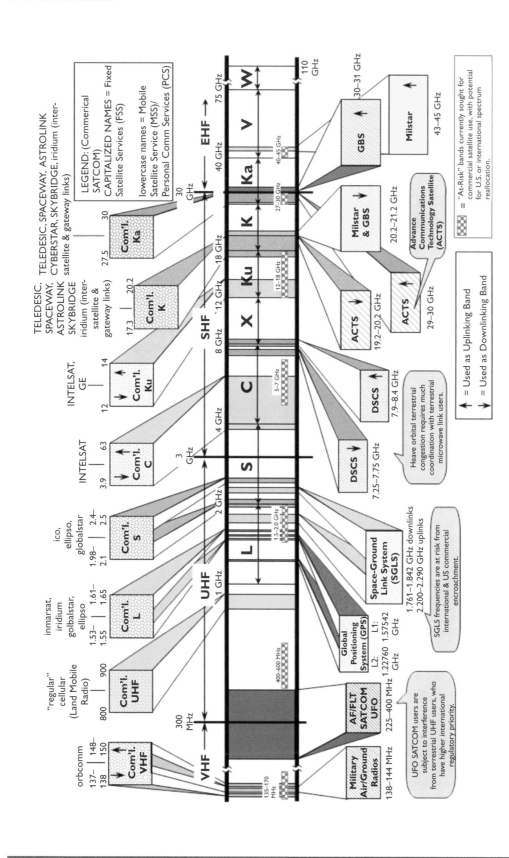

Figure 4.46   Satellite communications.

Figure 4.47   The Inmarsat satellite uses a broad beam pattern at the expense of data transfer speeds to cover large areas of the globe. Most of the earth's surface is covered by a constellation of only three satellites and as a result the pointing accuracy of the antenna is very forgiving.

Low frequency bands such as the Inmarsat L-band (UHF) system (Figure 4.48) have wide beam patterns and are very forgiving in their directional accuracy, while Ku-band (SHF) satcoms require much more accurate beam pointing often requiring fixed or gyro-stabilized platforms to ensure that they stay within the required pointing accuracy. Data rates, however, are significantly higher at higher frequency transmission rates (Figures 4.49 and 4.50).

Satellite orbits[5]

The orbit that is chosen for a satellite depends upon its application. Those used for direct broadcast television or for maintaining communications over a certain geographical region use a geostationary orbit [28].

Other satellites such as those used for satellite phones may use low earth orbits to keep path losses to a minimum. Similarly, satellite systems used for navigation such as Global Positioning (GPS) systems occupy a relatively low earth orbit to ensure consistent signal strength.

Satellites orbit the earth in one of two basic types of orbit (Figure 4.51). The most common is a *circular orbit* where the distance from the earth to the satellite remains constant. A second type of satellite orbit is an *elliptical orbit*, which is often used when only a particular part of the earth needs to be in

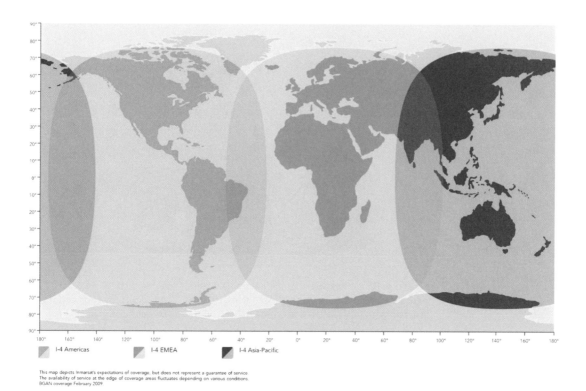

This map depicts Inmarsat's expectations of coverage, but does not represent a guarantee of service.
The availability of service at the edge of coverage areas fluctuates depending on various conditions.
BGAN coverage February 2009

Figure 4.48 The Inmarsat constellation uses the wide beam characteristics possible with lower frequency satellites to achieve near-global coverage with only four geostationary satellites. (Image courtesy of Inmarsat.)

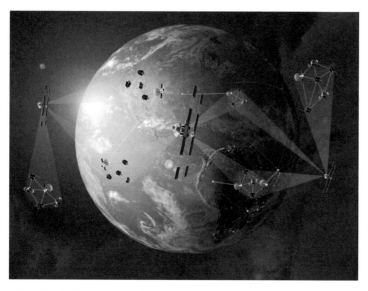

Figure 4.49 The DARPA Future, Fast, Flexible, Fractionated, Free-Flying Spacecraft United by Information Exchange (System F6) program intends to demonstrate that a traditional, large, monolithic satellite can be replaced by a group of smaller, individually launched, wirelessly networked, and cluster-flown spacecraft modules. (Image courtesy of DARPA.)

Figure 4.50   A multiband satellite-communications-capable tactical radio used for secure tactical communications near Al Asad Air Base, Iraq. The radio is used for contingency operations and can be used nearly anywhere on the earth's surface.

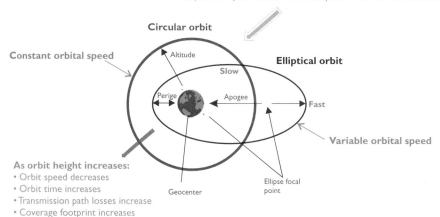

As the height of a satellite increases, the time for the satellite to orbit increases. At a height of 35,790 km, it takes 24 hours for the satellite to orbit. This type of orbit is known as a geosynchronous orbit, (i.e., it is synchronized with the speed of the earth's rotation.)

Circular orbit

Constant orbital speed

Altitude

Elliptical orbit

Slow

Perige   Apogee

Fast

Variable orbital speed

As orbit height increases:
• Orbit speed decreases
• Orbit time increases
• Transmission path losses increase
• Coverage footprint increases

Geocenter

Ellipse focal point

Figure 4.51   Satellites orbit the earth in one of two basic types of orbits.

close contact with the satellite for infrequent periods of time (Figure 4.52 and Table 4.5).

As the height of a satellite increases, so the time for the satellite to completion orbit increases. At a height of 35,790 km, it takes 24 hours for

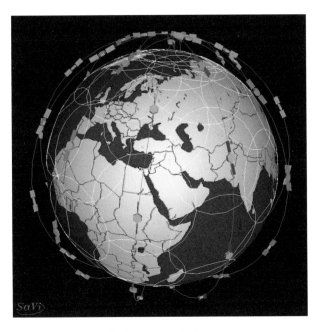

Iridium constellation
66 active satellites in 6 planes of 11.

Figure 4.52 The iridium communications satellite constellation comprises of 66 satellites in 6 planes of 11 satellites in low earth orbit. The coverage footprint for each satellite is indicated by the white circles on the earth's surface. (Image courtesy of L. Wood, P. Worfolk, et al. SaVi-Satellite constellation visualization software, http://savi.sf.net, 2010.)

Orbit name	Orbit initials	Orbit altitude (km above Earth's surface)	Details /comments
Low Earth Orbit	LEO	200–1,200	Provides for low signal attenuation.
Medium Earth Orbit	MEO	1,200–35,790	—
Geosynchronous Orbit	GSO	35,790	Orbits once a day, but not necessarily in the same direction as the rotation of the Earth—not necessarily stationary.
Geostationary Orbit	GEO	35,790	Orbits once a day and moves in the same direction as the Earth and therefore appears stationary above the same point on the Earth's surface. Can only be above the Equator.
High Earth Orbit	HEO	Above 35,790	

Table 4.5 Satellite Orbit at Varying Altitudes Depending on Their Function

the satellite to orbit. This type of orbit is known as a geosynchronous orbit (i.e., it is synchronized with the rotational speed of the earth).

One particular form of geosynchronous orbit is known as a *geostationary orbit*. In this type of orbit, the satellite rotates in the same direction as the rotation of the earth and has a 24-hour orbit period. This means that it revolves at the same angular velocity as the earth and in the same direction and therefore remains in the same position relative to the earth, which of course also spins through 360° every 24 hours. Geostationary orbits have the advantage that for fixed communications networks, the satellite remains in the same position, and no tracking is normally necessary.

In contrast, low earth orbits are just above the earth's atmosphere, typically between 160 and 2,100 km in altitude. Orbiting at this altitude, a satellite may only take about 90 minutes to completely circle the earth, traveling at around 27,000 km per hour. Low earth orbits are used by manned vehicles such as the space shuttle and the International Space Station. They are also used for weather, reconnaissance, electronic intelligence (ELINT), and remote sensing satellites. On a clear night it is usually possible to see with the naked eye several satellites in low earth orbit passing overheard.

Medium earth orbits are used by satellites requiring slow sweeps of large areas of the globe such as search and rescue satellites, which listen for distress beacons transmitting on fixed frequencies.

Data transmission delays

The time taken for a signal to reach a geostationary satellite is around 120 milliseconds depending upon the actual position of the ground station on the earth's surface which gives a round-trip time of about 0.25 second. This delay can make voice communications rather difficult, although data communications via satellite are well able to cope with such delays if the information is not sufficiently time-critical to interfere with its use. Over long distances, voice communication links often use fixed cables rather than satellites as the transmission delays incurred are far less.

In some applications high earth orbits may be required. For these applications the satellite will take longer than 24 hours to orbit the earth, and transmission path lengths may become very long, resulting in additional delays for the round trip from the earth to the satellite and back as well as increasing the levels of path loss.

Global Position System (GPS)

The GPS system is run by the U.S. Department of Defense and provides accurate positioning, navigation, and timing services across the globe on a continuous basis in all weather, day and night, to locations that have an unobstructed view of four or more GPS satellites.

The GPS constellation consists of 24 operational satellites comprising 6 orbital paths with 4 satellites in each path together with some spare satellites in orbit in case of catastrophic failures. The medium earth orbits are approximately 20,200 km above the surface of the earth and the satellites complete each orbit in about 12 hours.

The GPS satellites are named Navstar satellites and each one weighs around 850 kilograms. They are about 5.20 meters across with the solar panels extended, and they transmit about 50 watts through the GPS antennas, although the solar panels themselves generate around 700 watts.

The GPS satellites (Figure 4.53) also carry a set of nuclear detonation detectors consisting of an optical sensor, an X-ray sensor, a dosimeter, and an electromagnetic pulse (EMP) sensor, which form a major portion of the United States Nuclear Detonation Detection System [29, 30].

Ultrahigh frequency (UHF) satellites (Figure 4.54) are the workhorses for tactical ground, sea, and air forces, with services provided by both military and civilian satellite operators. They are limited to relatively narrow bandwidths and modest data transmissions rates and are typically used for tactical communications where heavy encryption is not required, thus avoiding the burden associated with high encryption and security requirements.

Figure 4.53   Artist's impression of a GPS Block II-F satellite in orbit.

Figure 4.54   The MUOS UHF satellite system will provide 3G-type services for military users from a constellation of geostationary satellites.

Figure 4.55   The MILSTAR satellite constellation is a U.S. satellite communications system with narrow 1° beams transmitted from a geostationary orbit. The effect of the narrow beam is that users can operate within 5 km of the enemy without being detected. Additional protection against jamming is provided from the use of a frequency-hopping waveform using several thousand hops per second in bands of 2 GHz.

UHF satellite users are often subject to interference in populated areas from civilian UHF communications systems such as TV broadcasts, which often have higher regulatory priority. UHF satellites tend to have a large geographical beam coverage, which is useful for civilian applications, but in a military environment this tends to increase their vulnerability to jamming. The large footprint, however, does permit UHF satcom ground antennas to be pointed in the approximate direction of a satellite and still capture a strong signal.

Super high frequency (SHF) satellites typically support long-distance secure communications requirements of military forces that cannot be met by fixed ground-based communications systems. SHF satellites are capable of providing high bandwidth and high data rate communications allowing suitable accommodation for encryption and security and are often used in joint operational environments to relay imagery and information from other data intense applications over long distances. SHF communications require large antennas to achieve high bandwidths, and their use is therefore limited to large platforms such as ships and ground stations.

Extra high frequency (EHF) satellites are used for special-purpose applications where a low probability of detection and interception, antijam performance, and the ability to operate even during the presence of electromagnetic pulses from nuclear weapons are important. They are often used in a narrowband application, which results in reduced antenna and installation in relatively small platforms.

EHF satellites have highly directional narrow beams, which make detection and interference by the enemy extremely difficult, but are not suitable for coverage of large areas due to the large number of beams required. EHF satellite communications form part of the U.S. Navy submarine high data-rate antenna program, which exploits the survivability of the EHF MILSTAR satellites (Figure 4.55) along with the narrow focus spot beam, which can communicate with submarines with relatively low risk of detection.

Jam resistance versus bandwidth

Civil satellite systems can transmit data at fairly high speeds, typically above 10 Mbps, and operate in the X- and Ku-bands. Most military satellite communications systems, however, trade bandwidth for jam resistance. Satellites such as Milstar, a U.S. DoD system, operate at higher Ka (EHF) frequencies, providing the ability to transmit at higher bandwidths, but this is traded for jam resistance and as such the data transmission rates are reduced to around 20 kbps. On the U.S. Milstar II satellite, these rates have been improved by using special waveforms above 1.5 Mbps.

Vulnerability to jamming (Figure 4.56) is a potential problem in commercial satcoms where noise jamming introduced within the wide footprint of the satellite uplink can be a highly effective means of degrading or jamming the satellite's performance. As most civil satellite communication satellites operate in a geostationary orbit, their footprints are intentionally large, and unlike military satellites where reception and transmission requirements are designed to be highly directional to minimize this problem, jamming waveforms can easily be introduced into the uplink, even through the sidelobes. Military satellites overcome this problem by having beams that are narrower and more directional in nature, making it more difficult to introduce jamming signals and easier to use antenna gain to blank out unwanted signals.

Future satellite communications networks will seek to minimize the terrestrial footprint through which jamming signals can be introduced by providing a higher number of smaller footprints on the battlefield by using relay networks of UAVs and other long-endurance aerial platforms. As UAVs become increasingly commonplace in the battlespace, the option will exist to improve the bandwidth available to commanders and troops on the ground through the proliferation of such networks, and the directional nature of multibeam footprints on which uplinks are based will enable an

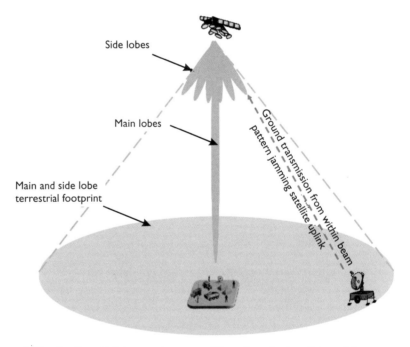

Figure 4.56 Satellite sidelobes can be susceptible to jamming by off-boresight transmissions from ground sources.

inherent improvement in jam resistance for local signal collection and broadcast. When combined with self-forming ephemeral networks as described earlier, this can produce an enduring high-speed communications network with high resistance to detection, interference, and jamming.

Satellite communications on the move

The ability for satcom systems to transmit on the move (Figure 4.57) is largely dependent on the beamwidth that needs to be captured and hence the pointing accuracy of the antenna. Even with gyro-stabilized systems, it can be very difficult for a vehicle-mounted satellite antenna to accurately lock on to the satellite. The use of lower-frequency UHF satellites is therefore preferable for vehicle mounted systems, although this will limit the maximum data speeds that can be achieved.

Where higher frequency systems such as EHF are used, data rates will be reduced while the vehicle is on the move and transmission rates can be increased with improved beam pointing accuracy typically by a factor of 8 or 10 when the vehicle is stationary.

Unlike commercial vehicle-based systems, military on-the-move satcom terminals must be capable of providing uninterrupted connectivity in all environment and terrain conditions. They must be capable of reacquiring the satellite if the lock is lost or the signal blocked by man-made objects, terrain, or foliage and to operate in all weather conditions.

Figure 4.57   A convoy of future on-the-move platforms for the U.S. Army Warfighter Information Network-Tactical (WIN-T) EHF satcom systems. The WIN-T network will be capable of providing 1-Mbps continuous data rates to other users while on the move.

Airborne platforms are inherently more stable and with the assistance of computer controlled and steered electronic or mechanical antenna can achieve highly accurate beam pointing and are therefore capable of effectively exploiting high frequency satcom systems.

Satellite data link applications and vulnerabilities

The use of satellites to extend data link ranges is now commonplace, and the advent of software defined digital radios and variable waveforms has enabled most data links to be carried over satellite links. Both Link-16 and the CDL data links have satellite datalink variants available, and virtually all data links are capable of being connected over long distances to support deployed forces.

The protection of space-based communications systems is becoming of increasing importance, given the accelerating capabilities of potential adversaries in the domain of antisatellite weaponry and capabilities. Given the criticality and importance of space-based communications, it is clear that disabling even one communications satellite could have a disastrous effect on the adversary's networked communications capability. Defensive systems are being planned to protect these vital assets starting with a comprehensive space-based situational awareness picture from which effective defensive actions may be planned along with redundancy at platform and network levels.

At a local level, defensive satellites may be employed to guard a constellation of unarmed or unprotected satellites, and satellites themselves may be fitted with defensive aids such as rocket launchers or defensive nets to fire at an attacking platform or kill vehicle. Among other defensive aids under consideration are the ability to be able to maneuver out of harm's way (and reposition once the threat has passed) and lightweight armor to shield more sensitive parts of the satellite [31]. Ground-based satellite infrastructure such as ground-control stations and a communications receiving station will also naturally become prime targets in future networked warfare.

Control of unmanned platforms by satellite data links

Long-range autonomous unmanned vehicles such as unmanned air vehicles (UAVs) and unmanned ground vehicles (UGVs) operating at long distances or where LOS communications are not possible such as in mountainous regions can be controlled by satellite data links, which may also serve to provide ISTAR imagery back to the ground station. There are various methods by which such vehicles can be controlled (Figure 4.58).

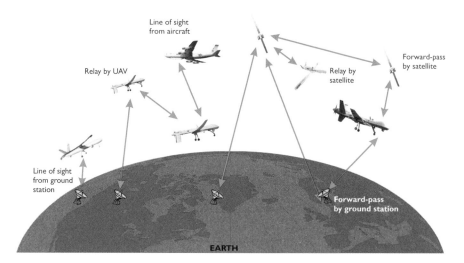

Figure 4.58 UAV and airborne platform networks with line-of-sight data links can be used to provide relatively jam-resistant communicant networks.

- **Line of sight:** The most common form of data link for tactical UAVs, but limits the vehicle to LOS operation, restricting the range of the system.
- **Satellite relay:** A UAV that is controlled via satellite may require several satellites to act as relays, but can operate virtually anywhere on the globe providing it is within the footprint of the final satellite's downlink. Its use is not inhibited by terrain or low-level atmospheric conditions.
- **Airborne relay:** UAVs may also be controlled via data links connected between multiple platforms, often another UAV of the same type. Ground stations need to be located much nearer to the operational theater as without satellite links they will need direct LOS to the first air vehicle in the chain. In addition to extending the range, this also provides the ability to control a UAV from higher altitude, enabling LOS control even in mountainous terrain.
- **Forward pass:** Similar to the airborne relay concept, but involves at least one relay from a ground station.

Similar principles can be applied to unmanned ground and maritime platforms.

Predator and Reaper UAVs (Figure 4.59) have two C2 data links; a low bandwidth telemetry data link that describes the status of the UAV on-board systems, and a high bandwidth data link to stream the imagery and telemetry from the UAV's surveillance and reconnaissance sensors (Figure 4.60).

Figure 4.59   A USAF MQ-9A Reaper taxis in preparation for a mission in support of Operation Enduring Freedom.

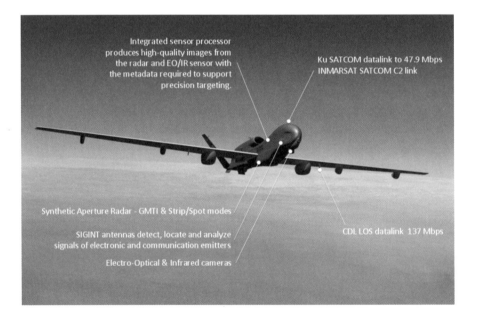

Figure 4.60   Northrop Grumman/EADS EuroHawk [32]. (Figure courtesy of Northrop Grumman.)

Predator UAVs have a 200-Kbps outbound channel for command and control and a 3.2-Mbps return channel for data dissemination, which will increase to 45 Mbps by 2015. The Global Hawk UAV, with operational data rates of up to 50 Mbps in the return channel, is expected to increase the capacity to 274 Mbps by 2015.

Typically during takeoff, the UAVs are controlled from the launch site (Figure 4.61), often using UHF, X-, or Ku-bands for line-of-sight communications. After takeoff, control can be handed off to mission controllers who fly the drone and operate an array of sensors as well as weapons until control is handed back to the ground crew just before landing [33].

(a)

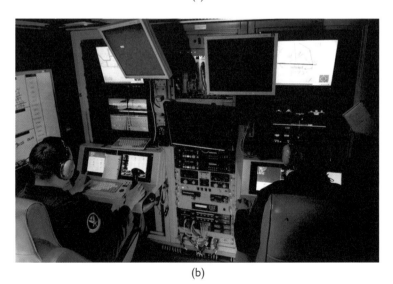

(b)

Figure 4.61 (a, b) Long-range UAVs will continue to be ideally suited for dull, difficult, and dangerous missions. Here an MQ-1 Predator patrols over Afghanistan. Once launched from the operational theater, command and control of the UAV is passed to operators at Creech AFB in Nevada. Data latency via the Ku-band satellite link is typically 1.5 seconds, between Nevada & Afghanistan.

Laser data links as airborne and space-based communications methods are feasible as a replacement for RF data links where high bandwidths communications are required. Current satellite technology has demonstrated the feasibility of data transmission rates of up to 10 Gbps (equivalent to the simultaneous transmission of 600 high-definition TV channels) and data rates of up to 100 Gbps are envisaged in the future. In addition to the very high-speed nature of the communications links, such a network is virtually jam-proof, as it is extremely difficult to effectively interfere with such a narrow directional beam operating at such high frequencies.

Laser data links may also be used for underwater communications where theoretically a high-powered laser may reach as far as 700m through water if it is tuned to the appropriate optical window in the green-blue part of the visual spectrum. Such systems, however, need significant power, and current research is examining the feasibility of using terrestrial-based lasers with space-based mirrors to reflect the beam to the submarine.

Underwater data links and communications

The provision of effective underwater communications significantly increases the utility of submarines and underwater platforms as an integral part of networked sensors and effects systems (Figure 4.62). As such, underwater communications and networks are critical to the domination of the battlespace. Missions such as search and tracking and covert surveillance of underwater and above water targets all require effective near real-time underwater communications. Historically, however, effective integration of underwater assets has been limited by the lack of communications capabilities, particularly at depth and at speed between the submerged vehicles and above-water platforms.

Communication with submarines is technically challenging because RF electromagnetic radiation used in terrestrial and airborne communications data links cannot easily penetrate dense substances such as water.

The degree to which electromagnetic waves penetrate through a conductor is described by the concept of *skin depth*. The skin depth is the depth at which an electromagnetic wave is attenuated to $1/e$ of its surface magnitude. Skin depth, d, is inversely proportional to the square root of the transmitted frequency:

$$d = \left[1/\left(\mu \pi f \delta \right) \right]^{1/2}$$

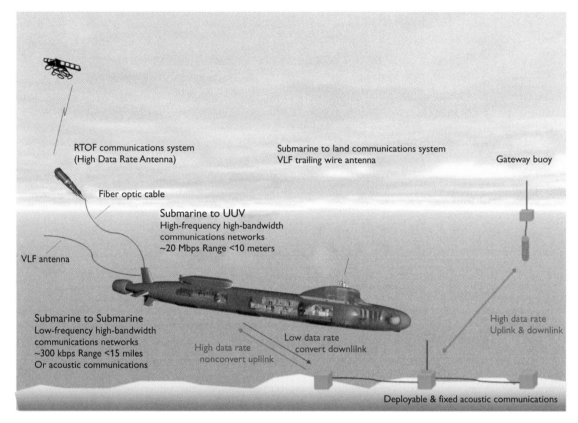

Figure 4.62 Submarine communications and data links.

where μ is magnetic permeability; e is the base of the natural logarithms ($e = 2.7183...$), f is frequency, and δ is electrical conductivity.

For sea water approximations using values of $\mu = 4\pi \times 10^{-7}$ and $\delta = 5$ (ohm-m)$^{-1}$, this gives a skin depth for 30 kHz of 1.3m and a skin depth of 4.0m at 3 kHz.

Although submarines can surface to communicate through normal RF frequencies, they jeopardize revealing their position, and strategic submarines in particular such as those involved in nuclear deterrence need to stay submerged for the duration of their mission—often up to 6 months—to ensure that their location is not revealed.

Submarine missions increasingly require significant improvements in communications bandwidth in the same way as other networked terrestrial and air platforms. The vision is to allow submarines and, increasingly, unmanned underwater vehicles (UUVs) to communicate without limitations imposed by depth, speed, or location and with sufficient bandwidth to enable real-time network connectivity to maximize the effectiveness of data and intelligence collected by the submarine, while ensuring that it plays a full role as an integrated networked effects system (Figure 4.63).

Communication with submerged submarines typically involves the following methods:

Laser communications

Laser communication between aircraft or satellites and submarines is an evolving technology of use for high-speed communications with shallow submerged submarines. Data link performance is limited by the adverse effect of solar illumination during the day, resulting in the need for high-power lasers during daylight hours to counter the sun's effect. Blue lasers tend to perform better in open ocean environment and green lasers are more suited to more turbulent littoral environments. To ensure that the location of the submarine is not compromised during the communications transmission, the aircraft or satellite broadcasts a laser inquiry message to a predesignated transmission area, which is then detected by the submarine, which in turn initiates its underwater laser transmission sequence to the airborne platform to confirm the link before duplex communications are established between the two platforms [34].

Figure 4.63 The U.S. Navy Virginia-class fast attack submarine USS *Virginia* (SSN 774) cruises through the Mediterranean.

Underwater acoustic transmission

Sound exhibits different characteristics to RF signals and travels well in water, although differing thermal layers that occur at varying depths can disrupt its path. Underwater microphones (hydrophones) can detect sound typically up to several hundred meters in range and it is well known that both the U.S. and Russian Navies have placed acoustic communications equipment on the seabed of areas frequently visited by passing submarines both to detect enemy submarines and communicate with friendly ones through acoustic signals (Figures 4.64 and 4.65).

Acoustic detection is particularly effective underwater as each submarine has a unique acoustic fingerprint, which enables individual submarines to be identified with a high degree of accuracy. Acoustic communications can be effected by the use of an underwater loudspeaker from these stations, although this system does have limitations as it will announce that a submarine is in communication with its underwater network, and tends only to be effective as a one-way communications channel from the base station to the submarine.

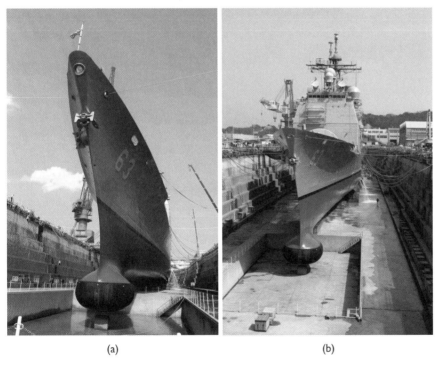

(a)   (b)

Figure 4.64   The Ticonderoga class guided missile cruiser USS *Cowpens* (CG 63) at the completion of its ship's repair force (SRF) dry dock period in Yokosuka, Japan. The bow sonar is clearly visible in these photos.

Figure 4.65   Sonar buoys are loaded into a P-3 Orion aircraft prior to a NATO training exercise in February 2006.

Underwater sound communication (Figure 4.66) is therefore quite effective where submarines and ships are operating together in relatively short-range environments typically up to 10–30 km to preserve data transmission speed and where the presence of a submarine is already known to the enemy.

The bandwidth of acoustic transmissions is, however, low, and as such, data messages are limited in their complexity and size.

Very low frequency (VLF)

Typically terrestrial and airborne communication transmissions occur in the HF-VHF-UHF-SHF frequency bands to maximize bandwidths, data rates, and reception quality in atmospheric conditions. Communications with submarines and underwater vehicles, however, require communications at the lower end of the electromagnetic spectrum in the VLF, and ELF frequency bands as higher frequency signals are incapable of penetrating water to any useful depth.

Figure 4.66   A U.S. Navy MH-53E Sea Dragon assigned to the "Vanguard" of Helicopter Mine Countermeasures Squadron Fifteen (HM-15) retrieves an AN/AQS-14A Side-Looking Sonar used to detect underwater mines.

In the VLF band (3–30 kHz), radio waves are capable of penetrating sea water to a depth of approximately 20m depending upon the salinity of the water. Transmission in the VLF bands, however, significantly reduces the data transfer (bandwidth) rates and requires the submarine to operate at relatively shallow depths. A submarine operating at greater depths may deploy a VLF trailing wire antenna attached to a buoy that rises to a few meters just below the surface, although careful design is needed to ensure that the buoy and cable remain undetected by sonar or other sensors. VLF antennas are typically 61m in length and the antenna tow line 500-m long [35].

Most VLF transmissions tend to use minimum shift keying (MSK) or continuous-phase frequency-shift keying (CPFSK) modulation techniques.

VLF systems are hindered by low data transmission rates and large antenna installations and are usually used for one-way communications to the submarine and as a backup for global communications during hostilities when nuclear explosions may disrupt higher frequency communications or satellites are destroyed by enemy action.

Extremely low frequency (ELF)

Electromagnetic waves in the ELF frequency range (30–300 Hz) can travel through the oceans and reach submarines at depths of over 300m, even under the polar ice cap, and are therefore typically used to contact more deeply submerged vehicles (Figure 4.67).

Data transmission rates can be very limited, typically in the region of a few bits per minutes, and to overcome this, most ELF transmissions use three-digit codes that are repeated continuously over a period of time to ensure that the signal is received in an environment where background noise is a significant adverse factor in reception and transmission.

Due to the very low bandwidth characteristics of ELF signals, it takes about 20 minutes to transmit such a short coded message via ELF frequencies typically instructing a submarine to come to the surface to receive further instructions (through a higher bandwidth system) [36].

Such codes would offer a predetermined set of messages, providing instructions such as "return to base" or "come to periscope depth to receive further orders." ELF communications are only one-way and do not permit for the submarine to acknowledge signals or communicate on the same

Figure 4.67   1982 aerial view of the Clam Lake, Wisconsin, ELF transmitter facility.

wavelength due to the impracticality of installing ELF transmitters on a submerged vehicle. ELF submarine antennas required to receive signals comprise long (often up to several kilometers) wires, which may be jettisoned from the submarine in an emergency.

U.S. Navy and Russian ELF transmitters typically work at frequencies between 75 Hz and 85 Hz, which equate to a wavelength of 4,000–3,500 km, equivalent to the distance from New York to Los Angeles. Ideally, a half-wavelength dipole antenna would be used, but clearly the construction of such a long antenna by traditional means would be impractical. As a substitute, an antenna is often constructed in an area of low electrical ground conductivity using two huge electrodes, buried in the ground, typically 60 km apart, and connected to a signal station equidistant between the two via wires on poles acting as antenna feed lines. Due to the poor ground conductivity, the current between the electrodes is able to penetrate deep into the earth, effectively using the earth as a radiating antenna. Such an arrangement, however, is very inefficient, and dedicated power stations are often constructed to power the antenna in order to produce a radiated power of only a few watts [37, 38].

Submarine surface communications

Compared to surface ship communications, submarines need to take into account the challenging subsurface environment to which communications equipment is exposed, physical limitations, and the need for extreme stealth when communicating.

Submarine antennas, are unique in design, shape, materials, and performance due to the stringent requirements associated with a submarine's space and weight limitations, extreme environmental conditions, and stealth considerations. In designing the communications antennas, RF absorbing materials are often used that will permit only selected frequencies to pass through the antenna, thereby limiting unwanted side lobe transmissions and ensuring that unwanted signals from enemy radars are absorbed rather than reflected.

While RF communication transmissions provide a high data rate, they require the submarine to expose a radio mast or buoy-mounted antenna, risking detection or alerting the enemy to its presence. However, a surfaced submarine or one operating just below the surface with a towed surface communications buoy can use standard radio-frequency communications to achieve high data rate exchanges with their command centers.

Submarine communications make extensive use of the HF band (3–30 MHz) for long-range, over-the-horizon communications (Figure 4.68). The most efficient transmissions require fairly large antennas and are therefore suited to communications with submarines from a land-based station. The primary drawback with HF transmissions is that they are highly susceptible to atmospheric and electrical changes in the ionosphere and therefore several parallel frequency channels are often used to ensure a reliable communications link to be established.

Today HF communications are generally being replaced by dedicated military communication satellites with highly directional encrypted low probablity of intercept (LPI) transmissions to minimize the risk of detection. The U.S. Navy's system called Submarine Satellite Information Exchange System (SSIXS) is a component of the U.S. Navy's UHF Satellite Communications System, and in the United Kingdom, Skynet 5 provides the secure military satellite link. (HF signals cannot be used for satellite communications as they are reflected by the ionosphere.)

Faster data rates will be achieved through the USN High Data Rate (HDR) Antenna program, which will permit connections to the Joint MILSTAR Satellite Program in the extremely high frequency (EHF) band and provide the ability to access time-critical tactical information from the Global Broadcast Service (GBS).

Figure 4.68   Future submarine communications may also exploit the use of airborne lasers which can penetrate shallow water enabling communications to submarines without the restrictions of trailing wires or buoys or enable communications at greater depths via a subsurface communications buoy, which would avoid the telltale wave signature caused by surface buoys. Here the Royal Australian Navy's Collins-class submarine, HMAS Rankin (SSK 78) cruises out to sea at periscope depth in July 2004.

The HDR antenna will also provide the capability to receive time-critical tactical information from the Global Broadcast Service (GBS) and access to the Defense Satellite Communications System (DSCS) in the super high frequency (SHF) frequency band.

Software-defined radios

It is becoming increasingly difficult to distinguish between radios and data links as virtually all modern radios transmit voice over digital data waveforms. Many data links carry data-only messages; however, most modern digital radios are also capable of transmitting voice and data in digital format.

Taking advantage of digital transmission capabilities, digital radios capable of transmitting voice and digital data have evolved in recent years to exploit the rapid evolution of emerging computing capabilities and the shift to introduce commercial-off-the-shelf (COTS) technology solutions. Rapid progress in these areas has led to similarly swift advances in meeting the needs of tactical data transmission requirements for secure voice, situational reporting, map data, still images, and limited video.

Most digital radios are software programmable, which enables them to be adapted to work with any standard waveforms, often across multiple data link types. Software defined radios (SDRs) can be dynamically reprogrammed to provide operating system upgrades and to facilitate the installation of software applications to radios in the field. Such flexibility also provides the ability to create network-specific radios with even the waveform being able to be reprogrammed to meet operational requirements.

A software-defined radio is defined as one in which some or all of the physical layer functions (see OSI model later in this chapter) are software-defined.[6] In practical terms this means that functions that have previously been implemented in hardware form (e.g., oscillators, noise filters, amplifiers, modulators, demodulators, and tuners) are instead implemented through software functions embedded in generic computers. Such a design produces a radio that can send and receive different waveforms or even transmit multiple waveforms through the use of embedded and easily changed software. This arrangement also facilities the updating of the embedded software to the latest required standard through software patches transmitted to the radio itself.

By programming specific waveforms into SDRs, virtually any SDR radio with suitable hardware and transmission characteristics can be programmed to communicate with existing generation wireless communications systems, as well as new and enhanced digital systems. One of the main advantages of SDRs is their ability to be programmed to provide connectivity to other communications systems, such as commercially available radios, cellular telephone systems, and standard telephone networks, rather than being limited to bespoke military systems.

Within the SDR family, *high capacity data radios* (HCDRs) are used across local networks to transfer data between network nodes. Many digital data radios, often euphemistically called high capacity, are now relatively limited in their ability to shift adequate amounts of data around the battlespace to meet current and future C2 requirements. Most high capacity data radios operate at speeds below 0.5 Mbps and tend to be limited to point-to-point data and voice applications or tactical broadcast applications extending through to limited short-range networking capabilities (Figures 4.69 and 4.70).

A *combat net radio* (CNR) is usually defined as a networked radio that provides a half-duplex transmission and uses either a single radio frequency or a discrete set of radio frequencies when in a frequency-hopping mode [39]. CNRs are primarily used for push-to-talk operated voice radio nets or in automated networks for data transmission supporting C2 functions, often forming the basic layer for tactical C2 from division to battalion and company level.

Figure 4.69   A communications engineer at a Combined Air Operations Center, uses a "Radio over Internet Protocol Routed Network" communication system console to monitor convoy operations in Iraq. March, 2006.

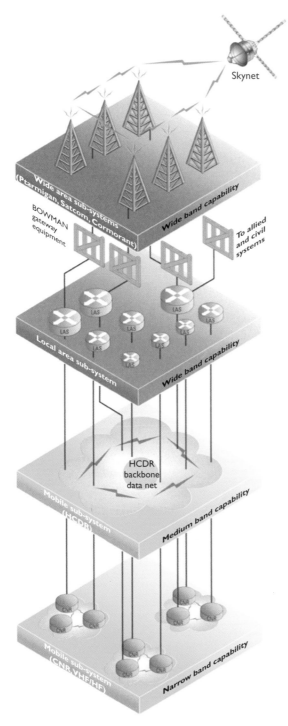

Skynet

Wide area sub-systems
(Ptarmigan, Satcom, Cormorant)

Wide band capability

BOWMAN
gateway
equipment

To allied
and civil
systems

LAS

Local area sub-system

Wide band capability

HCDR
backbone
data net

Mobile sub-system
(HCDR)

Medium band capability

CNR

Mobile sub-system
(CNR, VHF/HF)

Narrow band capability

Figure 4.70 Software-defined radios, high capacity data radios, and combat net radios form important elements of a multilayered network communications system connecting wideband strategic communications through to tactical narrowband data links. The U.K. Bowman communications network is shown in this illustration.

CNRs often provide advanced communications security (encryption) along with the ability to self-network and frequency hop for efficient spectrum utilization and electronic counter-countermeasures (ECCM). CNRs typically support voice and data communications with data transfer rates up to several hundred kbps.

Networked communications, broadcast systems, and data links

Networked systems provide multiple functions across the information gathering process from data collection (sensing) to data processing, data transmission, data and information association, and cataloging (classification) and data and information storage.

As analog data transmission systems are increasingly replaced by digital transmission systems and as data from any source can be digitized for transmission, it is becoming difficult to distinguish between voice and data communication systems. With the advent of digital communications, the ability to send high volumes of digital data or voice traffic across the same networks has become the norm and has transformed the ability of people and machines to operate as a truly integrated network.

Until the 1860s military communications were dominated by physical dispatch and then by the telegraph until about 1915 before being replaced by the telephone until relatively recently [41]. Modern C2 structures comprise multiple integrated tactical and strategic networks that support data, voice, and still and video images and operate over various information distribution systems such as multiple point-to-point links and network broadcast systems using information push-and-pull techniques. The goal of all these systems centers on the need to achieve information superiority and, in this regard, many systems are designed to achieve a key step in that process of shared situational awareness.

Without situational awareness and the means to direct forces, the outcome of any engagement is likely to be adversely affected either through prolonging the contact with the enemy and risking higher attrition rates or reducing the win probability of friendly forces through suboptimal decision-making. The need for secure communications across networked forces is therefore a time-critical component of any engagement strategy, and the threats to maintaining assured connectivity are much more diverse and represent proportionately greater opportunities to disrupt network-based effectiveness.

In addition to the complexities of network establishment described earlier, data links and digital voice communications also present implementation challenges:

- **Transmission standards:** A wide range of digital data links will be found in any network-enabled environment. Transmission standards and communications protocols for voice and data communications vary enormously and need to be standardized to enable sensors, effectors, and platforms to be networked to different data links and C2 networks with relative ease.

- **Information access:** Not all information will be authorized for release to all network participants. There is a fine balance, however, between limiting information access and restricting the shared situational awareness of participants. Even among those authorized to receive information and voice transmissions, managing and updating the security access protocols for authorized users can be a significant impediment to sharing data and information.

- **Information security:** Security can be provided by preventing detection of the signal using LPI waveforms and transmission techniques and then if the signal is detected, preventing its analysis through the use of advanced encryption techniques. It is important to conceal the presence of the data link transmitter so that its location cannot be detected, jammed, or attacked, or the transmission patterns (rather than the information itself) analyzed.

- **Data link integrity:** Data link and digital voice transmissions must be able to overcome natural and man-made obstacles such as weather or enemy jamming and remain functional. Careful design of transmission frequencies, modulation patterns, and error correction algorithms will help to minimize disruption. Data link signal integrity can be improved by the use of advanced waveform modulation, although this creates additional burdens by adding to the amount of data that needs to be transmitted.

- **Transmission capacity:** The size of the pipeline over which the data can be sent will determine the speed of data transmission and the functionality that a data link can provide. A situational awareness data link may be able to operate at around 100 kbps, while a real-time video data link will need to provide transmission speeds typically in the range of 1–10 Mbps. While data compression techniques continue to improve, there remains no real substitute for bandwidth.

Data link principles

Military data links employ a variety of secure communications principles operating at varying data transmission speeds, but all transmissions involve the exchange of digital information in a structured and secured manner.

Digital Radios

The following digital radio systems are typical of the planned and current digital radios in use within NATO countries.

Bowman

The U.K. Tri-Service HF, VHF, and UHF tactical communications system provides secure voice communications, data services, and situational awareness down to the section commander level. It provides the main tactical digitization network for the British Army, linking together units to support network-enabled capability and the integration of units in to the command and battlespace management approach. To support the situational awareness function, the Bowman radio has an embedded GPS system for automatic position location, navigation, and reporting.

The Bowman system comprises a family of networked voice and data radios with self-organizing, self-healing capabilities, obviating the need for a central coordinating node.

The Bowman HCDR provides a high-capacity data backbone of up to 500 kilobits per second, linking clusters of VHF radios at distances of up to 10 km, while the narrowband capability provides voice and data communications to troops at section leader level using secure data rates at speeds of 56 kbps using frequency-hopping spread spectrum and encryption techniques. HCDR data links can be connected via standard IP protocols to other networks.

Falcon II Radio

The AN/PRC-117F (Figure 4.71) is a man-portable, software-defined combat-net radio covering the 30–512-MHz frequency range. It employs FM, AM, PSK, CPM modulation, frequency-hopping voice and data and secure encryption. Transmission rates are up to 64 kbps depending on the type of encryption employed and the frequency band in use.

Joint Tactical Radio System (JTRS)

JTRS is modular family of software programmable digital voice and data communications data links providing the backbone for many current and future battlespace networks. Covering the spectrum from 2 to 2,000 MHz, the JTRS family of radios will be capable of transmitting simultaneous voice, video, and high-speed data across various military networks and will be compatible with a wide range of civil and military radios.

It is capable of transmitting voice, video, and data. JTRS uses advanced LPD/LPI waveforms and standard communication protocols to ensure interface compatibility with a wide range of NATO standard data links, although it is also capable of networking different network-specific waveforms to provide compatibility and interoperability between various networks. Waveforms currently planned to be covered by JTRS capabilities include the following, but will certainly be expanded in the future to include many others:

• Wideband networking waveform (WNW);

Figure 4.71 A member of the International Security Assistance Force (ISAF) Command Network (ICN) Team set up an AN/PRC-117F Radio in the tactical satellite (Tac SAT) role on TV Hill in September 2006. (Photo courtesy of U.K. MoD.)

- Soldier radio waveform (SRW);

- Joint Airborne Networking–Tactical Edge (JAN-TE);

- Mobile User Objective System (MUOS);

- Single Channel Ground to Air Radio System (SINCGARS);

- Link-16;

- Enhanced Position Locating Reporting System (EPLRS);

- High frequency (HF);

- UHF SATCOM demand assigned multiple access;

- Have Quick II.

The versatility of JTRS is designed to permit the automatic creation of wide area networks providing C2 and situation awareness to the front line.

The JTRS family of radios is split into several principle domain-specific groups.

Ground domain:

- *Ground mobile radio (GMR):* Includes the development of a wideband networking waveform (WNW), an IP v6 Internet protocol (IP)-based waveform designed to allow mobile ad hoc networking (MANET).

- *Handheld, manpack, and small form factor (HMS) radios:* Designed for small air and ground unmanned vehicles, soldier communications networks, unattended ground sensors, and intelligent munitions; typically up to a 10-km range and data rates up to 2 Mbps [40].

Airborne, maritime, fixed station:

- *JTRS Airborne Maritime and Fixed (AMF):* Designed for installation in more than 100 U.S. platform types including fixed and rotary wing platforms, surface ships, and submarines and fixed ground stations, with future compatibility planned for the U.K. Bowman digital radio system.

MIDS-JTRS:

- *The Multifunctional Information Distribution System:* JTRS program to provide compatibility with Link 16 waveforms.

Special radios:

- *JEM:* The JTRS JEM program adds JTRS capability to the existing handheld multiband inter/intrateam radio (MBITR) to create the JTRS Enhanced MBITR (JEM). In a program led by the U.S. Special Operations Command.

Networking enterprise domain (NED):

- *JTRS Networking Enterprise Domain (NED):* Develops the waveforms and network management software for the JTRS family of radios.

JTRS Enhanced MBITR (JEM)

The AN/PRC-148 JEM radio (Figure 4.72) provides a good example of modern software defined radio capable of voice and data through to top secret level. It operates at frequencies between 30 MHz and 512 MHz on one of 256 preset channels covering numerous waveforms and modes (current and planned) such as:

Figure 4.72   The AN/PRC-148 JEM radio.

- AM/FM

- HAVEQUICK I/II;

- MIL-STD-188-241-1/-2 (SINCGARS);

- MIL-STD-188-181B (56 kbps);

- MIL-STD-188-181C (SATCOM IW);

- Blue Force Tracking;

- Mobile Ad Hoc Networking (MANET);

- Retransmission;

- Civil air traffic and marine frequencies.

PR4G

The Poste Radio de 4eme Generation (PR4G) is a secure software programmable radio operating in the 30–88-MHz VHF band and capable of handling simultaneous digital voice and data communications in IP formats if required. Transmission security is provided by fast frequency hopping and free channel search modes or an automatically managed mix of the two as well as providing an interface for additional cryptographic equipment.

Single-Channel Ground-to-Air Radio System (SINCGARS)

The SINCGARS radio (Figure 4.73) is primarily a digital VHF voice radio, but can also transmit limited amounts of digital data at 16-kbps data transmission speeds. It has its own dedicated waveform that has been adopted by many other software programmable radios.

Figure 4.73 A U.S. Army soldier practices with the SINCGARS radio during field trials in 2008.

Talon V/UHF Radios

The Talon software defined radio offers various waveforms in addition to its own waveform including the SINCGARS and Saturn waveforms. The Talon waveform provides enhanced encryption capabilities and a country unique identification code (CUID).

Wideband Network Radio (WNR)

A tactical digital network providing connectivity applications for the U.S. Army and U.K. Army (via the Bowman radio system). It operates in the UHF frequency band at 225–450 MHz and uses self-organizing networking techniques to network multiple users.

The aggregate throughput of the WNR system can obtain data rates of over 2 Mbps, supporting multiple users at a data rate of 288 kbps per user, although these rates will be significantly reduced depending on the overlaid data and security protocols. The WNR covers a range of 35–40 km in open terrain.

Information can be transmitted across data links in either a raw format such as video or radar imagery or processed using varying degrees of data fusion from collaborating sensors (typically on the same platform) to provide for the continuous exchange of information concerning established tracks and targets.

The timely transmission of information is a key enabler in the C2 task and supports the ability to control, monitor, and evaluate operational activity across the battlespace, as well as forming an integral part of the linkages across the kill chain functions.

Data links are also an essential tool in the production of ISTAR information and a common real-time situational awareness picture, which relies on the integration of intelligence and sensor information from a wide range of sources (Figure 4.74).

Given the wide range of functions that data links are required to serve, no single data link can support every operational requirement, which will vary widely in data transmission rates and operating environment (Figure 4.75). Significant challenges also exist as a result of the variation in data link message formats, degree of data processing, and associated requirements, timing and synchronization, security protocols, and network protocols, all of which vary significantly across data link types.

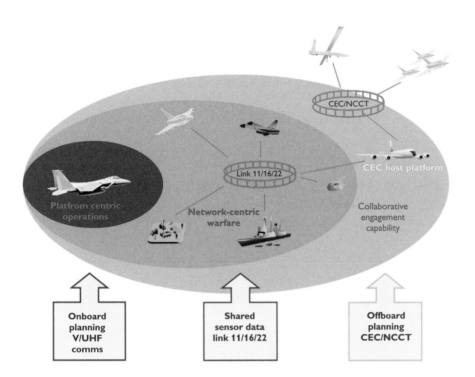

Figure 4.74 Evolution of integrated effects through data link improvements.

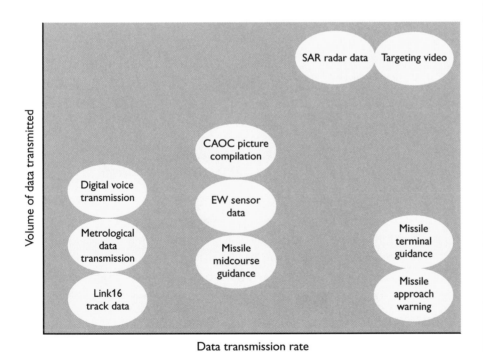

Figure 4.75 Data rate requirements will vary significantly depending on the application.

Data links serve a number of operational purposes (Figure 4.76) and can be considered either as an *uplink* where information is transmitted to the platform or *downlink* where information is transmitted from the platform. The provision of ISTAR data to the commander is:

- To support situational awareness and decision making;
- To support and coordinate crosscueing of platforms and sensors across the kill chain or defend chain (see Appendixes F and G);
- To enable tactical data to be exchanged between coalition forces;
- Data links on platforms such as UAVs and U-2 where sensor data is downloaded via the data links for processing compared to Rivet Joint and Nimrod MRA4, for example, where processing is done on board;
- On-board processed images such as SAR imagery, GMTI data, or track data to require significantly less transmission bandwidth than unprocessed data;
- C2 linkages to extend operational use of platform to include remotely command delivery of effects;
- Data links for increased tempo of operation to form the NEC picture;
- To extend situational awareness and operational effectiveness of collaborating platforms and their associated sensors;
- To assist in the combat ID task and the positive identification of friendly forces deployed in the battlespace.

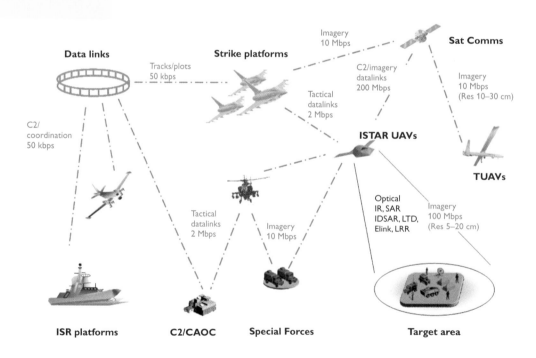

Figure 4.76 Typical data transfer requirements for joint strike operations (simplified for clarity).

The use of data links to provide an increase in data accessibility from sensors and networks is key to the improved tempo of decision-making and acceleration of the OODA loop, as described earlier, with the number and quality of decisions being heavily dependent upon available bandwidth and the capabilities of the attendant data links (Figure 4.77).

Data transmission rates will vary significantly depending on the function of the data link. At the low end of requirements are sensors that feed processed data into a network containing, for example, three-dimensional time and location-based track data. The transmission of processed data requires relatively limited bandwidth to operate effectively, and it is this principle that enables data links with relatively low transmission rates such as Link-16 to function effectively. When multiple tracks are detected and reported by the sensor, the bandwidth challenge increases rapidly, with the potential degradation of time accuracy caused by data queuing and delays in transmission increasingly degrading the accuracy of the information. At the other end of the scale, modern radar and E/O sensors, especially those involved in imaging functions such as video, SAR radar, and GMTI functions, need significantly increased bandwidth, with real-time functionality and utility often being defined by the available bandwidth.

Link type	Link 4A	Link 11	Link 16/22	TTNT	Satellite	Phased array
Data rate	5 Kbps	2.25 Kbps	54 Kbps	2 Mbps	Civil > 100 Mbps Mil > 3 Gbps	Tx 548 Mbps* Rx > 1 Gbps*

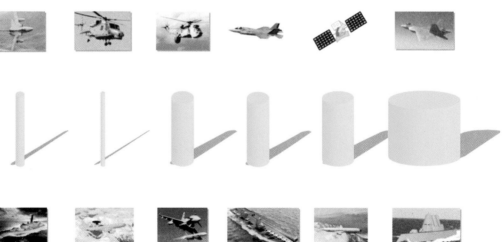

*Figures quoted are for F/A-22 Raptor APG-77 AESA radar. Larger Tx and Rx rates will be possible with larger AESA radars

Figure 4.77 Data transmission capabilities vary significantly between platforms and applications.

One way to overcome the limitations in bandwidth is for the platforms themselves to process the sensor and image information on-board and send only the processed data across the network. The limitation of this, however, is that the on-board processing power is usually limited compared with that on the ground, and if preformatted images are transmitted, it leaves fewer options for sensor fusion algorithms to use raw data as a basis for their intelligence analysis.

While directional data links tend to be difficult to intercept and jam, most RF and microwave networks are still vulnerable to noise jamming, where a randomly generated noise signal is transmitted on the same frequency or bandwidth as the network itself. Frequency-hopping networks and data links can still be defeated by broadband jamming techniques, which simply cover a larger bandwidth to blank out the signal over the frequency-hopping range. This problem is particularly prevalent in commercial satellite communication systems where noise jamming introduced within the footprint of the satellite uplink can be a highly effective means of degrading or jamming the satellite's performance.

One potential means of overcoming radio frequency jamming is to use highly directional laser uplinks and downlinks, which would need an enemy to in turn jam the satellite receiver using a laser on the same frequency. Although frequencies in the light part of the spectrum lose a lot of energy through propagation effects, there is a significant amount of bandwidth available at these high frequencies, so even with degraded performance, suitable data encoding can still ensure that high data transmission rates can be achieved.

Data communications structures

Data communications protocols tend to be structured around standardized functional models permitting designers to develop interfaces common to the communication needs of groups or user communities. Such nonproprietary models are usually referred to as open standards, enabling the construction of open systems architecture that can be developed by any third party.

Several international standards have been developed such as:

- The TCP/IP model as used across the Internet;
- The Open Systems Interconnection (OSI) standards model;
- The Global Information Grid (GIG) model.

The functional blocks of these models and the interactions between the model layers will vary depending on how information is required to be collected, processed, and disseminated across the network.

The TCP/IP model

Strictly speaking, the TCP/IP model comprises two transmission protocols: the Transmission Control Protocol (TCP) and the Internet Protocol (IP). The TCP/IP model is generally described as using only four layers[6] of application, transport, Internet network, and host (some references add physical layer[7]), which are not as rigidly structured as the seven-layer OSI model. However, each of these layers accommodates the functionality of the OSI layers.

In March 1982, the U.S. Department of Defense declared TCP/IP to be the standard communications protocol for all military computer networking [50].and the general trend across NATO communications systems is to move away from TDM-based systems to an *everything over IP approach,* including telephony services, voice, video, data, and chat.

The OSI model

The Open System Interconnection Reference Model (OSI Reference Model or OSI Model) (Table 4.6) describes a protocol for layered communications and computer network protocol design, developed as part of the Open Systems Interconnection (OSI) initiative [51].

OSI communications principles are structured around a framework comprising seven discrete functional layers. Each layer has a dedicated role of greater or lesser importance depending on the type of communications protocol involved; in fact, some protocols may combine the functions of several layers into a single functional layer. Layers are defined by a collection of conceptually similar functions that provide services to the layer above and request and receive service from the layer below.

	Data unit	Layer	Function
Host layers	Data	7. Application	Network process to application
		6. Presentation	Data representation and encryption
		5. Session	Interhost communication
	Segment	4. Transport	End-to-end connections and reliability
Media layers	Packet	3. Network	Path determination and logical addressing
	Frame	2. Data link	Physical addressing
	Bit	1. Physical	Media signal and binary transmission

Source: [52].

Table 4.6   OSI Model

- **Application layer:** The application layer provides standardized software applications for end users and also provides application-specific services to other programs. Applications are often provided from a variety of sources and as such their input and output syntax requirements may differ widely.

- **Presentation layer:** This layer, sometimes called the syntax layer, ensures that the data input and output from diverse software programmes and encryption standards in the application layer is presented in the correct format and syntax to suit the applications from which it is sent and received.

- **Session layer:** This layer controls the communication protocols between computers, establishing, managing, and terminating the connections between local and remote applications.

- **Transport layer:** This layer manages the provision of reliable data transfer across all layers to its end-users, ensuring flow control, error detection and correction, efficient bandwidth usage, and resource optimization.

- **Network layer:** The network layer coordinates and manages the transfer of variable length data messages across a network of networks or within an individual network.

- **Data link layer:** This layer provides functional point-to-point data transfer between two or more physically connected devices, enabling reliable data delivery between two physically connected devices.

- **Physical layer:** This layer defines the electrical and physical properties of the communications components. It defines and manages the transmission modulation and receptions standards, converts signals into either analog or digital signals, and may also manage the flow control, error detection, and correction processes.

The GIG model

The Global Information Grid (GIG) model is a hybrid of the OSI and TCP/IP reference models (Figure 4.78). It adds a mission layer and combines the presentation layer and session layer into a single service layer.

- **Mission layer:** The mission layer provides the specific aggregation of applications from the application layer necessary to perform a particular military mission.

- **Service layer:** The service layer acts as an interface between different applications ensuring data in a correct format is presented participating applications. This layer also provides a control structure for connections and structured interactions between applications.

The OSI Model and the Future Combat System/Army Brigade Combat Team Modernization Program

The U.S. Army's Future Combat System (FCS) program was formally cancelled in April 2009, to be replaced by the Army Brigade Combat Team Modernization program. FCS was a highly integrated network of task-oriented modular systems that would enable soldiers to comprehend, shape, and dominate the future battlefield by being able to see first, understand first, act first, and finish decisively [22]. Much of the network-enabled functionality and concepts of the FCS program will be preserved in the Army Brigade Combat Team Modernization program, and as such it is worthwhile mentioning the key network elements of the FCS program that have been built along the lines of the OSI model:

Communications and Computers (CC) Systems

The networked systems are connected to the command, control, communications, computers, intelligence, surveillance, and reconnaissance (C4ISR) network by a multilayered communications and computers (CC) network with a particular focus on range, capacity, and dependability over extended distances and complex terrain. The network will support advanced functionalities such as integrated network manaement, information assurance, and information dissemination management to ensure dissemination of critical information among sensors, processors, and warfighters.

Intelligence, Reconnaissance, and Surveillance (ISR)

A distributed and networked array of multispectral intelligence, reconnaissance and surveillance (ISR) sensors mounted on the various vehicles provides the network with the ability to see first and act first. Networked intelligence, reconnaissance, and surveillance (ISR) assets will provide timely and accurate situational awareness (SA), enhance survivability by avoiding enemy fires, enable precision networked fires, and maintain contact throughout engagement. The networked system processes real-time ISR data, outputs from survivability systems, situational awareness data, and target identification information to update the common operating picture (COP) containing information on friendly forces, battlespace objects (BSOs), BSO groupings, and their associated intent, threat potential, and vulnerabilities.

Network

The communications network enables soldiers to perceive, comprehend, shape, and dominate the future battlefield through the real-time distribution and dissemination of information and data. To achieve this, the network is reliant on robust, reliable, and high-capacity data links that are structured across five discrete layers: the standards layer, the transport, services layer, the applications layer, and the sensors layer, and the platforms layer.

The Standards Layer

The Standards Layer is the foundation of the network, providing the governance and protocols for which the other layers are shaped and formed.

Transport Layer

The Family-of-Systems (FoS) are connected to the C4ISR network by a multilayered transport layer. The transport layer provides secure, reliable access to information sources across the network. The transport layer comprises several heterogeneous communication systems including the Joint Tactical Radio System (JTRS) and Warfighter Information Network-Tactical (WIN-T). The use of all available networked resources provides a robust, layered communications network with multiple redundancy paths that seamlessly integrates ground, near ground, airborne, and space-borne assets for constant connectivity to the GIG.

The Network Management System will be utilized to manage the entire network including radios with different waveforms, platform routers, and local area networks (LANs), information assurance elements, and hosts. It provides a full spectrum of information capabilities required during all mission phases, including premission planning, rapid network configuration upon deployment in the area of operations, monitoring the network during mission execution, and dynamic adaptation of network policies in response to network performance and failure conditions.

Services Layer

Central to the network implementation is the services layer, commonly referred to as System-of-Systems Common Operating Environment (SOSCOE), which supports multiple mission-critical applications in real-time, near real-time, and nonreal-time frameworks.

The services layer framework allows for integration of critical interoperability services that translate Army, Joint, and coalition formats to common native, internal message formats, and access to common databases such as the situational awareness database.

The network software is supported by application-specific interoperability services that act as proxy agents for each Joint and Army system. The application layer can also access interoperability services that act as proxy agents for each Joint and Army system through applicable program interfaces.

Applications Layer

The applications layer is responsible for providing the integrated ability to assess, plan, and execute network-centric mission operations using a common interface and a set of nonoverlapping functional services such as:

- Integrated On-The-Move Common Operational Picture Real of the 4-D Battlespace;
- Real-time collaboration among dispersed warfighters;
- Automated deconfliction of Blue Forces, air/ground space, and fires;
- Automated planning and rehearsal decision making process;
- Dynamic sensor planning, tasking, and collection visualization to support Commanders Critical Information Requirements (CCIR) at all levels;
- Rapid battlefield damage assessment tied to networked fires;
- Full control and autonomy of unmanned systems and payloads.

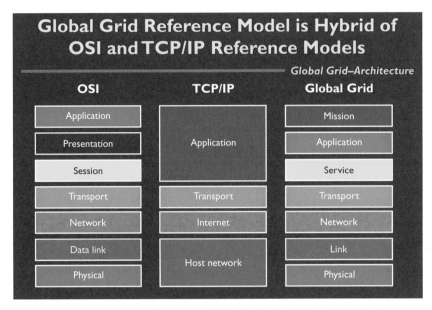

Figure 4.78 The global grid reference model is a hybrid of OSI and TCP/IP reference models [53].

- **Link layer:** Similar to the data link layer in the OSI model, the link layer manages the two-way communications protocols between the many users on the GIG using different data access techniques such as publish and subscribe and push/pull data requests.

Composition of a data link

Data link transmissions comprise a series of encrypted, digital message sets broadcast across a communications bearer at a range of speeds and frequencies depending on the type of data link in use. Transmissions may take place over a dedicated data link or alternatively over digital radio transmissions that serve as a means of secure voice and data transmissions. Message formats depend on the available bandwidth and also differ even at a basic level such as the message bit lengths.

The employment of data links requires careful planning to ensure that requisite platforms are not only equipped with the appropriate equipment, but are issued with the correct encryption and access codes. The proliferation and diversity of data links must also be integrated into the busy electromagnetic spectrum used by civil and military authorities to ensure that there are no conflicts in the spectrum frequency ranges assigned to each data link. Extending the use of data links over satellite networks requires additional planning and liaison activity to be undertaken with civil and military satellite operators.

Effective use of data links requires careful premission data link planning to evaluate the required information flow, data transmission rates, security, and connectivity to plan an effective communications system in advance to enable all participants to participate on the network (Figure 4.79).

Consideration is also required to consider how the network can function effectively if the configuration changes through assets joining or leaving the network, if the mission requirements change, or if elements are destroyed or degraded through enemy action.

Where point-to-point data links are used, formats can be bespoke to meet the specific needs of the system. However, multiple access techniques are required in networked data links to enable multiple data links to share data, and this requires a standardized approach to organizing the data for transmission.

All multiple access transmissions follow a standard protocol format, such as an IP format depending on the type of data link. Typically, a transmission sequence will comprise a standardized number of information bits divided into discrete sections aimed at synchronizing the message transmission to the network slot (Figure 4.80).

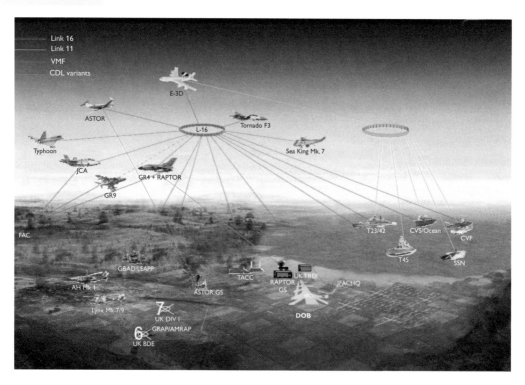

Figure 4.79   Multiple data link types are used to establish a C3-ISTAR network specific to mission requirements.

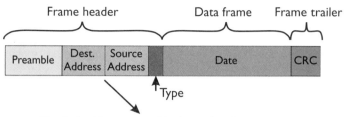

Figure 4.80 Typical data frame structure.

The preamble is used to synchronize timing including clock rates, frame and character synchronization and for error detection.

Point-to-point protocols

In point-to-point data links a single shared communication channel is used to exchange data between two nodes, with up to two or more simultaneous transmissions taking place without interference, at any one time. Specific individual frequencies can be nominated per link and because the point-to-point data links tend to be directional in nature they can achieve longer ranges as the energy is not spread directionally or across the spectrum.

Data link transmission techniques

There are several approaches that can be used in point-to-point and multiple access data links:

- **Simplex:** One-way transmission;
- **Half duplex:** Alternating two way transmission;
- **Full duplex:** Simultaneous two-way transmission.

Where dedicated links are required between permanent or semi-permanent systems, point-to point data links are often used. The directional nature of point-to-point data links also makes them ideally suited for covert operations where energy is focus in a narrow beam and is therefore difficult to detect by enemy sensors.

Unlike radar antennas, which are highly directional and optimized for sending and receiving signals in one direction, multiple access data links often use relatively unsophisticated omnidirectional antennas, which broadcast in all directions to provide wide spatial coverage. Such antennas require no mechanical or electronic steering or associated power supplies and are therefore cheap, small, and lightweight compared to directional antennas.

Typically, frequency hopping, encryption, burst transmissions, and directional antennas are used to minimize the risk of detection and interference with data link transmissions.

Digital data compression

Digital data compression is essential to ensure that the high volumes of data generated by sensor systems such as video imagery and C2 data is transmitted as fast as possible without having to queue for transmission and that the data latency is minimized to assist with accurate intelligence and situational awareness assessments.

Video compression ratios can typically be achieved in the region of 70:1, speech typically at a ratio of 6:1 at an acceptable quality level, and data typically at a 2:1 compression ratio.

Multiple access protocols

In networks where data links need to access multiple nodes or to receive their information from multiple nodes, standardized multiple access protocols are used at each transmit/receive point to build a single cohesive network. At each node, a distributed algorithm determines how stations share the channel and frequency allocations, determining when a station will transmit, on what frequency, and using what code and modulation pattern. Information about the protocols to be used is often sent across the network to participating nodes.

For all multiple access protocols, the key building block is that of timing and synchronization across the network so that data transmission can be precisely queued and received and transmitted.

There are two main types of data transmission used to ensure synchronization across the network.

- **Asynchronous transmission:** Each character is synchronized by a dedicated start bit.

- **Synchronous transmission:** A larger frame of data is transmitted as one contiguous block.

Multiple access protocols enable multiple data transmissions to be made across single channels. Key multiple access principles are as follows:

- **Channel partitioning:** Each channel is divided into small partitions such as time slots, codes, or frequencies, with each partition being allocated to a node for its specific use.
- **Random access:** Nodes are allowed to transmit as and when required. If data from multiple sources collides, algorithms assist in recovering the data stream from the collisions.
- **Taking turns:** Nodes transmit in a sequential pattern, which is tightly coordinated across the network to avoid collisions. Each network participant or node has a reserved slot during which time it can transmit.
- **Polling:** The network appoints a master node that invites subnodes nodes to transmit in turn.
- **Token passing:** Control, which can be imagined in the form of a token, is passed from one node to the next sequentially to form a token ring network (Figure 4.81).
- **Demand assigned:** A central node allocates a transmission slot to subnodes based on a request for transmission, which includes details of the transmission slot required and details of the transmission parameters.

Across military communications networks, the following (Figure 4.82) are among the more commonly used network protocols, most of which combine several of the techniques mentioned above. Code division multiple access (CDMA) employs spread-spectrum technology and a special coding scheme (where each transmitter is assigned a unique code) to allow multiple users to be multiplexed over the same channel. By contrast, time division multiple access (TDMA) divides network access by time, while frequency division multiple access (FDMA) divides it by allocating specific frequencies. Demand assigned multiple access (DAMA) allocates channel space depending on requests from users.

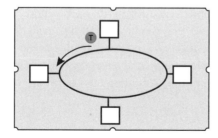

Figure 4.81 A token ring passes control of the network in a sequential order around the network.

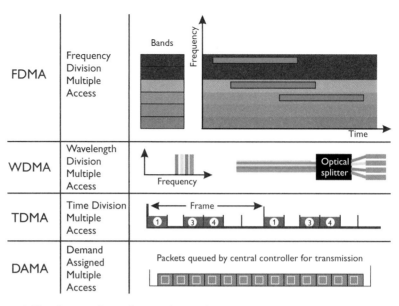

Figure 4.82 Commonly used network modulation protocols.

Time-division multiple access (TDMA)

Time-division multiple access (TDMA) is a channel partitioning digital transmission technique that allows several users to access and share a single frequency channel without interference from other users by dividing the signal into different time slots and allocating unique time slots to each user. The users on the TDMA net transmit in turn according to the time slot they have been allocated, thereby allowing multiple participants to share the same transmission frequency channel.

The preassigned time slots ensure that only one transmitter is active at any one time, and as such is an efficient means of sharing a limited bandwidth on the same frequency between multiple network participants.

Each station gets a fixed length slot, which is the time required to transmit the data packet in a standardized format in each round. The network will contain a number of slots that can be allocated to participants and each unused slot will go idle if there is no participant allocated to that slot. Each user is effectively therefore allocated a time gate that ensures that there will be no interference from other simultaneous transmissions on the same frequency.

TDMA networks are controlled by a net control station (NCS), which coordinates and allocates transmission slots and polls each participating unit

(PU), which in turn transmits its data prior to reverting to the receive mode. The transmission cycle continues until all PUs have transmitted once before cycling through participants on the network in turn once again. The time taken to poll all participants is called the *net cycle time*.

The NCS collects information from PUs, integrates it into a common data framework, and retransmits it to all participants on the network. To ensure data integrity, a checksum calculation is checked at the receiver. If an error is detected, the frame is simply dropped.

TDMA and CDMA techniques are always used in combination with FDMA so that a given frequency channel may be used independently of signals on other frequency channels and other networks.

TDMA is used in many TADIL-standard data links such as Link-16/JTIDS. In a JTIDS network, there are 64 frames per 12.8-minute epoch. Each frame is 12 seconds in duration and is composed of 1,536 time slots. There are 98,304 time slots in a single JTIDS net (Figure 4.83).

Dynamic TDMA (DTDMA)

In a dynamic TDMA network, transmission slots can be assigned on demand to participating transmitters as they join and leave the net. TDMA allows all users on the net to be informed or addressed with a common message set to ensure, for example, that all users have access to the same situational awareness picture.

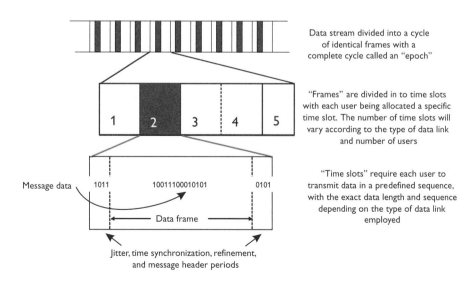

Figure 4.83   The Link 16 TDMA message structure.

In addition to the use of TDMA in the TADIL family of data links, TDMA principles are used by the U.S. Combat Net Radio (CNR) and in WiMax transmissions.

Extended time division multiple access (ETDMA)

ETDMA is a technique used in TDMA-based voice communications networks and uses the pauses contained in normal speech to send data (including other voice communications) to substantially improve upon the efficiency transmission principles. In exploiting these pauses, the system scans the transmission buffer of the participating devices, notices that the node has data to transmit, and allocates bandwidth accordingly. If a node has nothing to transmit, the sequence continues to the next node.

Carrier-sensing multiple access (CSMA)

In CSMA, transmitters share the same frequency, and the term *carrier sensing* describes the arrangements whereby collision detection processes arbitrate access to the desired channel. In some shorter-distance networks, it is possible to listen to the channel before transmitting, which is called *sensing the carrier,* and avoids interrupting other transmissions already taking place before transmitting on a shared physical medium, such as a hardwired cable or a band of the electromagnetic spectrum. If another signal is sensed, the node waits for the transmission in progress to finish before initiating its own transmission. Of course, a single channel transmission can fully utilize the channel without having to worry about collision avoidance.

Multiple access arrangements facilitate the ability for multiple nodes to send and receive on the same network and enable transmissions by one node to be received by all other nodes on the network if required.

Wi-Fi and wired Ethernet transmission use CSMA techniques for their transmission protocols.

Carrier sensing multiple access with collision detection (CSMA/CD)

Collision detection is used to improve CSMA performance by terminating transmission as soon as a collision is detected and reducing the probability of a second collision on retry. This system is used in Ethernet connections.

CSMA/CD protocols listen both before transmission and during transmission to detect possible or actual collisions, and if any are detected while transmitting, the transmission is aborted to free up the channel. The protocol then waits for a random time interval before trying to resend the data.

Code division multiple access (CDMA)

CDMA is a form of direct sequence spread spectrum communication, which allows multiple frequencies to be used, thereby facilitating the ability for many nodes to transmit and receive information simultaneously over a single communication channel (multiplexing).

CDMA codes every digital packet it sends with a unique key, known as a "chipping" sequence, which enables the target receiver to identify the signal by recognizing the unique key and demodulate the associated signal.

Wideband code division multiple access (WCDMA)

WCDMA is a wideband spread-spectrum communications protocol that uses asynchronous code division multiple access (ACDMA) to achieve higher speeds and accommodate more users compared to most time division multiple access (TDMA) schemes used today. ACDMA avoids the need for each user to transmit their assigned sequence at the same time (as in synchronous CDMA), thereby permitting significantly improved usage of the available bandwidth. WCDMA is a used in commercial wideband spread-spectrum 3G mobile telecommunication interfaces and is also the protocol to be used in the U.S. MUOS networked UHF satellite system.

Frequency-division multiple access (FDMA)

In FDMA, frequencies are allocated dynamically within a designated wideband channel so that each node accessing the network transmits on a different frequency. Transmission on slightly different frequencies prevents interference between radios operating on the same network.

Each node wanting to participate on the network is allocated a specific narrow frequency within the wideband channel on which to transmit, although this can be wasteful of the available bandwidth as the channel is assigned irrespective of whether or not it is transmitting. If the transmission path deteriorates, the controller automatically allocates another channel, which may be subject to less interference or noise.

Although this method is simple and reliable, it has its limitations. Each single-channel network or one-way link requires the allocation of two radio frequencies. A duplex (simultaneous two-way) link requires four separate frequency channels.

FDMA techniques are used in satcom and 3G mobile communications.

Demand assigned multiple access (DAMA)

DAMA is a technique that matches user demands to available channel bandwidth. DAMA assigns users variable time slots that match the user's information transmission requirements.

DAMA generally sets up channel assignment on a first-come, first-serve basis, although other types of formats such as prioritized cueing access and minimum percentage access can be used. The prioritized cueing access technique is suitable for command type networks, while the minimum percentage access approach is suitable for less time-critical support and logistic nets. Regardless of format, DAMA generally increases channel capability by 4 to 20 times over normal dedicated channel operation [54].

DAMA protocols are used by data communications systems such as satellite networks.

Error correction

There are several basic methods that can be applied to ensure accurate reception of transmitted data based on either forward error correction using error correction codes that avoid the need to retransmit or backwards error correction where erroneous frames are detected, discarded and retransmitted:

Backwards error correction

Backwards error correction techniques such as the automatic repeat-request (ARQ) protocol provides a means of improving the accuracy of the received data by sending the data along with an error detection code, which the receiver uses to check for errors, and if required, request retransmission of erroneous data. In many systems employing ARQ techniques, the request and confirmation process is built into the process and the receiver sends confirmation of having correctly received the data. If no acknowledgement is received within a predetermined period of time the data is re-sent.

This process has several drawbacks for battlefield systems. First, it requires significant additional bandwidth to ensure that data can be rebroadcast.

Second, it adds latency to the data transmission process, which may not be appropriate for time critical operations, and third, it requires the receiving platform to broadcast back to the transmitter, which may also be impractical for receivers that do not wish to reveal their position.

Forward error correction

Forward error correction (FEC) provides a means of improving the accuracy of the signal reception by adding redundancy to the transmitted information using a predetermined algorithm. This permits the receiver to detect and correct errors without the need to request additional data from the sender, thereby reducing the effective bandwidth requirements and improving data accuracy.

In environments where signals are competing against high background noise such as in high signal density environments or at long range or during deliberate jamming, a relatively high proportion of errors are likely to be generated in the signal reception. The ability to correct these errors means that the signals can be broadcast reliably even in relatively high noise environments. This enhancement translates into extended range, improved performance in a denied environment, higher data reception rates, and increased data accuracy.

The maximum percentage of errors that can be corrected is determined by the design of the algorithm, and different forward error-correcting codes are therefore designed to match specific transmission requirements.

The two main types of FEC are block coding and convolutional coding:

- Block codes work on fixed-size blocks (packets) of bits of a predetermined size. There are many types of block codes, but the most common is Reed-Solomon coding or error correction. Reed-Solomon error correction works by oversampling[8] a polynomial equation constructed from the transmitted data. The polynomial is sampled at multiple points to overaccurately determine its pattern so that the receiver can reconstruct the original polynomial even in the presence of a few bad points. *Reed-Solomon error correction* is used in a wide range of commercial applications, most notably in CDs and DVDs, but is also suited to noise reduction in transmitted signals and as such is often used in commercial data links such as satellite transmission error correction. Other FEC block code protocols include Golay, Reed-Muller, and Hamming codes, which are suited to varying network applications.
- Convolutional codes work on bit or symbol streams of arbitrary length. A convolutional code can be turned into a block code, if desired.

Convolutional codes are most often decoded with the Viterbi algorithm, though other algorithms are sometimes used.

FEC codes typically examine the patterns evident in the previous several hundred, received bits to determine how to decode the subsequent groups of bits. In a simple example, an oversampling of three times for every bit of transmitted data received could be used to reduce errors and determine the intended transmission pattern, with the correct output being determined by the most frequently occurring value in each group of oversampled data as shown below. Such a process will, however, result in higher bandwidth requirements.

The simplest form of error correction is for the receiver to assume the correct output is given by the most frequently occurring value in each group of digits such as groups of three as shown in Table 4.7.

Such an arrangement is data intensive, and other, more sophisticated FEC methods exist to minimize the overhead burden associated with this technique.

One further advantage of FEC is that it does not require handshaking protocols to establish connections between the source and the destination, and can therefore be used to broadcast data to multiple destinations simultaneously from a single source.

Double pulsing

Double pulses of data can be sent in order that they can be checked upon receipt for consistency. If errors are detected, depending on the type of protocol used, either the data can be rejected or a request can be issued for a repeat transmission. Double pulsing is also useful in jammed environments to ensure a higher degree of integrity of transmitted data.

Code received	Interpreted as
000	0
001	0
010	0
100	0
111	1
110	1
101	1
011	1

Table 4.7 Forward Error Correction Codes Can Add a Significant Burden to Bandwidth Requirements, but Result in High Data Transmission Integrity

Interleaving techniques

Interleaving is used in digital data transmission technology to protect the transmission against burst errors. These errors overwrite a lot of bits in a row, so a typical error correction scheme that expects errors to be more uniformly distributed can be overwhelmed. Interleaving is used to help stop this from happening.

Data is often transmitted with error control bits that enable the receiver to correct a certain number of errors that occur during transmission. If a burst error occurs, too many errors can be made in one code word, and that code word cannot be correctly decoded. To reduce the effect of such burst errors, the bits of a number of code words are interleaved before being transmitted. This way, a burst error affects only a correctable number of bits in each code word, and the decoder can decode the code words correctly.

This method is popular because it is a less complex and cheaper way to handle burst errors than directly increasing the power of the error correction scheme.

Whatever the type of data link employed, a means of ensuring collaboration between participating units is required to ensure accurate and timely collection and dissemination of data and a standardized approach to enable data links to join and leave the network without disruption to other participants. Various multiple access protocols are used for that purpose.

All data links make use of techniques that permit digital data to be combined into one signal over a shared medium. Such a process is known as *multiplexing* and permits multiple signals to be sent and received simultaneously over the same data link.

Multiplexing uses a structured protocol to combine multiple signals on to a single communication channel by creating several subchannels, each one able to transmit discrete data. Multiplexed signals are unscrambled using the reverse process once the signal has been received, known as demultiplexing, which takes place in the receiver.

A device that performs multiplexing is called a multiplexer, and a device that performs the reverse process is called a demultiplexer.

The main types of multiplexing used in network data links are:

- *Frequency division multiplexing* (FDM), where several information signals are transmitted simultaneously over the same shared medium, using slightly different frequencies;

- *Wavelength division multiplexing* (WDM), where multiple signals are sent using slightly different wavelengths;
- *Time division multiplexing* (TDM), where signals are sent at slightly different times on the same wavelength and same frequency to avoid them overlapping;
- *Code division multiplexing* (CDM), where signals are sent at the same time using a different code understood by the receiver to distinguish one signal from another.

Data link and radio modulation techniques

Digital data is sent in bits, individual "1" or "0" signals. The digital bit stream needs to be formatted to enable a signal to be sent and decoded. This process is called *modulation* and involves varying the amplitude, frequency, or phase of the signal (see Appendix D for details).

Data link modulation techniques use advanced and complex forms of modulation to provide low probability of intercept (LPI) waveforms and secure data transmission ensuring a high probability of the signal getting through to the destination even in dense RF environments or in the presence of enemy jamming.

Low probability of intercept waveforms

A combination of techniques will typically be used together in a data link to reduce the probability of detection, interception, and analysis and is typically referred to as LPI waveforms. Even if a data link transmission is detected, the use of spread spectrum techniques will make it very difficult to rapidly piece together the signal and even if this is achieved, encryption will present another hurdle to understanding its contents. Of course, once the presence of a transmitter has been established much can be gleaned from transmission patterns even without decryption. Routine transmission patterns can be distinguished from those used in higher alert states and other transmitters responding to a transmission may also become active as they participate in transmission protocols such as acknowledging C2 instructions.

LPI transmission techniques include burst transmissions and spread spectrum transmissions including frequency-hopping patterns.

Burst communications

Burst communications involve compressing the data and transmitting it in short bursts at unpredictable times and in an irregular time pattern so that it is difficult to detect and intercept. If the transmission duty cycle is low, the signals will be particularly difficult to detect by receivers sweeping transmission bands; in addition, the higher data rates spread the spectrum, making it difficult to detect transmission waveforms.

Burst communications techniques are often used in radios and transmission equipment for Special Forces and security services where relatively small amounts of information need to be transmitted with minimal risk of detection.

Frequency hopping and spread spectrum transmission techniques

The term *spread spectrum* refers to the modulation technique of spreading the carrier signals over the full bandwidth or spectrum of a device's transmitting frequency. The result appears very similar to background static or white noise and is therefore particularly difficult to recognize as a deliberate transmission and even more difficult to detect multiple transmissions to form a coherent data sequence if decryption is required. Spread spectrum signals typically cover a much wider band than the information they are carrying to make them blend more into background noise patterns.

The transmission pattern of a spread spectrum signal (Figure 4.84) has a symmetrical pattern centered on the carrier frequency, similar to a normal AM transmission; however, the added noise causes the spectrum coverage to be much broader than that of a standard AM transmission.

Spread spectrum systems transmit on similar power levels to conventional narrowband transmitters, but because the signal covers a broader frequency range, they transmit at a much lower spectral power density (measured in watts per hertz). The lower power density of the transmission helps to ensure that the transmitted signal does not stand out against the background noise.

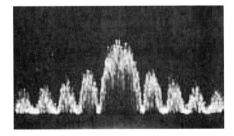

Direct Sequence (DS) Spread Spectrum signal

Figure 4.84   The waveform of a typical spread spectrum transmission.

Spread-spectrum transmissions offer several advantages over fixed-frequency narrowband transmissions:

- Spread-spectrum signals are difficult to detect, as they are usually combined with frequency-hopping techniques, which will make the transmission appear as background noise. Spread spectrum waveforms have no discernable patterns as they follow pseudo-random frequency hops and may include variable time intervals between transmissions.

- Spread-spectrum signals are difficult to decode even if detected, as the pseudorandom transmission sequence will need to be decoded in order to sequence data strings together for decryption. Even if decryption is attempted, it will not produce real-time results and will therefore be of limited use considering the time-sensitive nature of the type of information transmitted in typical tactical data links.

- Spread-spectrum signals are can be used in effectively in an environment where noise, interference, and complex signal environments are present. The process of reconstructing a spread signal helps to distinguish it from background noise and interference, while the frequency diversity in typical transmission patterns provides a degree of immunity from effects such as multipath interference.

- Spread spectrum signals are difficult to jam. Spot jammers will find it virtually impossible to follow the transmission signals in real time and jam them, while broadband noise jammers may also interfere with transmissions across frequencies used by the same side, even assuming that it is possible to detect the presence of the transmissions in the first place.

- Multiple simultaneous transmissions can be made by spread spectrum systems operating across the same bandwidth if different spreading codes are used, and as a result, bandwidth can be utilized more efficiently (Figure 4.85).

There are two main spread spectrum transmission techniques: direct sequence spread spectrum (DSSS) and frequency hopping spread spectrum (FHSS).

Direct sequence spread spectrum

Direct-sequence spread spectrum (DSSS) transmissions use pulse amplitude modulation techniques to break the data being transmitted into a higher frequency pattern by multiply the original data being transmitted by a pseudorandom sequence of binary (1 and -1) values, which appear as a random noise pattern. The pseudorandom noise pattern is overlaid across the data signal at a much higher frequency than that of the original data signal, thereby spreading the energy of the transmission across a much wider bandwidth.

Figure 4.85 Communications interception systems are tuned to detect and decrypt a range of different frequencies and communications signals.

DSSS waveforms are used for spread spectrum multiple access (SSMA), a technique that uses a wideband signal which may be many hundreds of megahertz wide, and allows multiple users to share a single wideband channel.

The bandwidth of the DSSS waveform depends on the frequency of the pseudorandom code sequence used directly to modulate the data-bearing carrier. Binary code sequences as short as 11 bits or as long as $[2^{89} - 1]$ have been employed for this purpose, at code rates from under a bit per second to several hundred megabits per second [55].

Direct-sequence (DS) transmissions use a system whereby the receiver knows the pseudorandom code used by the transmitter and is able to synchronize the transmit and receive patterns. The receiver can then use the same code to decipher the overlaid pseudorandom noise pattern in order to reconstruct the data signal. Typically, the transmitter and receiver use fixed hopping patterns across the frequency slots so that once synchronized the transmitter and receiver can maintain communication by using the table. Such tables will be changed frequently to ensure data security.

Synchronized DS transmissions are also used to good effect in satellite-based global positioning systems where the relative synchronizations between multiple transmitters can be used to determine the timing of signal transmissions, which, in turn, can be used to calculate the receiver's three-dimensional position if the transmitters' positions are known.

Frequency-hopping spread spectrum

Frequency-hopping signals rapidly "hop" from frequency to frequency over a frequency-wide band. The specific sequence of transmission and the frequencies used are determined by a pseudorandom coded sequence known to both transmitter and receiver.

The rate of hopping from one frequency to another is a function of the information transmission rate required. Hop frequencies vary depending on the system in use, but hope rates of over 10,000 hops per second are not uncommon

Unlike a DSSS transmission pattern, which occupies a Sin^2-shaped envelope, the transmitted waveform of a frequency-hopping signal is flat over the band of frequencies employed. The bandwidth of a frequency-hopping signal is simply a function the number of frequency slots available, multiplied by the bandwidth of each slot (Figure 4.86).

Frequency-hopping transmissions are resistant to spot jamming by an opponent who does not know the hop sequence over which the hopper is operating and may therefore have to jam across a broader frequency range. Although hoppers can provide a large jamming margin, they offer little protection against detection as the transmission waveform can be clearly identified against background noise.

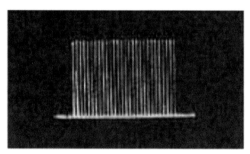

Frequency Hop (FH) Spread Spectrum signal

Figure 4.86   The output spectrum of a frequency hopping system is flat over the band of frequencies employed.

Adaptive frequency-hopping spread spectrum (AFH) transmissions such as those used in Bluetooth systems place their hopping sequence in parts of the spectrum that are less dense thereby improving their resistance to interference from other transmission sources. Such adaptive modulation is easier to implement with FHSS than with DSSS as the frequency bands are more clearly defined and therefore easier to place in an FHSS system.

Further modulation techniques are outlined in Appendix C.

Data link networks

Within network-enabled systems, virtually all nodes will be linked to one or more networks through a series of data links. Typical data link applications include:

- Transmission of digital voice communications;
- Transmission of C2 message sets to networked units;
- Transmission and receipt of information from remote sensors;
- Transmission and receipt of information from remote sensors;
- Transmission of target coordinates to network participants from on-board sensors using standard message formats;
- Transmission of situational awareness information to assets on the network who cannot transmit on their own sensors or who need to remain stealthy;
- Mid-course updates for missile guidance systems;
- Telemetry from weapons systems such as the transmission of preimpact missile seeker imagery for guidance and damage assessment;
- Transmission of sensor video imagery to C2 centers from various manned and unmanned platforms.

Data link networks have typically relied on star-shaped networks where one node is designated as a hub (often on a large platform such as an aircraft or ship). The hub node coordinates the timing of data transmissions from collaborating spoke nodes. Network signals are broadcast and received through an omnidirectional antenna to ensure 360° coverage and spoke nodes transmit to the hub preferably using a directional antenna to minimize the risk of interception and reduce the susceptibility to jamming.

Spoke nodes cannot communicate directly with each other and need to communicate via the hub. Such arrangements are typical of Link-11, Link-16, and collaborative engagement networks.

Future networks will rely more on ad hoc networking and Internet protocols where nodes with join, participate on the network, and depart without the need for strict coordination.

Data link transmission ranges are determined by the frequency band of the transmissions, the transmission power, the waveforms in use and the network arrangements. Many tactical data links are carried on UHF bearers and are thus restricted by radio horizon and line of sight (LOS) limitations, resulting in ranges typically of up to 500 nm for air-air transmission, although figures of 300 nm are more realistic for most data links when the waveforms required to ensure adequate error correction and jam resistance are taken into account.

Surface-to-surface and surface-to-air data link ranges will be much shorter given the LOS limitations although data links such as Link-16 can relay transmissions between nodes to ensure that the extended network shares the same situational awareness picture.

Data links such as Link-11 operating in the HF band will be able to operate over longer ranges, but typically at slower speeds (Figure 4.87).

Figure 4.87 A variety of data links enable the E-8 JSTARS to rapidly assimilate and process intelligence and targeting data.

Typically data links can be structured to form unique members-only clubs where specific missions, operational requirements, or data sets can be managed and accessed by authorized participants only. Such groups are usually referred to as *network participation groups* (NPGs) and typically operate on slightly different frequencies across the same network (Figures 4.88 through 4.90).

Tactical data links

The ability of a participant to operate effectively in a networked environment is determined by the communications capabilities and interoperability provided to and from that participant. Common data interchange standards, information management, and network services protocols are needed to enable the many networks and their functions to interact seamlessly and with minimal data latency.

Typically, many C2 functions such as ISTAR data dissemination, target and track information, and intelligence dissemination, along with processes such as e-mail, chat rooms, and the creation of working groups, are supported by tactical data links such as:

- Mission-specific data links (HIDL for UAVs and JSTARS SCDL);
- Common situational awareness data links (GBS, IBS, SADL);

Figure 4.88 The F-35 Joint Strike Fighter will be equipped with multiple data links including a multimode software radio, LPI.L and PD intraflight data link, and comprehensive JTIDS/VMF implementation.

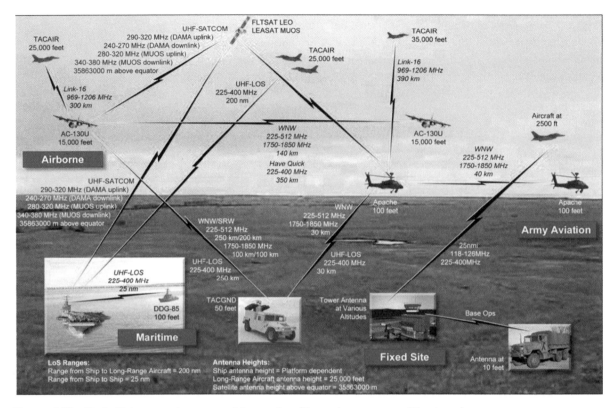

Figure 4.89   Typical joint operations network integration. (Illustration courtesy of Raytheon.)

- Force control data links (JTIDS and AFATDS or equivalents);
- Engagement networks for time sensitive targets (CEC/NCCT);
- Weapons control (weapons data links);
- Administration and logistics (DISN, NIPRNET, SIPRNET, or equivalents).

Typically, many different networks will exist in the same battlespace, and individual platforms will be connected to multiple data and voice networks specific to the mission or task to which they are assigned.

NATO data link connectivity

At a strategic level, other data links also exist to coordinate interactions between national-level C2 centers with deployed headquarters between multinational control centers and to interface with government facilities (Figure 4.91).

Figure 4.90 Aircraft such as the NATO AWACS can act as a network node for multiple different data links, collecting, fusing, and disseminating data to collaborating platforms and ground stations.

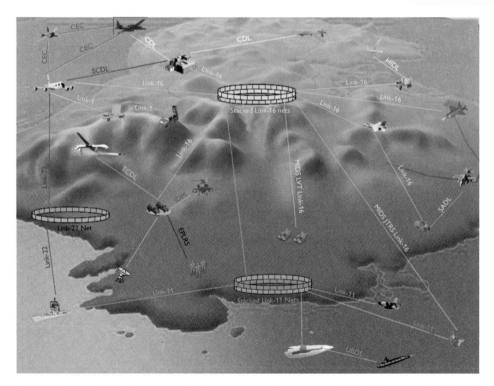

Figure 4.91 NATO forces within the modern battlespace use a diverse range of data links to enable C2 and coordinate ISTAR assets across the battlespace and to achieve a common situational awareness picture. (The multiple gateway and processing systems that interface data transmissions are not shown here for simplicity.)

NATO Data Link Message Formats and Waveforms

A *message format* is the structure of the bit transmission sequence and transmission protocols associated with digital data communications.

A *waveform* is the radio and/or communications functions that occur from the user input to the radio frequency output and vice versa.

Advanced Field Artillery Tactical Data System (AFATDS)

AFATDS employs a unique data exchange standard to enable the networking of five participating countries the United States, the United Kingdom, France, Italy, and Germany. Data formats enable fire support coordination and planning and NBC reporting using a series of standard messages sets (Figure 4.92).

Figure 4.92   Soldiers from 4th Battalion, 11th Field Artillery Regiment, fire their howitzer to obtain proper calibration, near Mosul, Iraq.

Internet protocols (IP)

Many data links use Internet Protocol-based formats and open architecture structures overlaid with suitable security protocols. Such an approach also enables modifications to networks to be made relatively quickly when compared with earlier data link-specific protocol standards.

In addition to the benefits of improved interfaces through the use of common message format standards, modern IP-based networks are generally able to adapt better to changes in the network caused by units joining or leaving or being degraded through enemy action through their ability to reroute signals through other parts of the network.

Common Message Format (CMF)

Integrated Broadcast Service (IBS) is standardizing all communications into a new message format called the common message format (CMF). CMF consists of XML tags marking a data item or document, rendered through a translation layer to a specific output document format. The data is then passed to IBS servers where it is disseminated to subscribers on the IBS network. The CMF will use a new standardized waveform called the common interactive broadcast waveform, which will be broadcast from the military UHF satcom network.

Bandwidth Efficient Advanced Modulation (BEAM)

The BEAM waveform provides a high data rate of up to 80 kb/s over a standard 25 kHz UHF SATCOM channel; a significant improvement on previous satcom waveforms.

F and F/J series messages

F-Series message sets are used by the Link-22 network and contain 72 bits per message.

F/J messages are J messages with 2 additional bits of overhead data and are able to be sent over a Link 22 network.

Interim Joint Message Standard (IJMS)

The Interim JTIDS Message Standard was developed as a message format because the hardware for JTIDS was available before the U.S. TADIL-J (Link-16) message standard was fully defined. IJMS was developed to provide an initial JTIDS operating capability and was implemented in the NATO E-3A and associated ground infrastructure, U.K. E-3D and French E-3F.

IJMS is based on modified Link-11 messages structured to fit into the JTIDS architecture and consequently has a greater capacity and ECM-resistance than Link-11. The IJMS format, however, cannot be used with the JTIDS architecture and data speeds are not as fast as the final JTIDS standard as a result.

Internet Protocol Version 6

IP v6 is the latest Internet Layer protocol standard for packet-switched Internet systems. IPv6 uses a 128-bit address, whereas IPv4 uses only 32 bits, resulting in a significantly larger address space than IPv4.

IPv6 also implements new features that simplify autoconfiguration and router message handling along with significant improvements in Internet Protocol Security (IPsec).

J-Series Message Standards

The J-series messages sets as used by Link-16, Link-11, and Link-22 data links group messages according to 13 different categories and identify messages by the format of Jx.x. For example, the message J3.2 is the Air Track message and J12.0 is the Mission Assignment message.

A more complete listing is shown in Appendix B.

JTRS Message Standard

IP-based encrypted message standard used for the JTRS series of data radios. The JTRS radio is capable of handling numerous waveforms.

M-Series Messages

Used for Link-11A connections. The exact message format depends on the mode in use.

High Performance Waveform

The high performance waveform (HPW) facilitates the secure transmission and reception of e-mail and larger files over satcom and line-of-sight (LOS) AM/FM nets by adapting to varying channel conditions. HPW ensures error-free data delivery using high-speed, over-the-air data rates up to 64 Kbps on LOS nets and up to 56 Kbps on wideband satcom nets.

Mobile User Objective System (MUOS)

MUOS will use commercial third generation (3G) wideband code division multiple access (WCDMA) cellular phone technology applied to a military UHF (300-MHz–3-GHz frequency range) SATCOM radio system using geosynchronous satellites. MUOS will enable JTRS radios to support the transmission of waveforms across cellular mobile networks through the Mobile User Objective System (MUOS) waveforms (Figure 4.93).

Figure 4.93 The Mobile User Objective System (MUOS) located at Naval Computer and Telecommunications Area Master Station Pacific, Wahiawa, Hawaii, is a next generation narrowband tactical satellite communications system intended to significantly improve ground communications for U.S. forces.

Saturn Waveform

Second generation antijam Tactical UHF Radio for NATO (SATURN) (Figure 4.94), operating at 225–400 MHz and using phase shift keying (PSK) modulation with antijam waveforms.

Figure 4.94   The Lynx Mk 8 was the first platform in the United Kingdom to enter service with the Saturn radio upgrade along which was accompanied by the installation of a new SIFF system, avionics upgrades and improved mission avionics.

Soldier Radio Waveform (SRW)

The soldier radio waveform (SRW) will operate in three channels at frequencies between 350 MHz and 2,700 MHz. In the U.S. Future Warrior program, it is envisaged that a single channel will be used for intra-team networking while the network leader's radio will support multiple channels and serve as a node to connect with C2 networks. SRW radios are designed to deliver an instantaneous burst data rate ranging from high-speed bursts of 450 Kbps to 1.2 Mbps and low-speed bursts of 2 Kbps to 23.4 Kbps, while using a low-probability of intercept stealth mode to limit the probability of interception or interference by enemy electronic surveillance systems [22]. JTRS will see LPD/LPI waveform characteristics added in later increments

Variable Message Format (VMF)

Variable Message Format (VMF) is a digital message format that can be broadcast across radio frequency point-to-point systems or networks. It is not a data link in its own right, but is a message standard that harmonizes variable message formats used in NATO data links under a standard referred to as the K Series Message Standard [56].

VMF messages consist of variable length messages and provide a common means of exchanging digital data between operational units with varying needs for volume and detail of information making use of the same core common Data Element Dictionary used by Link-16, Link-22, and the common message format (CMF). The use of structured message sets is also particularly useful when operating across networks with limited bandwidth for message transmission and reception.

The VMF message standard forms, or will form, the backbone of tactical data exchange requirements for the U.S. Army, in addition to many future NATO and coalition forces, predominantly in the ground environment.

Wideband Network Waveform (WNW)

A JTRS Increment 1 waveform includes orthogonal frequency division multiplexing and antijam characteristics. Future WNW waveforms will include LPI/LPD characteristics and bandwidth efficient advanced modulation (BEAM) techniques. It is typically used for networked real-time video distribution between collaborating platforms and ground stations.

A wide range of force and mission-specific tactical data links, networking initiatives, and gateways are employed by armed forces around the world. Some of the more common systems in use by NATO forces are outlined next [56].

Some of the more common data links used by NATO partners are described here [57].

Battlefield universal gateway equipment (BUG-E)

Many interface systems have been developed to enable different data link standards to communicate with each other. The BUG-E system enables communications between a range of different data links, allowing, for example, Link-16 equipped aircraft to communicate with the situational awareness data link (SADL) used by the U.S. Army and USAF CAS aircraft such as the A-10 Thunderbolt II (Figure 4.95) and the F/A-16 Fighting Falcon. Forward air controllers (FACs) can relay target coordinates via the BUG-E gateway to AOCs or directly to CAS aircraft to enable the engagement of time-critical targets.

Improved data modem (IDM)

The improved data modem (IDM) is a U.S. data link modem using existing VHF/UHF airborne radios to pass near real-time targeting data between

Figure 4.95 An A-10 Thunderbolt II ground attack aircraft pulls up sharply out of a low-level strafing run during a combat search and rescue demonstration. The digitally upgraded A-10 is equipped with satellite-guided precision weaponry and advanced communications data links such as the SADL system for transferring information with ground-based warfighters.

joint services air and ground weapons platforms in support of suppression of enemy air defense (SEAD), close air support (CAS), and forward air control (FAC) missions. It is used by many NATO and coalition forces and is currently installed in more than 3,000 platforms [58].

The IDM provides the pilot with a nine-line message that can be displayed in the pilot's head-up display (HUD) or multifunction display (MFD). It also provides the location, altitude, and track data of up to four collaborating flight members.

The IDM also acts as an interface between networks enabling ground units such as forward air controllers (FACs) equipped with suitable data radios to pass and receive threat data and images to and from aircraft and helicopters.

The IDM provides four half-duplexed radio channels, with one communications port, data rate, and link format operational per channel at any given time (Figures 4.96 and 4.97). Each radio channel is capable of being configured into one of three different communication ports: analog, digital, and secure digital. The IDM can operate using CPFSK and FSK analog modulation achieving data rates of up to 2.4 kbps or through digital radios using ASK digital modulation achieving data rates of up to 16 kbps. Over VHF radios the IDM achieves LOS ranges in the 30–88-MHz FM band or in the 116–153-MHz AM band. UHF radios using the IDM system also achieve LOS ranges using the 22–400-MHz band.

Figure 4.96 The U.S. Army uses the IDM to transmit Longbow Radar digital target data from Longbow-equipped AH-64Ds to collaborating air and ground platforms, passing up to 256 targets for target classification and assignment.

Figure 4.97 An F-16 Fighting Falcon aircraft returns to the fight after receiving fuel in June 2008, during a mission over Iraq. F-16s are equipped with the IDM modem for CAS missions.

Coalition Enterprise Regional Information Exchange (CENTRIXS)

CENTRIXS provides a ship-to-shore satcom IP link enabling ship-to-shore Web services, secure e-mail, and chat communications with coalition partners to compliment existing ship-to-ship HF e-mail capabilities.

Common data link (CDL)

The common data link (CDL) standard (Table 4.8) aims to harmonize interchange and waveform standards for unprocessed broadband, imagery, and signals intelligence data for assets used in the primary collection and processing of ISR information. The CDL provides a secure, jam-resistant, duplex point-to-point data link, operating at a range of speeds from 10.7 Mbps to 274.2 Mbps, and defines standard interface requirements and processing techniques to ensure interoperability across the CDL network. Its interface standards have been applied to a family of data links including the tactical common data link (TCDL) and high integrity data link (HIDL). Where the CDL is networked rather than used in a point-to-point mode, it is referred to as a networked CDL (N-CDL).

The CDL family covers both line-of-sight (LOS) and, using satellite relay, beyond-line-of-sight (BLOS) capabilities, permitting the remote operation and management of ISTAR sensors carried by CDL-equipped platforms positioned virtually anywhere on the globe. There are two types of Defense Satellite Communication Systems satellites in use to support CDL operations: Senior Span, operating in the I-band, and Senior Spur, operating in the Ku-band in order to improve bandwidth.

The A-CDL data link is used as a high-speed trunk data relay in point-to-point airborne links installed on manned and unmanned aircraft. It operates at higher bandwidths than the N-CDL data link providing an ability for duplex transmission with send and receive transmissions operating at the same data rates of up to 274.176 Mbps. Airborne platforms equipped with the A-CDL data link are able to establish multiple connections to other A-CDL data links

CDL	Common Data Link
A-CDL	Advanced CDL
MP-CDL	Multi-Platform CDL
N-CDL	Networked CDL
P-CDL	Pulsed CDL
R-CDL	Radar-based CDL
SE-CDL	SatCom Extension to CDL

Table 4.8   Common Data Link Variants and Descriptors

enabling multiple A-CDL networks to be established over ranges of several hundred km. Modulation waveforms using by the CDL are principally QPSK, O-QPSK formats (see Appendix C).

The SE-CDL (Figure 4.98) uses a CDL waveform broadcast and received by military satellites, which can either act as a direct node or as a network extension to a CDL system.

Enhanced Position Location Reporting System (EPLRS)

The EPLRS is the U.S. Army's and U.S. Marine Corps current digital data networking system providing high-speed, automated data exchange for static and on-the-move platforms up to LOS ranges. EPLRS is the backbone of the Army Battlespace Command System (ABCS) and Force Battle Command Brigade and Below (FBCB2) networks and is also used by the U.S. Navy for ship-to-ship communications and for the coordination of amphibious operations, and for USAF CAS missions.

The EPLRS system operates between 420–450 MHz using spread spectrum techniques to reduce the probability of detection and enhance security of transmission.

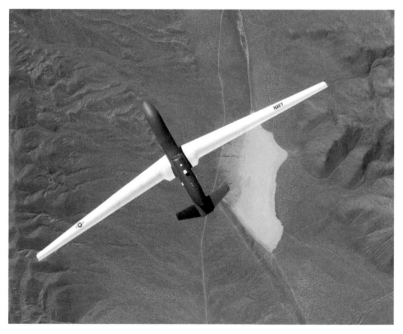

Figure 4.98   RQ-4 Global Hawk UAV carries the airborne signals intelligence payload, which will increase battlefield signal collection capabilities and the SE-CDL data link for control and information relay.

EPLRS (Figure 4.99) provides a range of C2 functions including data communication and position location reporting, navigation to waypoints and targets, C2 information such as artillery fire support requests, and the exchange of air track data. It is a key data link for users requiring a near real-time situational awareness in a joint operational environment. EPLRS is compatible and interoperable with other NATO standard data links such as SADL and is capable of integrating data between the two systems during close air support operations. It will also form the principle data link for future army interoperability programs such as the U.S. Land Warrior programs.

The EPLRS system can also be configured for collecting data from widely dispersed networked systems across the battlespace and sending the information back to the C2 centers for processing and onwards distribution (Figure 4.100). The EPLRS network automatically reconfigures itself to overcome the UHF line-of-sight limitations or to improve data transmission integrity in dense RF environments or where jamming is taking place, using automatic forwarding between links where necessary.

EPLRS uses TDMA, spread spectrum, frequency hopping, and forward-error correction to provide a robust self-relaying digital network. It operates at UHF frequencies of 225–450 MHz with a data transmission rate between 57 Kbps and 486 Kbps, but with the potential to increase rates to 1 Mbps [59].

Figure 4.99 U.S. Army M1A1 Abrams tanks are fitted with the EPLRS data link to exchange tactical situational awareness information.

Figure 4.100   Soldiers from the U.S. 1st Cavalry Division on patrol in Kahn Bani Sahd, Iraq.

High integrity data link (HIDL)

The HIDL data link is a NATO compatible, duplex narrowband jam-resistant data link designed for the guidance and control of multiple UAVs (Figure 4.101). Consisting of airborne and surface-based terminals, HIDL will operate at 225–400 MHz (UHF) and variable data rates between 3-kbps to 20-Mbps bandwidth to control UAVs as platform sensors out to line-of-sight (LOS) ranges of 200 km. It will operate with the wideband TCDL to transmit UAV sensor imagery to ground stations. Antijam features are provided by time and frequency diversity within the waveform data messages.

Joint composite tracking network (JCTN)

JCTN uses omnidirectional beacon antennas in a point-to-point network configuration to achieve data transfer rates of up to 10 Mbps. The JCTN system is used in the CEC network to coordinate and disseminate fused data tracks related to time-sensitive targets.

Situational awareness data link (SADL)

The situational awareness data link (SADL) is a networked data link based on a modified Enhanced Position Location and Reporting System (EPLRS) Ground Radio, which, as the name suggests, is able to report the position

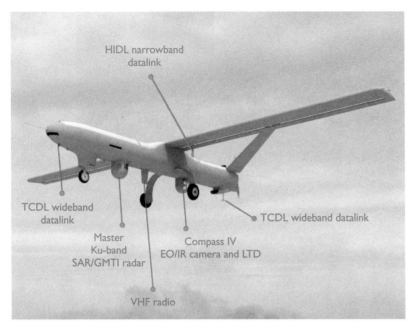

HIDL narrowband datalink

TCDL wideband datalink

TCDL wideband datalink

Master Ku-band SAR/GMTI radar

Compass IV EO/IR camera and LTD

VHF radio

Figure 4.101 The Watchkeeper Tactical UAV is fitted with the HIDL and TCDL data links.

and identity of the host platform on a network and as such is particularly useful for situation awareness and in supporting air-to-ground combat identification. The SADL integrates U.S. Air Force close air support aircraft such as the F/A-16 Fighting Falcon and A-10 Thunderbolt II with the digitized battlefield providing fighter-to-fighter, air-to-ground, and ground-to-air data communications. The SADL can synchronize with specific EPLRS ground networks to enable the display of ground targets while also independently sharing fighter-to-fighter data from other SADL-equipped aircraft as well as showing positions of EPLRS-equipped aircraft and ground units.

The SADL (Figures 4.102 and 4.103) is a networked data link system operating in the UHF band between 425 to 447 MHz and uses a TDMA architecture to integrate network participants. It provides a message set similar to Link-16 through a data rate of 2.5 kbps.

Surveillance and control data link (SCDL)

The SCDL provides a secure, all-weather dedicated link between both the U.S. E-8C JSTARS aircraft and the U.K. Astor aircraft to multiple ground stations. The JSTARS (Figure 4.104) and Astor GMTI/SAR (Figure 4.105) radar system can detect, classify, and track the movements of enemy and

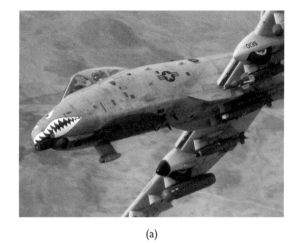

(a)

(b)

Figure 4.102 (a, b) The most significant change to the newly modified A-10C Thunderbolt II CAS aircraft is the addition of the situational awareness data link. With SADL, the A-10C joins a massive Internet-like network of land, air, and sea systems.

friendly ground forces, including tracked and wheeled vehicles as well as slow and low-flying aircraft or helicopters and maritime targets.

The SCDL relays information between air and ground terminals [known as the as common ground station (CGS)] across multiple networks.

The SCDL uses a TDMA protocol and a fast frequency hopping, spread spectrum waveform, with error correction coding techniques to mitigate jamming. The frequency hopping agility enables the SCDL to operate in shared spectral bands to avoid areas of potential interference from competing

Figure 4.103 U.S. Air Force F-16 Fighting Falcon aircraft with 77th Fighter Squadron fly over Shaw Air Force Base, S.C., during Operation Iron Thunder, February 7, 2007. On the F16 the SADL displays EPLRS-equipped units within 500 meters of a target that the pilot plans to engage. The five closest EPLRS units are displayed, irrespective of the target proximity.

transmissions and to provide a high degree of assurance of transmitted data reaching its intended recipient.

The SCDL transmission modules provide for selectable or adaptive data transmission rates, which exploit the maximum bandwidth available within the TDMA network. Transmission speeds in the Ku-band (12.4–18 GHz) of up to 1.9 Mbps can be achieved, although typical data rates are much less.

Tactical common data link (TCDL)

The CDL-compliant tactical common data link (TCDL) provides a family of secure digital, point-to-point data links for use with both manned and unmanned airborne platforms to transmit radar data, electro-optical sensor imagery, video, and other sensor information to ground terminals. The TCDL program is aimed at developing a family of CDL-compatible, low-cost, lightweight, near real-time, digital data links providing interoperability between multiple TCDL terminals and common data link (CDL) networks (Figure 4.106).

Duplex spread data transmission rates of 200-Kbps transmit and 10.71-Mbps receive, at Ku-band frequencies over airborne ranges of 200 km and beyond, are possible with current TCDL standards, making the TCDL the data link of

(a)

(b)

Figure 4.104   (a, b) The USAF E-8C Joint Surveillance Target Attack Radar System (Joint STARS) has a crew of 21 Air Force and Army personnel to detect, track, report, and target enemy ground movement while employing various radar modes, data processing systems, and extensive communications suites including a wide range of data links. Data links such as Link-16 and the surveillance control data link (SCDL) are used to pass SAR and GMTI radar information to the ground station modules (GSMs), which are the Army component for the Joint STARS program.

choice for air-to-ground high frame rate video relay. Signal reliability is achieved with spread spectrum Ku-band transmissions in the 14.40–14.83-GHz band, and reception in the 15.15–15.35-GHz band [60].

(a)

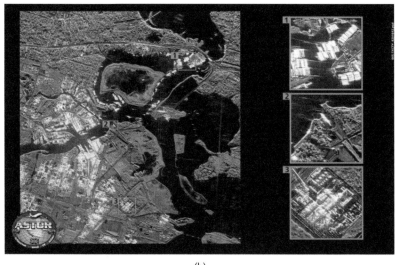

(b)

Figure 4.105 (a) The United Kingdom's ASTOR aircraft is fitted with the SCDL data link for high-speed transfer of GMTI and SAR radar imagery. (b) A typical SAR radar imagery from the RAF Sentinel ASTOR Aircraft, which is fitted with an upgraded version of the Raytheon ASARS-2 radar as used on the U-2 aircraft. It is capable of providing high-resolution images of the battlefield at ranges of several hundred kilometers. The SAR mode provides high-quality radar images of the area surveyed, while the MTI mode will detect moving vehicles operating in the area. The SAR can be operated in spot mode to produce high-resolution imagery over relatively small areas of fixed location. The SAR swath mode can collect lower-resolution imagery broadside to the aircraft as it proceeds but at a much greater width than SAR spot. Multiple passes using SAR swath mode can effectively provide wide area surveillance of fixed and static targets. These images can be exploited by the airborne mission crew, or downlinked from the aircraft in near real time to the ASTOR Ground Station, to generate intelligence reports for commanders. (Images courtesy of Raytheon.)

(a)

(b)

Figure 4.106   (a, b) A USAF F-16 fitted with a Raptor DB-110 reconnaissance pod. The images received by the pod can be transmitted via a real-time data-link system to image analysts at a ground station or can be displayed in the cockpit during flight. The imagery can also be recorded for postflight analysis. The RAPTOR system can create images of hundreds of separate targets in one sortie; it is capable of autonomous operation against preplanned targets, or it can be retasked manually for targets of opportunity or to select a different route to the target. The stand-off range of the sensors allows the aircraft to remain outside heavily defended areas to minimize the aircraft's exposure to enemy air-defense systems. CDL and TCDL data links will be used in the ASTOR aircraft and Reconnaissance Airborne Pod for Tornado (RAPTOR) programs. Inset photo shows Raptor imagery typically achievable at ranges over 40 km. (Images courtesy of Goodrich Corporation.)

Multifunction advanced data link (MADL)

The multifunction advanced data link (MADL) (Figure 4.107) is a directional data link selected by the U.S. DoD to provide broadband communications between future stealth platforms and to replace narrowband Link-16 communications technology available today.

Tactical targeting networking technology (TTNT)

The tactical targeting networking technology (TTNT) is an MIDS-JTRS based system, which operates using IP-based data transmission principles and can transmit data at 2 MBPS over a 100-nm range. This particular data link is focused on high tempo, fast jet operations where high data rates combined with low data latency are required to provide tactical aircraft with the most current intelligence from bandwidth-hungry sensors.

The TTNT system provides the ability to share imagery and data relevant to a particular target from any sensor or intelligence source that has information relevant to the particular location through connection to the Global Information Grid (GIG). When a tactical aircraft identifies or is tasked to investigate a target of interest, the TTNT system, via operators typically based in a CAOC, can access the most relevant information from the GIG that will support ISTAR operations and assessment of the target of interest.

Using an IP-based approach, TTNT data transmissions are automatically connected to the network within 5 seconds of registration and can access and share its sensor data to support targeting operations in the area of interest for up to 1,000 users on a single network.

As with most modern data links, the TTNT data link is optimized for LPI and LPD features. Unlike Link systems such as Link-16, whose data latency can be 10–12 seconds, the TTNT data latency over ranges of up to 150 km is engineered to be no more than 2 milliseconds. This is a significant improvement over previous data links and provides the backbone for concepts such as CEC and NCCT where information must be provided to other platforms and data fusion engines in a virtually real-time environment. Link networks are useful in providing surveillance data, but data latency becomes increasingly important when that information is required for targeting information against high-speed, fleeing, or transient targets.

Warfighter information network–tactical (WIN-T)

The WIN-T network enables soldiers in deployed environments to communicate, plan, and collaborate from multiple locations across the

Figure 4.107   Due to communications security considerations, the intelligence and situational awareness picture generated by F-22 Raptors cannot be transferred to collaborating platforms. As stealth aircraft, F-22s are not equipped with conventional data links such as Link-16, which can be easily detected by enemy SIGINT sensors, but today rely on a platform-specific stealthy, narrow-beam intraflight data link (IFDL) designed to relay data and to synchronize the situational picture only among F-22 Raptors. The IFDL data link is incompatible with other data links, and as such F-22 Raptors cannot communicate with other collaborating aircraft or platforms. Future F-22 Raptors along with B-2s and F-35s will be equipped with the MADL data link, which will allow communications between limited types of collaborating stealth platforms.

battlespace and provides on-demand access to local situational awareness information, linking troops to small groups of networks. It is based on an intelligent self-forming and self-healing collaborative network system. Forming a key element of future soldier networking capabilities, initial operating capability is planned for 2013.

WIN-T is designed to provide the backbone of the tactical network, continuous and full communications-on-the-move capability at all echelons, joint and coalition voice and data services to all command posts, and a more survivable and less complex network. WIN-T's single integrated network will provide multilevel classified Joint and coalition voice and data services to all command posts. Unlike most current military networks, WIN-T will also offer seamless interoperability with other networks, including legacy, joint, coalition, and, where necessary, commercial networks.

As an integral component of future army tactical networks, WIN-T is a critical element in the U.S. Army's transition to robust network-based operations. WIN-T provides the key capability for on-the-move communications through a three-tiered architecture (ground, airborne, and space) that enables continuous network connectivity.

The overall system will comprise standard elements such as wireless networks supporting voice over IP (VoIP) communications, over IPv4 and IPv6. The system will also integrate the Joint Tactical Radio System (JTRS), personal communication devices, and small satellite links.

The ground layer will equip soldiers, sensors, platforms, command posts, and access nodes (signal shelters) with integrated transmission (radio) systems, switching, and routing capabilities that will serve as WIN-T points of presence (Figure 4.108). The airborne layer will serve as an access node and relay by positioning transmission, switching, and routing capabilities onto airborne platforms. The space layer will serve as an access node and relay by leveraging the transmission, switching and routing capabilities provided on the satellite [61].

Figure 4.108   The vehicles and highly mobile equipment of the Army's Warfighter Information Network–Tactical (WIN-T), during field tests at Ft. Huachuca, Arizona, in December 2008.

Link-series tactical digital information link (TADIL)

A tactical digital information link (referred to as a TADIL in the United States) is a standardized communication link suitable for transmission of digital information operating to defined NATO interoperability standards.

In U.S. Joint Services parlance, Link-16 is the same as the TADIL-J standard, Link-11 is to TADIL-A, and Link-4A is to TADIL-C. The letter designation refers to the message set standards employed by the data link; Link-16, for example, uses J-series message sets. The term TADIL is now largely being replaced by TDL (tactical data link).

The NATO Link series of data links has been through the following number of evolutions.

Link-1

Link-1 is a nonsecure point-to-point digital data link developed in the late 1950s for coordinating air defense assets through target engagement and fighter allocation within the NATO Air Defense Ground Environment (NADGE). Link-1 provides a limited air surveillance, air defense, and link management message set between ground-based air surveillance and control system (ASACS) units such as control and reporting centers (CRCs), combined air operation centers (CAOCs), and sector operation centers (SOCs).

It has, by modern standards, a relatively low data rate of up to 2.4 kbps in multiples of 600 kps and tracks are shared according to priority. As a result, the display of friendly tracks can be several minutes behind the real-time picture. Air surveillance units equipped with the Link-1 system are allocated geographical areas of surveillance, called track production areas (TPA), whose boundaries are overlapped by track continuity areas (TCA) with data being shared automatically with adjacent units.

Link 4, 4A, and 4C

Link-4 is a networked TDMA-based, nonsecure digital data link used specifically for air defense coordination and aircraft control and was developed in the late 1950s (Figure 4.109). It operates at a very low data rate of 5 kbps, using vector-based commands to pass intercept instructions to airborne platforms. Link 4 has evolved into two separate systems, Link 4A and Link 4C.

Figure 4.109 An F-14D Tomcat assigned to the "Bounty Hunters" of Fighter Squadron Two (VF-2) comes in for a landing aboard the aircraft carrier USS *Constellation* (CV-64) after completing aerial combat maneuvers (ACM) training in May 2003.

Link-4A (also known as TADIL-C by the U.S. DoD) was originally designed to improve the C2 of tactical aircraft by replacing voice-based communications with a digital communications link. While Link 4A demonstrated the utility of digital data links, its nonsecure nature and the need to operate in increasingly complex and hostile electromagnetic environments have resulted in more advanced waveforms with greater security and resilience.

Link 4C, along with the F-14 Tomcat for which the data link was originally designed, has been phased out of service. It was designed to enable aircraft carriers from which F-14s operated to control up to four aircraft at a time.

Link-11A and 11B

Link-11A (also known as TDL-A or TADIL-A) is a secure, but not ECM-resistant, computer-to-computer, digital radio communications data link supporting the comprehensive exchange of target tracks through the use of a standard message format. It was developed in the 1960s to overcome the limitations of earlier data links and to extend functionality to surface and subsurface tracks.

Within NATO, Link-11 is primarily used as a maritime data link; however, it is being adapted to support theater ballistic missile defense information exchange requirements, and consequently ground-based (surface-to-air missile) SAM systems are increasingly being fitted with Link-11 systems. Link-11 is widely fitted in NATO ships, airborne surveillance, and some ground facilities (Figure 4.110). Within the United Kingdom Link-11 is employed by the Royal Navy, Royal Marines, and Royal Air Force in its ships, ship-shore-ship-buffers (SSSBs), E-3D AEW, Nimrod, and tactical air control center (TACC).

Link-11A employs netted communication techniques using standard message formats. Data is exchanged using the Conventional Link Eleven Waveform (CLEW) over a PSK modulated data link operating at 1,364 bps (in the HF/UHF band) or 2,250 bps (UHF band). Participating nodes are synchronized on the network so that the end of the transmission of one node signals the start of the transmission of the next node in the network sequence.

Link-11A is susceptible to interference from ECM in the HF/UHF bands in particular and the Single Tone Link Eleven Waveform (SLEW), which employs data interleaving techniques to improve bit error rates and improve security which uses more robust encoding to achieve a greater ECM resistance but at lower data rates of 1,800 bps.

In the high frequency (HF) band, Link-11A has a beyond line of sight (BLOS) capability, delivering a theoretical range of around 300 nm. In the UHF band its range is limited to line-of-sight (LOS) ranges of around 25 nm in the surface-to-surface role or 150 nm in the surface-to-air role.

Figure 4.110   HMS Daring Type 45 Anti-Air Warfare Destroyer is equipped with a Link-11 system for coordination of naval assets. (Photo courtesy of BAE Systems.)

The Link-11A data set includes specific data such as EW information and the transmission of orders, alerts and commands, along with air track management (Table 4.9). Link-11 waveforms may also be broadcast over satellite communications systems.

The Link-11A protocol establishes a common, network-wide, composite track picture, based on a compilation of data from collaborating sensors and platforms that is shared between collaborating assets.

Link-11 uses a central platform-based hub called a net control station (NCS) to coordinate the transmission and reception of data, broadcasting and receiving in standard message formats, and polling each participant in turn for their data (Figure 4.111). Link-11 cannot transmit and receive data simultaneously, but does so in an alternating sequence using a polling system, where a Link-11 NCS polls each network participant in turn for their data. Link-11 may also be operated in a broadcast mode in which a single data transmission or a series of single data transmissions are made by one participant. Each Link-11 platform is equipped with a data terminal set (DTS) that transforms digital data into audio tones for transmission by radio.

Link-11A has six operating modes as follows:

- **Net synchronization:** Synchronizes the timing and participation of collaborating units;
- **Net test:** Tests the connectivity and data exchange protocols of collaborating units;
- **Roll call:** The normal roll-call mode of operation where each participating unit is invited to transmit its data in a sequential pattern;
- **Broadcast:** Allows a single unit to repeatedly transmit tactical data to all other collaborating units;
- **Short broadcast:** Allows a Link-11 unit to broadcast its tactical data once every time the transmit button is pressed;
- **Radio silence:** Allows a unit to receive but not transmit Link-11 data.

Track Number
Position
Course
Speed
Height / Depth
Identify indicator
EW data
Unit ID

Table 4.9   Typical Link-11 Data Sets

(a)

(b)

Figure 4.111 (a) Link-11 networks are coordinated through a designated network node such as the E-2 Hawkeye acting as a network control station (NCS). (b) The multiple electrically steered beams found on AESA radars enable them to operate in multiple simultaneous modes transforming their utility when compared to previous mechanically scanned radars that were limited to one mode at a time.

Link-11B

Link-11B (also known as TDL-B or TADIL-B) is a duplex, point-to-point, digital data link. It uses a time sequenced protocol and either standard message formats or binary digits in serial transmission data frames. The data link operates at a standard rate of 1.2 kbps with optional speeds available in multiples of 1.2 kbps up to 4.8 kbps. Units that exchange data via Link-11B are called a reporting unit (RU) or, when in relay mode, a forwarding reporting unit (FRU).

Link-11 will be supported until 2015 and during that time will increasingly be replaced by Link-16 and Link 22 networks.

Link-14

Link-14 is a nonencrypted, nonreal-time, long-range maritime data link operating in the HF, VHF, or UHF bands at a relatively slow speed of 75 bps and is designed to share situational awareness information with platforms not equipped with the U.S. Global Command & Control System (GCCS) and not able to receive Link-11 or Link-16 transmissions. Due to its slow speed, it was used primarily to provide a high interest track broadcast (HITB) to supplement other data links and voice reports. Link-14 was formally discontinued from NATO service in 2000.

Link-16

Link-16 (also known as TDL-J or TADIL-J) (Table 4.10) is a secure and encrypted NATO standard, digital broadcast data link used by land, air, and sea platforms to provide access to a single integrated situational awareness picture across collaborating units. It is the most common and widely used of all the current NATO Link-series data links and is used for the transmission of voice, data, C2, or timing messages.

Track Number
Position
Course
Speed
Height/Depth
Identify indicator
Special interest indicator (including EW data)
Platform type /specific type
Activity type
IFF Modes I, II, III, IV, SIFF
ETA and ETD to/from station
Force/Tell emergency indicator
Equipment status
Exact ordnance inventory (including specific missile type ie TLAM or SM2)
Radar and missile channels
Fuel available for transfer
Gun rounds remaining

Table 4.10   Typical Link-16 Message Sets

Transmission of processed and unprocessed data

Processed data

Link 16 data link will be the DoD primary tactical data link for all Services and Defense Agency Command and Control (C2), Intelligence (I), and where practical, weapon system applications . . .
. . . all processed information will be disseminated through Link 16. . .
. . . all processed information is now also permitted in CDL format. . .

Unprocessed data

Unprocessed broadband, imagery and signals intelligence data are required to be provided through the Common Data Link (CDL)

—DoD C3I Tactical Data Link Policy Letter, dated 18 Oct 94

Operating at one of three data transmission speeds (26.88 kbps, 53.76 kbps, or 107.52 kbps), Link-16 builds on the data exchange protocols established by Link 4A and Link-11. Multiple user access in Link-16 networks is enabled by TDMA-based protocols, with multiple networks in the same theater of operation being established by the use of FDMA techniques to form a series of stacked nets.

Link-16 data formats provide a significant improvement from earlier data links in the number of messages and data sets that can be exchanged, including a secure voice capability and the transmission of the precise participant location and identification (PPLI) by each Link-16 equipped platform.

Link-16 information is arranged in a series of predefined, fixed length binary data words called J-series messages, which are defined in the Joint Tactical Data Link Management Plan (JTDLMP).

The network is built around network participation groups (NPGs), each of which serve as a virtual network in particular functional areas, any or all of which can be accessed by participating units. NPGs can be considered building blocks of the Link-16 network. Because an NPG is defined by its function, the types of messages transmitted on it are also defined. The most common NPGs are:

- PPLI;
- Surveillance;
- Electronic warfare;
- Command;
- Mission management;
- Weapons coordination;
- Aircraft control;
- Fighter-to-fighter;
- Secure voice.

By dividing the net into NPGs, each unit can participate to the network groups that support its specific mission. Access control to each NPG is defined in advance of the mission with access available to all nets with the exception of the fighter-to-fighter NPG. Multiple simultaneous networks can be created or stacked to create a multinet architecture with the data transmitted on different frequencies (FDMA) in each net. Although it is theoretically possible to define a network containing 127 nets, studies have shown that operating more than 20 nets simultaneously in the same geographical area will result in degradation of communications. This is due to interference in the frequency division-multiplexing domain [62].

Link-16 uses a frequency-hopping technique that transmits data over 51 different frequencies. This technique allows the network to have multiple nets that can be stacked for use by different NPGs. Using a predetermined pseudorandom pattern, the frequency is changed every 13 microseconds or approximately 600 times during each time slot. There are a total of 128 possible stacked nets numbered 0–127 (each assigned a particular hopping pattern), with the number 127 reserved to indicate a stacked net configuration [63] (Figure 4.112).

Participation in a Link-16 network can be expanded by any asset using a standard Link-16 terminal to connect it as a node on a Link-16 network. Unlike other networks such as Link-11A, no central coordinating hub or node is required. Either the first participant sets the clock using a net time reference (NTR) signal and others synchronize accordingly as they join, or the network is synchronized by reference to GPS clock signals.

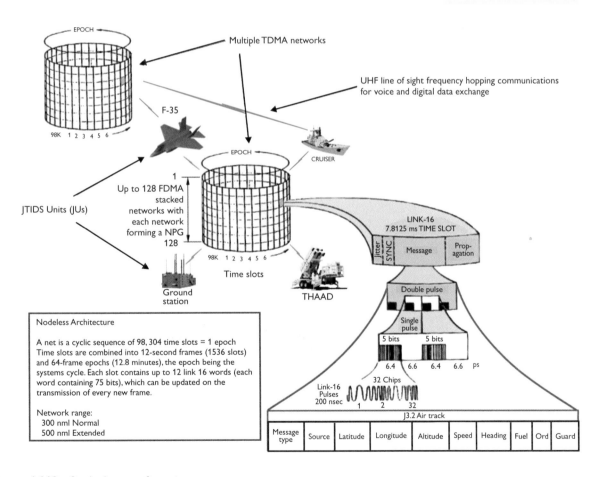

Figure 4.112 Stacked net configuration.

Data latency across the network is typically 10–12 seconds depending on the information reported and the number of participating units.

To extend the operational range of Link-16, the U.S. Navy (USN) is investing in satellite data links between Link-16 terminals to allow BLOS networking. The USN system, known as S-TADIL-J (or S-TDL-J), uses existing military UHF communications satellites and supports all Link-16 mission areas except for control (air intercept and fighter-to-fighter), voice coordination, system information, and net management. S-TADIL-J uses the 25-kHz demand assigned multiple access (DAMA) system working a roll-call protocol to provide network access to suitably equipped participants [64].

Link-16 networks (Figure 4.113) also have good resistance to electromagnetic interference and therefore the ability to operate in a hostile electromagnetic environment. The Link-16 waveform was developed to provide significant performance enhancements against optimized, band-matched jammers. To preclude jamming by a narrowband jammer, the transmission frequency of the terminal is changed for each pulse (77,000 hops per second) through 51 discrete UHF frequencies. The frequency-hopping pattern is pseudorandom and is determined by the Transmission Security Procedures [65].

Link-16 JTIDS and MIDS

The Multifunction Information Distribution System (MIDS) is a NATO standard, jam-resistant data, and voice communications link designed to

Figure 4.113 Pacific Ocean, October 10, 2005. A Link-16 equipped F/A-18C Hornet, assigned to the "Lancers" of Marine Fighter Attack Squadron Two One Two (VMFA-212), receives fuel from a KC-130 Hercules, assigned to Marine Aerial Refueller Transport Squadron One Five Two (VMGR-152) over the Pacific Ocean near Okinawa, Japan. Both tanker and F/A-18C position will be visible to users of the Link-16 network.

operate across the entire land, sea, and air battlespace. First developed in the 1970s, the requirement was originally fulfilled by a system called Joint Tactical Information Distribution System (JTIDS), but has been superseded by smaller and lighter terminals developed in the 1980s called MIDS. Like other Link series data links, it is designed to share situational awareness data such as air, land, surface, and subsurface target tracks, navigation information, standard orders, and alerts in near real time.

MIDS/JTIDS can support four message standards: Link-16 (J-Series), Interim Joint Message Standard (IJMS), Variable Message Format (VMF), and JTRS. MIDS communications ranges are typically up to 300 nm.

The MIDS/JTIDS terminals can support three TDMA message standards: Link-16, Interim Joint Message Standard (IJMS), and Variable Message Format (VMF). In addition, MIDS/Link-16 terminals provide the network for precise participant location and identification (PPLI) messages that enable secure positive identification to other similarly equipped platforms at ranges well beyond current IFF systems through exploitation of multihop data transmissions between terminals.

The MIDS-JTRS system is used, for example, in the USAF Tactical Targeting Networking Technology (TTNT) system and is capable of providing a wideband (2 Mbps), line-of-sight self-configuring encrypted IP-based data link, among recipients up to 100 nautical miles away [22].

The MIDS/JTIDS system (Figure 4.114) shares the same frequency band of some civil and military air traffic management systems, notably DME, IFF, GPS, TACAN, and consequently coordination is required between civil and military ATM authorities in peacetime when it is in operation.

Dynamic, multihop, beyond line-of-sight services will be provided through network routing functions supported in routers, the Joint Tactical Radio System (JTRS), and satellites. Automatic routing and relaying will take place on platforms using JTRS running the wideband network waveform (WNW). The WNW will support network services based on mobile ad hoc networking technologies. These technologies will permit all JTRS-equipped aircraft (manned or unmanned), all JTRS-equipped ground platforms, and other platforms to automatically become members of JTRS networks and provide adaptive, self-managed communication relaying services [65].

Given the increasing demand for increased bandwidth, Link-16 is being modified to take advantage of a bandwidth enhancement protocol termed the enhanced throughput (ET) protocol (Figure 4.115). Using the existing Link-16 waveforms and spectral characteristics, the ET protocol increases the selectable data rates from 2.5 to 10 times the current Link-16 data rate

Figure 4.114   July 10, 2007: An F-16 Fighting Falcon takes fuel from a KC-10 Extender over Iraq. F-16s along with F-15Es and E-2Cs are being equipped with the Multifunction Information Distribution System-Joint Tactical Radio System (MIDS-JTRS) to bring tactical aircraft into the networked battlespace.

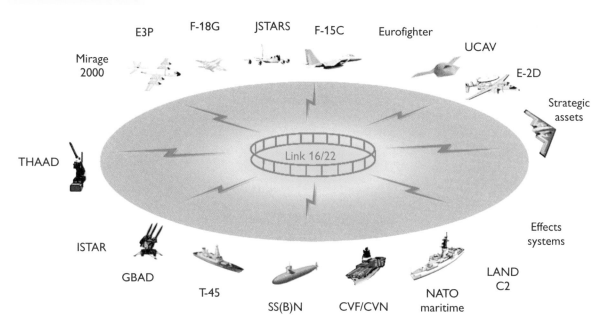

Figure 4.115   Link 16 and Link 22 will equip virtually all NATO networked platforms in the years ahead.

for coded messages. The ET capability will be introduced into existing MIDS and the JTIDS terminals.

Link-22

Link-22 is an advanced, high-speed, secure, encrypted digital data link utilized by land, air, and sea platforms. Link-22 uses TDMA or dynamic TDMA (DTDMA) multiple access protocols to provide increased flexibility and reduced net management data requirements and operates in the HF and UHF bands to provide BLOS capabilities.

Link 22 was originally conceived as the NILE (NATO Improved Link Eleven) program, and is something of a hybrid between MIDS Link-16 and Link-11, retaining many of the Link-11 protocols while benefiting from increased data transmission speeds and improved transmission security and fault tolerance, enabling improved data quality in adverse environments. It will in due course replace Link-11 systems and, as with current Link-11 systems, is principally designed for maritime applications.

Link 22 is designed to retain compatibility with the Link-16 family of data links through the use of common data structures and as such is compatible with other Link-16 family data links such as the STDL, S-TADIL J, Link-16, and VMF. As such, the 72-bit word message standard found in the Link-16 family can carry embedded J-Series Link-16 messages, known as the FJ-series, as well as the Link 22 F-Series messages [66].

Messages carry one of four priority statuses and may also request acknowledgement by the participant that the message has been received.

Five types of broadcast are available to Link-22 participants:

- **Totalcast:** Automatic broadcast to all link participating Link-22 units;
- **Neighborcast:** All radio frequency (RF) neighbors on each NILE network on which the NILE unit operates;
- **Mission area subnetwork (MASN):** A predefined logical group of units;
- **Dynamic list:** A list of two to five units specified in the request that may change as needs require;
- **Point-to-point:** A single participating Link-22 unit that is specified in the message.

A Link-22 unit may operate up to four networks simultaneously with the provision of mission area subnetworks without the need for a network control station. Messages are passed across the network through the use of automatic routing and relays between the network nodes, which benefit from being able to dynamically react to altering loads and conditions, including dynamic reconfiguration of the network and the addition and removal of network participants. All network units will be capable of operating on a single network and may also be configured to operate on up to four mission area subnetworks (MASN) simultaneously. A set of interconnected Link-22 networks is known as a *super network*.

Link-22 units automatically rebroadcast messages to ensure the widest possible reach of the situational awareness picture. The system network controller (SNC) automatically calculates whether onwards transmission is necessary based on its knowledge of the network and the intended destination of the message.

Participants wishing to join a Link-22 network as an active participant may do so through initiating a protocol called late network entry (LNE). Other units may also join as "receive-only units" where they do not transmit data to the network, but receive the tactical situational awareness picture without the need to transmit (Table 4.11).

AESA radar data links

Early radars used mechanically steered antennas (Figure 4.116) to point a single radar beam in the direction of the target. Scan patterns were created by having the antenna follow predefined search patterns to illuminate the area of interest. The ability of these radars to generate a single radar beam limited the ability of the radar to do anything other than what it was intended to do—search for and track targets.

The advent of active electronically scanned arrays (AESA) (Figures 4.117 and 4.118) has revolutionized the utility of the radar through its ability to be able to create multiple independent steered radar beams emanating from the same common antenna face. Electronically steered beams can be steered around 1,000 times faster than those created by a mechanically steered antenna, and by miniaturizing the design of the radiating antenna and assembling them as a large group of typically up to more than 1,000 cells, a phased array (or active electronically scanned antenna) can be created.

Through varying the phase and amplitude of the signal created by each individual radiating antenna, almost any number of beams can be generated and assigned specific tasks such as searching, tracking, or communicating. In the communications mode, dedicated beams can be created to provide high-bandwidth data transfer to enable the rapid transmission of radar imagery, video, FLIR, and EW data to participating network units. Each beam may also be programmed with different waveforms suited to the purpose of its creation.

	Link 11	MIDS/JTIDS	Link 16	Link 22
Alternative Name	TADIL-A or TDL-A for airborne applications. TADIL-B or TDL-B for shipborne applications.	Communications component of Link 16	TADIL-J or TDL-J	NATO Improved Link 11 (NILE)
Message Format	M-series messages	J-series messages	J-series messages	FJ Series messages or F-Series messages, which are part of the J-series message family
Frequency	HF/UHF Fixed frequency	UHF 960–1,215 MHz	UHF 960–1,215 MHz	BLOS HF (3–30 MHz) and/or LOS UHF (225–400 MHz) Fixed freq / freq. agile
Data Rate (Nominal)	Link-11A: 2.25 kbps (reduced with SLEW message format) Link 11B: standard rate of 1.2kbps with optional speeds available in multiples of 1.2kbps up to 4.8kbps	57.6 kbps (at moderate ECM resistance)	26.88 kbps, 53.76 kbps or 107.52 kbps	HF Data Rate 500 to 2,200 bps UHF Rate 12.7 kbps
Voice Capable	No	Yes	Yes	Yes
Max. No. of Participants	56 (theoretical) active PUs. Unlimited on receive only	Practical limit 240	Practical limit: 15 transmit, 30 receive	120 active PUs
Throughput Measure (Nominal)	20 tracks per 10 sec net cycle	75 tracks per 12 sec (with ECM protection)	30 tracks per 12 seconds	30 tracks per 10 seconds per user
Coverage–Range	HF: variable, 300 nm nominal UHF: LOS	LOS up to 500 nm in Air-Air mode but typically ~300 nm for ground-based transmissions. Range can be extended by data relay between collaborating nodes with up to 5 dedicated hops.	Near-world-wide satellite footprint	HF: typically over 300 nm HF. Relay may increase geographic coverage of HF > 1,000 nm and UHF > 300 nm
Max Track Position Accuracy	±500 yds	± 4 ft	±32 ft	± 60 ft
Max Track Velocity Accuracy	±2.8 m/h		±0.2 m/h	

Table 4.11   Summary of the NATO Link Series Data Link Characteristics

Figure 4.116 An early generation AN/APG-73 radar on a Boeing F/A-18C showing the mechanically steered antenna.

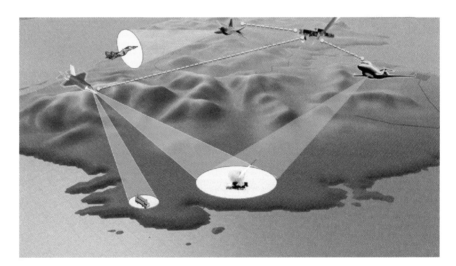

Figure 4.117 AESA radar data links are capable of the rapid transmission of high volumes of data between nodes, permitting sharing of near real-time imagery to other participating assets and C2 nodes.

Such flexibility has enable AESA radars to serve as highly advanced communication nodes able to do such tasks as creating detailed maps and transmitting them to other airborne or ground users with unprecedented speed, while continuing to perform their full range of traditional radar functions. This allows, for example, an aircraft to beam a SAR radar picture

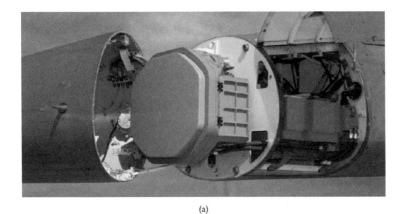

(a)

(b)

Figure 4.118   (a) The Raytheon Advanced Combat Radar, shown here on a Lockheed Martin F-16 fighter, is typical of modern generation AESA radars where multiple independent radar beams can be steered electronically, thus obviating the need for slow mechanically based beam pointing. By comparison, electronic beam steering is almost instantaneous. (b) The multiple electronically steered beams found on AESA radars enable them to operate in multiple simultaneous modes transforming their utility when compared to previous mechanically scanned radars that were limited to one mode at a time.

to a ground station and have it rapidly returned in an annotated format to show potential targets and other areas of interest (Figures 4.119 through 4.122).

Experimental work on an AN/APG-77 radar fitted to a U-2 aircraft has demonstrated the transfer of a 72-Mb synthetic aperture radar image in 3.5 seconds at a data rate of 274 Mbps. By comparison, the transfer would take 48 minutes on today's standard Link-16. High data transmission rates for the AN/APG-77 radar have been demonstrated in laboratory conditions of 548-Mbps transmit and up to 1 gigabyte per second on receive [67].

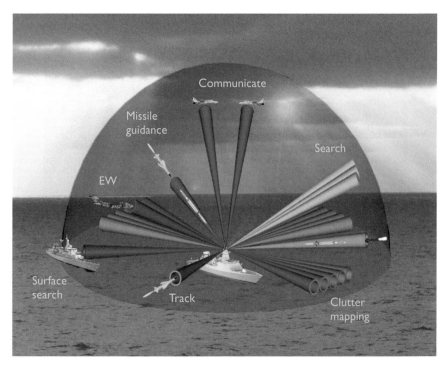

Figure 4.119   Electronically scanned radars may use multiple beams for multiple simultaneous functions. (Image courtesy of Thales.)

Figure 4.120   Republic of Singapore Navy frigate RSS *Formidable* (FFG 68) steams alongside the Indian Navy frigate INS *Brahmaputra* (F 31) in the Bay of Bengal. The four-faced phased array radar is clearly visible on the forward mast of RSS *Formidable*, as are the two mechanically scanned radars on the masts of INS *Brahmaputra*.

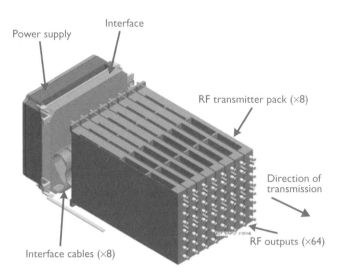

Figure 4.121   A phased array building block—in this case containing 64 radiating elements. Typically, phased array radars consist of more than 1,000 radiating elements. (Image courtesy of Thales.)

Figure 4.122   The AN/APG-77 radar coupled with the CDL modem has already proven its experimental use in the U-2 spy plane program to provide high-speed LOS data communications for both air-to-air and ground-to-ground applications.

Phased array antennas that can handle both civil and military satellite communications and CDL data link operations have also been demonstrated using frequencies between 2.2–15.35 GHz.

Such advances in radar technology will enable suitably equipped aircraft, ships, and ground vehicles to transmit and receive large, uncompressed data packages, such as synthetic aperture radar images, electro-optical images, and other data within seconds of it having been captured, providing a huge benefit to the OODA loop and kill chain decision-making processes.

Video data links

The transmission of imagery is increasingly being achieved through high-speed digital data links rather than dedicated video data links seen in earlier generations of equipment. With an emphasis on commonality of approach with other data links, the CDL series of data links is emerging as the principle data link for imagery transmission including EO, IR, and SAR radar imagery.

Other digital radio systems have also been adapted for the transmission of digital video. Also in widespread use is the airborne IDM application known as the Prism-IDM, which uses the integrated data modem a gateway for the transmission of video sources between various platforms such as those from FLIR or day/night television, which can be transmitted in JPEG format over existing onboard digital radio systems originally designed for voice transmissions.

Within existing Link series data links, a new J-Series message will be also be added to the current family of message standards to enable video images to be handled and accessed by suitably equipped platforms.

Weapons data links

Data links for the guidance of precision weapons systems are becoming increasingly important to ensure precision strike capability, especially where the weapons are launched from a long range. Data links between the launch platform and the missile can be provided through fiber-optic or wire cables, lasers, or RF data links. Data links can either be simplex (one-way), typically between the ground and the missile to enable it to be steered to the target, or duplex (two-way), providing information back to the operator from the missile sensor such as the TV or IR imagery to enable the correct target to be identified or reselected during flight.

As well as providing a high degree of survivability of the weapon operator, weapons data link applications generally provide the following advantages over autonomous or unguided munitions (Figures 4.123 and 4.124):

Figure 4.123   An AGM-154C JSOW-C seen an instant before impact against a test target. The AGM-154C (JSOW C) uses an infrared seeker for terminal guidance, while the Block III variant adds a Link-16 weapon data link and moving maritime target capability. (Photo courtesy of Boeing.)

Weapon in-flight target update:

- Updated guidance during engagement of moving and mobile targets;
- Key enabler for engagement of time-sensitive-targets;
- Improves precision in targeting hidden or concealed targets;
- Weapon can be retargeted while in flight;
- Avoidance of fratricide and collateral damage;
- Target acquisition and selection.

Weapon in-flight tracking:

- Strike deconfliction and synchronization;
- Health, arming status, location;
- Air-space coordination and deconfliction;

Weapon impact assessment:

- Poststrike assessment and restrike decisions;
- Recording of event.

Figure 4.124 Data links on electro-optical guided missiles and bombs enable the pilot to update the targeting of the missile in flight and remain outside the range of the local air defense systems. KAB-500Kr-E guided bomb on a Russian Air Force Mig29.

Weapon fusing selection:

- Collateral damage and fratricide avoidance;
- Selection of warhead and/or fusing mode during flight;
- Disengagement if target strike window closes;
- Force proportionality;
- Mission abort (if required).

Typically, weapons data links feature highly secure waveforms, such as direct sequence spread spectrum, interleaving, and deinterleaving techniques, frequency hopping, error detection, and correction codes. Data latency is minimized to give real-time control, and advanced compression techniques offer real-time video signals to be relayed from the missile during its flight. Depending on the frequency used, data link ranges may extend up to several hundred kilometers.

Instant messaging and chat systems

Recent operational campaigns have seen the rapid adoption and widespread use of chat rooms and instant messaging systems within C2 networks. Instant messaging and chat services were originally developed for commercial use and were easily adapted to military networks using TCP/IP protocols. Unlike commercial applications, however, users in military chat networks tend to have functional addresses rather than private addresses.

Chat rooms such as those provided by NATO's J-Chat system enable groups of users to share messages that can be seen by all participating group members. One-on-one messages may also be sent between users that cannot be seen by other participants; this is often referred to as "whisper mode" chat and instant message systems will also generally show the availability status of other users and whether they are on the network or not.

Instant messaging and chat rooms have the advantage that they can increase the tempo of communications between network participants, unlike e-mails, where users are watching out for messages and will respond by return.

Military Uses of Instant Messaging and Chat Rooms [68]

One-to-one chat

- Short message transfer complementing voice and network communications;

- Useful when voice is not practical or allowed;

- Near instant communication with specific user (node) on the network;

- Can be done in whisper mode, enabling message to be kept private;

Multiuser chat

Widely used for sharing information in real-time, including:

- Decisions to engage (field and HQ involvement);

- Allocation of targets;

- Intelligence updates.

Presence

- Sharing status information;

- Geolocation and other extended presence information.

Instant messaging systems are not without problems. In particular [69]:

- Group formation: How do members form a secure group for their specific mission or requirements?
- How do potential members learn that a group exists?
- In coalition operations or those involving the exchange of sensitive information, how are the security and privacy of messages and user presence assured?
- Is the communication robust enough to cope with unreliable, error-prone communication channels and denied environments?
- How can connections be formed to other users outside the immediate communications network?

Collaborative networks

Increasingly, the establishment of collaborative networks between different branches of armed forces and among collaborating nations has seen the adoption of COTS solutions to provide standardized data input protocols and to take advantage of well thought-out user interfaces. This has delivered significant advantages across the entire operational spectrum, not just on the time-sensitive intelligence and decision-making cycles, but also extending the benefits to logistics and operational planning areas.

Users are increasingly adopting common or at least interoperable interface standards to establish a wide area common mission network, which extends across bespoke national planning tools to link even the outer edge of the network. Units deployed in forward operating bases (FOBs) or patrol bases (PBs), can now access and integrate with other operational units in two-way near real time collaboration. This may include not only traditional tasks such as planning and intelligence updates, but also the ability to access and update databases and logistics systems with the latest operational requirements and situation reports.

Currently, the United Kingdom and NATO use Oracle Wise Web on their networks, while the United States uses Microsoft SharePoint as the foundation for interactive applications. Such software tools allow the shared generation of orders and plan and avoid the delays and confusion often in evidence through traditional stovepipe planning tools [70].

The ease with which users can construct specific applications using commercially available software programs such as these has advantages and drawbacks. While information can be captured and processed to meet the needs of small groups or node-specific requirements, the diversity of applications and the information fed into them can also provide a challenge when trying to assimilate the vast amount of data fed into a network, in particular the challenge of determining actionable intelligence from the vast amount of data and making links between data inputs to create actionable intelligence.

Global Information Grid (GIG)

The U.S. DoD defines its Global Information Grid (GIG) (Figures 4.125 and 126) as a "globally interconnected, end-to-end set of information capabilities, associated processes, and personnel for collecting, processing, storing, disseminating, and managing information." The idealized vision for the GIG is a publish–and-subscribe, plug-and-play network in which any application can be plugged into the network anywhere, at any time, to help achieve warfighting objectives [71].

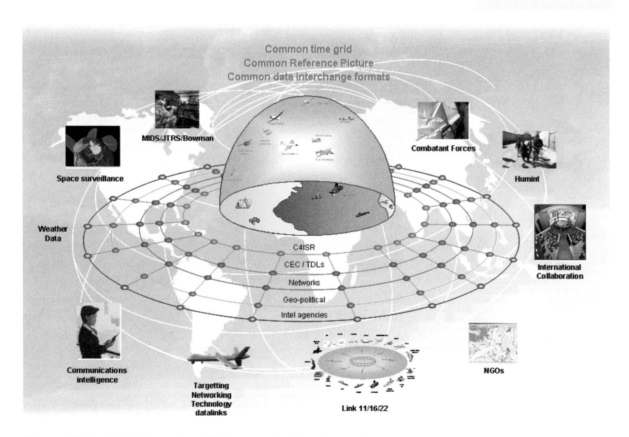

Figure 4.125   Global Information Grid showing typical data input and users.

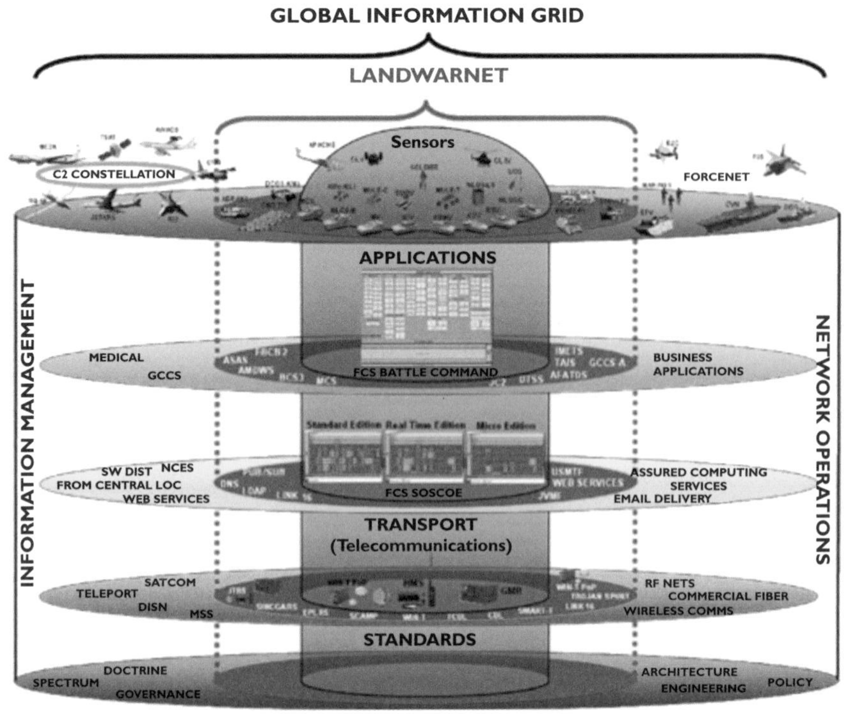

Figure 4.126 The U.S. Army LandWarNet program provides seamless data storage and information sharing to network participants and similar in concept to the U.S. Air Force C2C ConstellationNet and the U.S. Navy's Forcenet programs, all three of which are connected and integrated through the GIG. LandWarNet's key network elements consist of connectivity to the GIG, the WIN-T network, JTRS data radios, the Transformational Communications System, and the Network Centric Enterprise Services (NCES).

The target GIG allows all DoD users[9] (and their external mission partners[10]) to find and share the information they need, when they need it, in a form they can understand, use, and act on with confidence, and protects information from those who should not have it [72]. Information posted to the GIG becomes available to all relevant users from the grid (subject to security considerations), allowing them to pull information according to their need or have it pushed to them from the many applications connected to the GIG, based on predefined user criteria. The applications will provide the user with the information that they require, based on the highest quality data sources from across the network, to enable end-user decision superiority. For systems and users across the GIG, the goal is to get the data from the sensors or databases posted on the network for use by all participants at the first point where it becomes available in a consumable format.

The goal of the GIG is to deliver a universal network that comprises all U.S. DoD communications and computing systems and services, software (including applications), data, security services, and other associated services necessary to achieve information superiority. The GIG supports all U.S. DoD, national security, and related intelligence community missions and functions (strategic, operational, tactical, and business) in war and in peace. It provides capabilities from all operating locations (bases, posts, camps, stations, facilities, mobile platforms, and deployed sites) and interfaces to coalition, allied, and non-DoD users and systems [73]. The underlying backbone of the GIG infrastructure is the convergence over Internet-based protocols (IP) of all types of information such as telephony, multimedia services, video, and data.

The GIG will be structured to support and enable highly responsive, agile, adaptable, and information-centric operations characterized by [72]:

- An increased ability to share information;
- Greatly expanded sources and forms of information and related expertise to support rapid, collaborative decision-making;
- Highly flexible, dynamic, and interoperable communications, computing, and information infrastructures that are responsive to rapidly changing operational needs;
- Assurance and trust that the right information to accomplish assigned tasks is available when and where it is needed, that the information is correct, and that the infrastructure is available and protected.

The GIG relies on a Service Oriented Architecture (SOA)–based approach structured through common IP network and data interchange standards to improve access to information in the form of voice, video, and data between the DoD's numerous information, intelligence, and effects systems and those of its allies, particularly with a view to achieving a more network-enabled approach to fighting wars and achieving information superiority over adversaries—much the same way as the Internet has transformed industry and society on a global scale [74]. In effect, the GIG represents the integration of network-aware applications and systems that can exploit the information available to and from each node through adapting to the dynamically changing communications environment in which they are expected to operate.

GIG protocols are designed to provide multilayer seamless connectivity with sufficient bandwidth to support all the providers and receivers of intelligence information across a global network. The GIG consists of a constantly changing number of nodes and links designed to transform the vast increase in accessible information into an increased and timely combat capability. It represents a globally connected, end-to-end set of information capabilities, associated processes, and personnel for collecting, processing, storing, disseminating, and managing information on demand to warfighters, policy makers, and support personnel [75].

The GIG's role is to create an environment in which users can access data on demand at any location without having to rely on (and wait for) organizations in charge of data collection to process and disseminate the information. Data could emanate from a variety of sources, including weapon systems belonging to other military services, space-based intelligence, surveillance, and reconnaissance satellites, and DOD logistics, financial, and other systems that carry out business operations. Ultimately, DOD expects that most of these systems will become part of the GIG [74] (Figure 4.127).

Ultimately, the functionality of the GIG is designed to support network-enabled operations in creating a decisive warfighting advantage. In this regard, the DoD expects that lethality and survivability of equipment and personnel will be increased, the scale of deployed forces could be scaled down in favor of smaller more agile connected units, and the entire logistics and support footprint scaled down accordingly.

The GIG represents a vast network of interconnected systems and while all will in theory be accessible (subject to suitable security clearance), the specific information required to fulfill a mission will very much be subject to the capabilities and requirements of the stakeholders across the network who both supply and demand information. Through understanding the influences of stakeholders [76] and network cliques [77] across network structures, we can identify unique systems within the GIG that have specific roles across the network in supporting the achievement of information superiority. Studying the nature of stakeholders and cliques and the patterns of the links between them [78] gives us a functional pattern of information systems within the GIG that contribute to overall information systems performance for specific operational requirements [79].

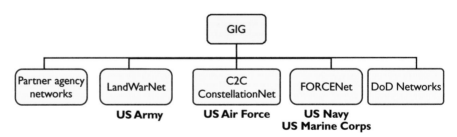

Figure 4.127   The GIG integrates and exchanges intelligence feeds from across the U.S. military community, DoD networks, and nonmilitary sources.

The GIG structure (Figure 4.128) will enable a migration towards sensor systems that take advantage of standardized IP-based multilayer protocols and shared infrastructure typically found in system-of-systems layering both to immediately post their information products for fusion with other intelligence, surveillance, and reconnaissance (ISR) sensors, navigation systems, or special community databases, and to receive information gathering tasking in near real time.

One key challenge in this multilayer concept is the ability to not only have true interoperability between national multilayer and multiagency networks, but for these networks to be able to operate between and across coalition and partner countries' networks.

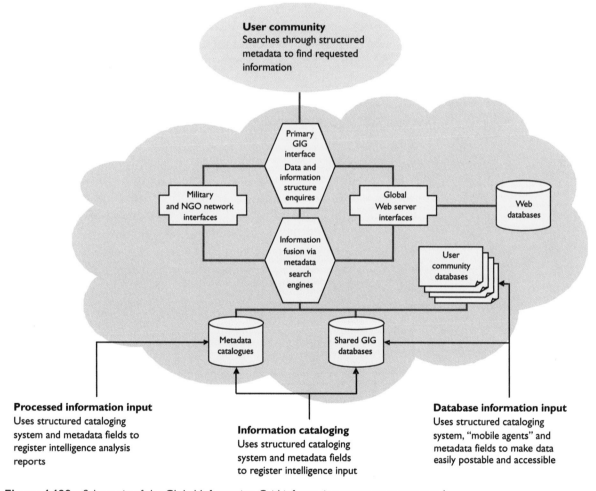

Figure 4.128 Schematic of the Global Information Grid information management protocol.

In order to ensure that this goal is achieved, common layered architecture, information exchange protocols and data standards using Extensible Markup Language (XML) and the Universal Description, Discovery, and Integration (UDDI [80]) protocol, operating in an IP environment need to be built into networked C4ISTAR systems.[11]

Where the integration of legacy systems or new systems is required that do not use IP formatting, a gateway system that effectively acts as a translator to place data on the network in an IP format will be required to ensure compatibility with data exchange standards.

Information transfer across the GIG network will involve a range of different transport mediums, preferably using open architecture public domain systems with suitable encryption ranging from JTRS and military data links to IP-based public domain networks. Standard interface criteria between compatible systems will ensure compatibility across multiple networks, and while any external interface is not considered part of the GIG, if it wishes to interface with the GIG network, it must meet GIG interface criteria (Figures 4.129 and 4.130).

Network broadcast systems

Integrated Broadcast Service (IBS)

The Integrated Broadcast Service or Intelligence Broadcast Service (IBS) is the worldwide DoD standard network for transmitting tactical and strategic

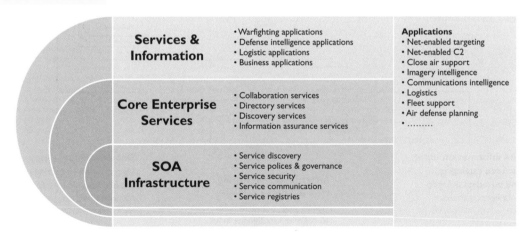

Figure 4.129   The GIG Operational Architecture core enterprise services [72]. The U.S. DoD intends to integrate virtually all command, control, and communications systems into the GIG.

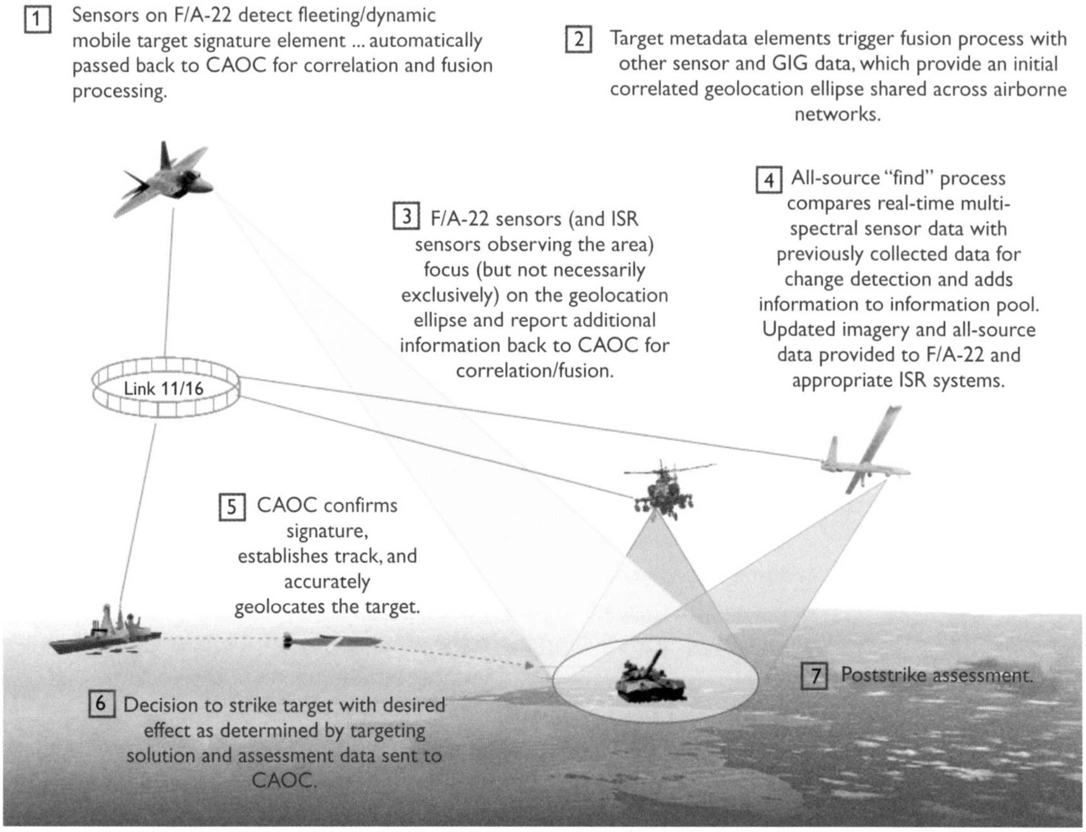

Figure 4.130 Typical engagement sequence showing an F/A-22 engaging a ground moving target through sensor and database integration with the GIG network [2].

intelligence and targeting data within a common format using the common interactive broadcast (CIB) waveform. IBS provides an ability to take disparate data fields and translate them into a common message standard, then translate them back into the message standard that the user requires.

The end result is a data format that appears on a user's display processor, either as images on a map or scrolling data that, when clicked on, conveys a more detailed track of information [81].

The IBS provides an interactive link to a virtual library of fused and correlated SIGINT intelligence data collected by multiple intelligence sources and agencies to provide engagement-quality targeting data to tactical and operational commanders and planning staff. It is designed to standardize the transmission formats of multiple intelligence broadcast systems into a common format and reduce the bandwidth requirements of multiple competing systems.

The IBS will broadcast as a push system over UHF satellites for high priority messages such as priority targeting information and threat warning and avoidance and SHF satellite broadcasts for routine message exchange where data latency is less critical. The IBS is also designed to operate over existing communications channels, such as SHF and EHF data link networks and the Global Broadcast System (GBS) using J-series and common message format standards.

The IBS system will receive multiple intelligence feeds in a variety of formats, and process, integrate, reformat, and transmit tactical data to suitably equipped platforms according to preset user-defined criteria. In particular, it will transmit data pertaining to situational awareness, target status and priority target updates, force protection, and indications and warnings to tactical C2, sensor, weapon, and ISTAR systems for direct use or for further processing and integration. Using the system's track correlation, repeated inputs of the same target or object of interest and multiple sensor sources can be detected and correlated to ensure that multiple tracks are not presented to subscribers.

The IBS system will replace three U.S. legacy UHF military broadcast systems: TIBS, TDDS, and TRIXS.

Global Broadcast System (GBS)

The Global Broadcast System (GBS) uses the services of three UHF follow-on (UFO) satellites to provide a continuous, high-speed, high-volume, one-way flow of information across a worldwide network.

The GBS rapidly broadcasts large volumes of intelligence information between rear echelon and deployed forces, particularly to the forward edges of a military force where communications capacities to support information dissemination have generally been limited. Typically, end users will include expeditionary forces, in transit or deployed forces, including those with limited priority or nonmission critical needs who need to be kept up to date with the current intelligence picture. Virtually all data formats can be disseminated including video, high-resolution imagery, and other large data files such as weather maps and situational awareness pictures, intelligence briefs, and operational planning data, all supported by an Internet-style user interface. The type of information is only limited by that which the user community decides to broadcast, and virtually any format can be accommodated.

Each GBS satellite in a global constellation will be served by a primary uplink site, technically called a *primary injection point* (PIP), where encrypted data is sent via RF transmission to the satellite for onwards broadcast to its designated geographical footprint covering the operational area. The deployed reception terminal is similar in concept to a set-top box and requires a small satellite antenna of 18 inches or 24 inches in length, permitting it to receive broadcasts of up to 23 Mbps. The GBS will also have the capability to broadcast intelligence feeds directly from deployed forces and to accommodate a capability for the users to request or pull specific pieces of information to meet mission requirements (thereby avoiding information overload of connected units), subject, of course, to appropriate levels of security clearance. These requests will be processed and prioritized by an information management center before the desired information is then broadcast back across the GBS satellite network (Figures 4.131 and 4.132).

Figure 4.131 The GBS network provides information to disadvantaged users with limited or no data access. GBS is a highly agile, extensible and scalable COTS-based system that provides full flexibility to broadcast mission critical data over both military and commercial satellites. (Photograph courtesy of Raytheon.)

Figure 4.132 The Predator UAV uses the GBS network to broadcast its video imagery to multiple users across the network. Here an MQ-1 Predator UAV, armed with AGM-114 Hellfire missiles, flies a combat mission over southern Afghanistan in the summer of 2009. The MQ-1 is deployed in Operation Enduring Freedom providing interdiction and armed reconnaissance against critical, time-sensitive targets.

References

[1] U.S. Department of Defense (DoD), Joint Directors of Laboratories, 1995.

[2] United States Air Force Scientific Advisory Board, *Report on Domain Integration*, Executive Summary and Annotated Brief, SAB-TR-05-03, July 2005.

[3] *An Introduction to Data Fusion*, Signal and Image Centre (SIC), Electrical Engineering Department of the Faculty of Applied Sciences of the Royal Military Academy, Belgium, 2010.

[4] Wang, Q., et al., "Multisensor Data Fusion in Distributed Sensor Networks Using Mobile Agents."

[5] White, Jr., Franklin E., "Data Fusion Sub Panel Report," *Proc. Fifth Joint Service Data Fusion Symp.*, October 1991, Vol. I, pp. 335–361.

[6] Joint Directors of Laboratories (JDL), Data fusion model, 1995.

[7] Steinberg, Alan N., and Christopher L. Bowman, "Rethinking the JDL Data Fusion Levels," *International Data Fusion Conference*, July, 2004

[8] "Transforming Data Into Knowledge: The Use of Advanced Visualization and Cognitive Aids to Support Improved Level 2 and Level 3 Data Fusion," *Technology Today*, Raytheon Publication, Issue 1, 2009.

[9] Daso, D., "New World Vistas," *Air Power Journal*, Winter 1999

[10] Thiele, Antje, et al., "Building Recognition from Multi-Aspect High-Resolution InSAR Data in Urban Areas," *IEEE Transactions on Geoscience and Remote Sensing*, Vol. 45, No. 11, 2007.

[11] Michaelsen, E., and U. Soergel, U. Thoennessen, *Potential of Building Extraction from Multi-Aspect High-Resolution Amplitude SAR Data*, FGAN-FOM Research Institute for Optronics and Pattern Recognition, Ettlingen, Germany, 2005.

[12] Open Geospatial Consortium, www.opengeospatial.org.

[13] Wesson, R., et al., "Network Structures for Distributed Situation Assessment," *IEEE Transactions on Systems, Man, and Cybernetics*, Vol. SMC-11, No. 1, January 1981, pp. 5–23.

[14] Jayasimha, D. N., S. S. Iyengar, and R. L. Kashyap, "Information Integration and Synchronization in Distributed Sensor Networks," *IEEE Transactions on Systems, Man, and Cybernetics*, Vol. SMC-21, No. 21, September/October 1991, pp. 1032–1043.

[15] Prasad, L., et al., "Functional Characterization of Sensor Integration in Distributed Sensor Networks," *IEEE Transactions on Systems, Man, and Cybernetics*, Vol. SMC-21, September/October 1991.

[16] Iyengar, S. S., D. N. Jayasimha, and D. Nadig, "A Versatile Architecture for the Distributed Sensor Integration Problem," *IEEE Transactions on Computers*, Vol. 43, No. 2, February 1994, pp. 175–185.

[17] A. Knoll and J. Meinkoehn. "Data Fusion Using Large Multiagent Networks: An Analysis of Network Structure and Performance," *IEEE Proceedings of the International Conference on Multisensor Fusion and Integration for Intelligent Systems (MFI)*, Las Vegas, NV, October 2–5, 1994, pp. 113–120.

[18] Qi, Hairong, S. Sitharama Iyengar, and Krishnendu Chakrabarty, *Distributed Multi-Resolution Data Integration Using Mobile Agents*, IEEE, 2001.

[19] Marshall, Preston, Program Manager, Strategic Technology Office, DARPATech, "Tactical Networking: It's More Than Just Command and Control," *DARPA's 25th Systems and Technology Symposium*, Anaheim, CA, August 8, 2007.

[20] Bell, Michael, *Introduction to Service Oriented Modeling: Service Oriented Modelling, Service Analysis, Design and Architecture*, New York, John Wiley & Sons, 2008.

[21] *C4ISR for Network-Oriented Defense*, March 2006, White Paper, 284 23-3064 Uen Rev A © Ericsson Microwave Systems AB 2006.

[22] http://www.army.mil/fcs/network.html.

[23] Miller, Gary M., and Jeffrey S. Beasley, *Modern Electronic Communication*, 7th ed., Upper Saddle River, NJ: Pearson, 2002.

[24] Poole, Ian, *Radio Propagation Principles and Practice*, Radio Society of Great Britain, Oxford Press, 2004.

[25] Fulghum, David A., "Duck the Soup," *Aviation Week & Space Technology*, January 18, 2010.

[26] www.wikipedia.com/vhf.

[27] Melia, Alan, "Understanding LF Propagation," *Radcom,* Bedford, U.K.: Radio Society of Great Britain, Vol. 85, No. 9, 2009, p. 32.

[28] http://www.radio-electronics.com/.

[29] Sandia National Laboratory's Nonproliferation programs and arms control technology, 2003.

[30] McCrady, Dennis D., "The GPS Burst Detector W-Sensor," Sandia National Laboratories, December 2009, http://www.osti.gov/bridge/servlets/purl/10176800-S2tU7w/native/10176800.pdf.

[31] Lennox, Duncan, "Protection of Satellites Is the New Race in Space," *Janes Defence Weekly*, October 28, 2009.

[32] http://www.emporia.edu/earthsci/student/graves1/GHWK2.jpg.

[33] *C4ISR Magazine*, August 2009.

[34] Scott, Richard, "DARPA Plans 'RIMPAC 2012' Demo for Submarine Laser Communications," *Jane's International Defence Review*, Vol. 43, March 2010.

[35] Sullivan, W., "How Huge Antenna Can Broadcast into the Silence of the Sea," *New York Times*, October 31, 1981.

[36] Deming, David, "The Hum: An Anomalous Sound Heard Around the World," *Journal of Scientific Exploration*, Vol. 18, No. 4, 2004, pp. 571–595.

[37] http://www.vlf.it/zevs/zevs.htm.

[38] www.fas.org.

[39] http://en.wikipedia.org/wiki/Combat-net_radio.

[40] JTRS HMS AFCEA presentation, "JTRS Handheld, Manpack, and Small Form Fit (HMS) Ground Domain," Program Manager COL Daniel Hughes, Product Manager LTC Richard Housewright, May 2007.

[41] *Security Engineering: A Guide to Building Dependable Distributed Systems*, Chapter 16: Electronic and Information Warfare, 16.3 Communications Systems, 2008, p. 323.

[42] IETF, RFC 1122, pp. 7–8.

[43] Tanenbaum, Andrew, *Computer Networks*, section 1.4.3, 2002.

[44] Dye, Mark A., *Network Fundamentals: CCNA Exploration Companion Guide*, 2007.

[45] Kurose, James F., and Keith W. Ross, *Computer Networking: A Top-Down Approach*, Addison-Wesley, 2010.

[46] Forouzan, Behrouz A., *Data Communications and Networking*, McGraw-Hill, 2006.

[47] Comer, Douglas E., *Internetworking with TCP/IP: Principles, Protocols and Architecture*, Upper Saddle River, NJ: Pearson Prentice Hall, 2005.

[48] Kozierok, Charles M., *The TCP/IP Guide*, No Starch Press, 2005.

[49] Stallings, William, *Data and Computer Communications*, Upper Saddle River, NJ: Prentice-Hall 2006.

[50] Hauben, Ronda, "From the ARPANET to the Internet," *TCP Digest (UUCP)*, http://www.columbia.edu/~rh120/other/tcpdigest_paper.txt, 1998.

[51] X.200: Information technology—Open Systems Interconnection—Basic Reference Model: The basic model, 2009, http://www.itu.int/rec/T-REC-X.200-199407-I/en.

[49] http://en.wikipedia.org/wiki/Osi_model.

[52] http://www.army.mil/fcs/network.html

[53] White, B. E., "Layered Communications Architecture for the Global Grid," The MITRE Corporation, supported by the U.S. Air Force Electronic System Center under contract number F19628-99-C-0001, March 2001.

[54] *U.S. Army Field Manual*, FM-24-11, Chapter 1-4, Transmission Techniques, 2005.

[55] The ABCs of Spread Spectrum—A Tutorial, http://sss-mag.com/ss.html.

[56] http://www.lm-isgs.co.uk/defence/data links/variable_message_format.htm.

[57] *Digital Battlespace Handbook,* The Shephard Press, 2009.

[58] Williams, David, "VMF—Operational Considerations," *IDLSoc UK*, Chapter 18, April 2006.

[59] Dewey, John K., and Robert E. Wilson, "Track #6: Extending GIG Connectivity to the Warfighter (3D)," *2007 LandWarNet Conference*, Session 7, August 23, 2007.

[60] Minges, Mark, Senior Engineer, Information Directorate, U.S. Air Force Research Laboratory, "Survey of Current Air Force Tactical Data Links and Policy," June 13, 2001.

[61] U.S. Army Modernization Strategy, Appendix A, 2008.

[62] Northrop Grumman Corporation, *Understanding Link-16: A Guidebook for New Users*, Northrop Grumman Corporation, Information Technology Communication and Information Systems Division, April 1994.

[63] U.K. Tactical Systems Reference Guide, http://www.synthesys.co.uk.

[64] U.S. Army, Marine Corps, Navy, Air Force Introduction to Tactical Digital Information Link J and Quick Reference Guide, FM 6-24.8, MCWP 3-25C, NWP 6-02.5, AFTTP(I) 3-2.27, June 2000.

[65] Capstone Requirements Document (CRD), *Global Information Grid (GIG)*, March 28, 2001, http://www.dfas.mil/technology/pal/regs/gigcrdflaglevelreview.pdf, April 1, 2003.

[66] Lockheed Martin UK, http://www.lm-isgs.co.uk/defence/data links/link_22.htm.

[67] "1 Joint Warfighting Science & Technology Plans Quiet Progress," *Aviation Week & Space Technology*, January 9, 2006, p. 24.

[68] Kille, Steve, CEO Isode Ltd., "Running XMPP over HF Radio," *HFIA Conference*, San Diego, CA, February 2009.

[69] Tölle, Jens, Tobias Ginzler, and Philipp Steinmetz, *Challenges of Instant Messaging in Tactical Environments—Concepts and Practical Implementation*, Research Establishment for Applied Sciences—FGAN Research Institute for Communication, Information Technology and Ergonomics, NATO Report RTO-MP-IST-083, 2008.

[70] "Coalition Warfare by Network," *Janes Defence Weekly*, January 27, 2010.

[71] Cruz, Charlie I., MSgt, USAF, "Netwars Based Study of a Joint STARS Link-16 Network," AFIT/GCS/ENG/04-06, Department of the Air Force, Air University, Air Force Institute of Technology, 2004.

[72] Department of Defense CIO, *Global Information Grid Architectural Vision, Vision for a Net-Centric, Service-Oriented DoD Enterprise*, Version 1.0, June 2007.

[73]   Capstone Requirements Document, *Global Information Grid*, March 2001.

[74]   United States Government Accountability Office, *Report to Subcommittee on Terrorism, Unconventional Threats, and Capabilities*, Committee on Armed Services, House of Representatives, July 2004 Defense Acquisitions, The Global Information Grid and Challenges Facing Its Implementation, GAO-04.

[75]   Department of Defense Directive Number 8320.02, "Data Sharing in a Net-Centric Department of Defense," December 2, 2004, certified current as of April 23, 2007.

[76]   Rowley 1977.

[77]   Provan and Sebastian 1998.

[78]   Granovetter 1973.

[79]   Iatrou, S. J., "Distributed Holistic Bounding," Naval Postgraduate School, Monterey, CA, June 22, 2005.

[80]   www.uddi.org.

[81]   Morales, Monica D., "IBS Capabilities Allow for More Seamless Data Analysis," *ESC Public Affairs*, July 2006, http://integrator.hanscom.af.mil/2006/July/07132006/07132006-04.htm.

[82]   "Hispars Project-Eucllid RTP6.2, Signal and Image Centre (SIC)," Electrical Engineering Department of the Faculty of Applied Sciences of the Royal Military Academy, Belgium.

Endnotes

1.   *Fuzzy logic* is a form of reasoning, derived from an approach whereby a logical state need not be exactly zero (false) or one (true), but can also be represented by any value in between.

2.   *Neural network* in this context is a real or virtual computer system designed to emulate the brain in its ability to process variable and imprecise data and to "learn" from inputs and patterns to improve processing accuracy.

3.   I am indebted to Ian Poole for his permission to take extracts from [24].

4.   Provision 2.1 of the International Telecommunication Union (ITU) Radio Regulations.

5.   I am indebted to Ian Poole for his permission to take extracts from [28].

6.   For descriptions of four-layer Internet protocol, see [42–44].

7.   For descriptions of five-layer Internet protocol, see [45–49].

8.   *Oversampling* is a process of sampling a signal with a sampling frequency significantly higher than twice the bandwidth or highest frequency of the signal being sampled.

9.   DoD users include information providers and (anticipated/unanticipated) information consumers, whether fixed or on the move, deployed or at fixed installation, human or software/hardware.

10. Mission partners generally participate through a secure gateway. These gateways permit members to be authenticated, produce and consume information services, and collaborate. However, the GIG and associated services also must allow unclassified information to be exchanged with uncleared civil or military partners outside the boundaries of the DoD Enterprise.

11. UDDI is a software protocol for retrieving Web-based data and services within XML-based Web service registries and defines a standard approach to Web service architectures.

Sailors and Marines conduct a foreign object damage walkdown on the flight deck of the aircraft carrier USS Harry S. Truman (CVN 75). The Harry S. Truman Carrier Strike Group is deployed supporting maritime security operations and theater security cooperation in June 2010. (U.S. Navy photo by Mass Communication Specialist 2nd Class Kilho Park.)

Future Trends in Network-Enabled Capabilities

Advances in modern warfare are increasingly centered on the utility of a network-enabled doctrine. The need to ensure that all assets—front line to rear echelon—benefit from this concept will further shift the balance of future military investment towards increased network integration, data fusion, and information management and exploitation rather than adding to the number of sensors and range of effects.

Military planners increasingly recognize that a huge amount of latent capability lies in the information and data available through existing underexploited systems, much of which are not currently shared or accessible to collaborating forces. Even within accessible networks, information is often not fused in real time, nor links made between data to provide the enhanced situational awareness.

Clearly, the efficient utilization of the vast amount of data and information that can be generated from networked sensors and systems will provide the next great evolutionary step in military warfare. Where investment in sensors does take place, it will generally be in specific areas such as those outlined below.

Data collection and sensing

Sensors will continue to provide the key means of data input into network-enabled capabilities and the need for increasingly real-time inputs will see further advances in capabilities and applications (Figure 5.1), with enhanced sensor discrimination against *difficult, stealthy, and fleeting targets* becoming a priority.

While sensor developments continue to address the challenge of such targets, platforms and equipment used in the battlespace will benefit from designs and materials to improve *cloaking, shielding, and invisibility*. Such stealth techniques will be applied not only to platforms, but also to increasingly small assets down to and including individual soldiers.

At an *operational level*, network-enabled capability (NEC) will further improve *remote and persistent surveillance* that can be proliferated throughout the battlespace, offering near real-time reporting and accuracy of target identification. Accurate and persistent real-time situational awareness will

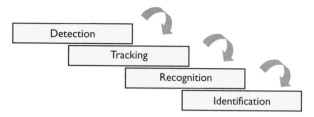

Figure 5.1   Functions required to deliver increasingly advanced sensor capability.

enhance the ability to ensure that appropriate military effects will be matched to suit the target type and desired outcome.

Current sensor research includes:

- Multisensor data fusion;
- Real-time situational awareness;
- Acceleration of detection, tracking, and identification functions;
- Improved performance in complex signal environments such as clutter and propagation;
- Improved performance in jamming and deception environments;
- Slow, small, manuevering; surface and air objects;
- Real-time sensor integration with weapon systems;
- Networked, nonintrusive microsensors;
- 3D mapping;
- Target identification, in particular of moving and fleeting targets;
- Damage/kill assessment;
- Improvements in sensor range;
- Geolocation accuracy;
- Detection of improvised explosive devices (IEDs) (Figure 5.2);
- Improved performance in the littoral environment;
- Hyperspectral sensors.

For all areas of research, the link with accurate *geolocation* capabilities pervades all areas of research into network-enabled capabilities. As such, geolocation capabilities extending to the underwater battlespace and space environments will provide increased accuracy for targeting and ISTAR applications. These capabilities will extend further to improve the accuracy of *positioning, navigation, and timing* of systems embedded in battlespace platforms.

Improvements in *locating and identifying friendly forces* will result in data becoming more widely shared across collaborating networks to greatly reduce the incidents of friendly fire. Similarly, knowledge about hostile

Figure 5.2 Detection of IEDs is an increasing priority in asymmetric warfare. Here a U.S. Army Stryker vehicle lies on its side after surviving a buried IED blast in 2007.

forces and their intentions will be made widely and quickly available to collaborating forces. Sources of data will be able to be automatically woven together through the use of standard geospatial reference grids.

Platforms, sensors, effectors, and personnel will all have the possibility of being allocated to *individual IP-based addresses* to enable comprehensive geolocation, data fusion, near real-time C2, and progress reporting across extended networks.

Detailed and automated *target identification* capabilities will become one of the greatest challenges we face and it is widely recognized that with asymmetric warfare, the identification and location of adversaries in the battlespace of the future is far more complex than has been the case with traditional mechanized warfare.

The *detection, identification, and tracking of difficult targets* will see further improvements, enhancing the ability to achieve clear *discrimination* between real and false targets and the ability to *identify decoys* and *overcome active deception techniques*.

In urban environments, the automated *tracking of specific individuals* through a crowded scene and plotting precise geolocation coordinates for each place the subject visits will continue to remain a priority for further investment. *Image and communications surveillance* data can be used in real time or stored for later analysis of suspicious patterns of enemy movements and determination of networks and adversary *centers of activity*.

Persistent, autonomous, *unmanned sensors* will proliferate in underwater, land, air, and space applications, enabling improved target identification and *target tracking*, together with day/night, all-weather *identification and discrimination* between friendly and enemy forces, particularly in adverse conditions.

The utility and effectiveness of *space-based sensors* will continue across multispectral domains for Elint and multispectrum EO surveillance and hyperspectral imaging (Figure 5.3).

Data transportation and networking

Network-enabled warfare relies on the flexibility of networks and their ability to adapt to meet the emerging threat scenario, the emphasis being on networking rather than networks in their own right. The backbone of effective networking is the provision of robust and secure interconnectivity and the ability for significant flexibility in forming and repairing network connections.

As users and networked applications demand increasing flexibility and unstructured patterns of use, future networks will see the continued

Figure 5.3 Further evolution in the concept of data collection will be seen in the concept of the "stand in" nature of sensors so that many more close in sensors will produce a much better accuracy of position, direction, identification, and intent than remote sensors, but, will individually have less coverage. Expendable UAVs with low-cost sensors are the obvious example, but, this will extend into areas of infrastructure such as urban environments and so forth with "throw away" sensors. It makes for simpler sensors, but, results in a need for much greater thought in the federation or centralization of their data and control.

development of *self-forming ephemeral networks* along with command and control structures that are capable of configuring and disbanding to meet specific mission requirements. These ephemeral structures will operate across and interface with multiple networks. *Common operating standards* and *security protocols* will evolve to facilitate interoperability between platforms and operating forces, and it is often the *interoperability standards* and security limitations rather than the capability of the network itself that constrain the full operational exploitation of NEC capabilities. Interoperable *multilevel* security will evolve between collaborating forces, permitting multinational forces to operate with relative efficiency compared to today's limitations.

Current network research includes:

- Multisensor data fusion and data standards;
- Data fusion across collaborating networks and associated interface standards;
- Real-time situational awareness;
- Collaborative engagement capabilities;
- Collaborative targeting;
- Multinational coalition frameworks;
- Coalition interoperability;
- Channel multiplexing;
- Data compression;
- Internet Protocol (IP) data formats;
- Security and encryption;
- Self-forming and self-healing ephemeral networks;
- Cooperative interaction between nodes and functionalities;
- Wireless networks;
- Dynamic spectrum allocation and resource management;
- Network attack.

Wireless *networks* will advance to enable the automated creation of self-forming networks suiting the limitations of the network paths, sensors, effectors, and communications protocols used across the network. Sign-in protocols used in sensors, effectors, and ISTAR assets will permit nodes to easily connect and disband to suit the mission requirements. Network wireless transmission will evolve to *cognitive radio* standards where radios will themselves determine the optimal part of the radio frequency spectrum over which to transmit rather relying on fixed protocols. Such arrangements provide for more efficient use of the spectrum through dynamic use of the available spectrum through unrestricted roaming between networks and services while achieving regulatory compliance based on the location where the radio is operating.

Fuzzy scheduling will enable network routing decisions to be made in the presence of uncertainty about available or optimal paths. *Policy cognitive operation*, moving intelligence into the network to make the "best" choices on delivery, will enhance routing decisions, and *deferred/hierarchical address binding* will enable network to deliver traffic without end-to-end address and routing information [1].

The current approach to open systems architecture will also expand to address the issues of *open network architectures*, standardizing collaborative data exchanges through network IP standards to enable input and output interfaces to be made with relative ease. Increasingly, networks will make use of a wider range of sensor and database information, with input data taken from a range of sensors and data input devices from mainframe databases, and enterprise architectures to proprietary application-specific packages such as the current J2EE and .NET architectures. This broadening in *data-handling* requirements will drive a standardized approach to linking data structures required to handle an increasingly wide variety of vector, raster, and nonspatial data sources.

Current limitations on capabilities driven by restrictions in data transmission speeds, data storage *capacities*, or data access speeds will soon be overcome. Data transfer *speeds* will be capable of exceeding hundreds of gigabits per second and will provide unlimited data transmission rates across the globe; storage database volumes will be measured in millions of terabits, and short access times for large volumes of data will enable faster *fusion* of real-time and database information.

Network and data transmission and *storage security* will remain a priority, with developments in redundancy, routing, and backup methods permitting effective operation even when the network is under attack or has been physically disabled.

Network security will see significant enhancements in security to repel attacks, detect unauthorized access, and repair damaged networks, while permitting networks to function unhindered for authorized users.

Data analysis and interpretation

Data fusion will see further advances with an increasing ability to fuse inconsistent geospatial data sources into common databases and to display that data without the actual source being visible to the user. Automated fusion between multiple sensors will be used to improve the accuracy of the target identification, especially at long ranges and in adverse conditions (Figure 5.4).

Fused and unfused data will benefit from automated *storage, indexing, analysis, correlation, search, and retrieval of multimedia and a wide range of databases.*

Networked resources such as the Global Information Grid will be further integrated into near real-time data to enable *correlation, pattern analysis and matching*, and enhanced intelligence analysis. Databases will move towards *common interface and interrogation standards*, permitting fusion and query techniques to quickly access data or task sensors to collect further data. Push and pull information systems will enable information to be integrated, shared, and accessed much more effectively than is the case today, enhanced by mass storage devices with built-in processing for indexing, crossreferencing, and retrieval.

As sensor technology improves, so will advances in *mapping technology*, where fusion of a range of sensor measurements such as temperature, color, and three-dimensional images, combined with situational awareness pictures and huge advances in storage capabilities, will permit the presentation of mapping information to warfighters allowing familiarity with territory they have never seen with their own eyes. Improvements in sensor types such as Light Detection and Ranging (LIDAR) will permit mapping of the entire

Figure 5.4 Modern sensors such as the Searchwater radar on the U.K. Royal Navy's Sea King helicopters can detect moving targets over an area of 50,000 km^2. The challenge is processing the data that results from coverage of such a vast area and identifying from it contacts that merit further investigation.

surface of the earth to unprecedented levels of accuracy. In addition, *change detection* across a much wider range of parameters along with pattern analysis and *artificial intelligence* will permit an almost instant knowledge of how the shape of the battlefield has changed [2].

Intelligence analysis will benefit from hypothesis formation based on fuzzy logic and intelligent agents for distributed threat prioritization and assessment.

Computing systems will become more adept at learning and reasoning to improve the output of *cognitive systems*, which seek to reason to explore alternatives and identify the best solution for a given situation. To support these advances, other areas of technology such as *expert systems, scene understanding, and knowledge-based systems*, often using *biologically-based processes*, will need to progress further [3].

Enhanced and increasingly accurate time-sensitive situational awareness will contribute to further advances in *process automation*. Automated processes, such as automatic change detection, and movement indication will enhance the detection and tracking of targets in real time. The ability to reason within the context of dynamically changing spatial data will support advanced features such as tracking of individuals or vehicles in dynamic and complex environments, permitting the analysis of movement and the automated *interpretation* of enemy actions and associated intentions to help to prioritize target allocation.

Language processing will become increasingly accurate and automated to facilitate the interception, analysis, and interpretation of voice and text-based intelligence gleaned from foreign languages and to enable the mass analysis of communication media in a domestic environment.

Information displays (Figure 5.5) will become increasingly refined to fuse a wide variety of sensors and integrate near real-time data with database and intelligence sources to improve accuracy and to refine the interpretation of increasing data sources and updates. Similarly, the display of increasing volumes of information will be refined to display factors relevant to understanding situational awareness and to enhance the effectiveness of the OODA loop process.

Networked decision making and C2

Further advances in decision making and C2 capabilities will focus on improving the speed and accuracy of decision-making for the warfighter by

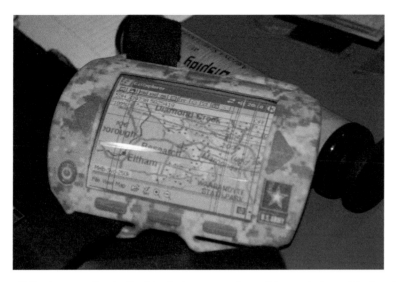

Figure 5.5 An example of a flexible display screen that will increasingly enable situational awareness to be shared down to the lowest tactical levels—or in NCW terms—to every node on the network.

recognizing that *information exploitation* can provide a decisive advantage to the outcome (Figure 5.6).

At a C2 level, further improvements in the effectiveness of *force employment* capabilities will be possible through wider dissemination and accommodation

Figure 5.6 Sharing command intent, particularly at the edges of the network, will be greatly enhanced by accessible information displays, enabling near real-time situational awareness and reporting.

of the command intent, particularly at a local level. Improved automated situational awareness, threat assessment and decision assessment tools for *local decision making*, and enhanced information inputs and frequent refreshes will improve the ability of commanders at all levels to effectively exploit ambiguous intelligence information and to tailor force and mission plans to improved engagement outcomes.

Information warfare will become increasingly important as engagement outcomes become more determined by access to and the manipulation of information on both sides.

The *tempo* of warfare will continue to increase, enhanced by the increased ability to fuse near real-time data to achieve a shared situational awareness picture. Common data fusion standards will enhance cross-sensor data fusion for multispectral sensor data fusion, and will enable near real-time sensor tasking to reflect priorities across the network rather than for an individual platform.

Building on these improvements, battlespace *situational awareness* capabilities will be enhanced by the accurate and continuous analysis of friendly and enemy forces, their moves, and the likely and actual outcomes of engagements, even at a local level, and also by providing mission-specific information from multiple networked sources to enhance mission effectiveness. Aided by advances in computing, database structures and database access situational awareness picture compilation will become much faster and will extend across the *kill chain* (see Appendix E for an explanation of the kill chain concept) to enhance the postengagement kill assessment.

In addition to presenting the current operational picture, improvements in artificial intelligence and processing speeds will enable situational awareness data and feeds from multiple intelligence sources to be integrated and analyzed to enable better *prediction of outcomes and intentions*. These technologies will form the foundation of advanced *simulation* techniques to model outcomes and *improve decision-making* processes.

Effects systems

Future effects (Figure 5.7) will become increasingly accurate in nature and will see a huge increase in the range of *nonlethal effects* available. *Increased accuracies* delivered through laser guidance and satellite guidance will enable a reduction in the size of weapons as the effect can be delivered precisely on target and does not need to cover a large radius (Figures 5.8 and 5.9).

Figure 5.7   A bow view of the experimental U.S. Navy stealth ship *Sea Shadow* underway in San Francisco Bay.

Figure 5.8   Precision guidance systems—here laser guidance modules on freefall bombs—will rapidly become commonplace as unguided freefall weapons are phased out of use.

A greater focus will be placed on soft kill rather than on hard kill effects where the use of high power microwaves, laser, and other RF devices will seek to disable rather than destroy adversary capabilities.

Figure 5.9 GPS-guided munitions such as those seen here on an MQ-9 Reaper will become increasingly commonplace.

Information warfare will become a priority as military planners recognize the importance of information in winning wars and its potential to massively disrupt the adversary's infrastructure. Offensive and defensive capabilities will continue to receive significant funding as alternatives to more traditional effects systems (Figure 5.10).

Platforms and hardware

Network-enabled *equipment* and sensors will continue to reduce in *size* and see lower *power consumption* requirements and will be more accessible and practical for soldiers in the field.

Sensors and the platforms on which they are mounted will become more autonomous and be capable of acting with increasing degrees of intelligence in their movements and collection of data. Autonomous platforms will form nodes on networks, enabling the network to be fought as an entity rather than each individual platform (Figures 5.11 through 5.14).

Figure 5.10 A U.S. Marine Corps Cobra gunship fires precision-guided rockets during a training mission, off the coast of Okinawa, Japan, in 2002.

Figure 5.11 Future unmanned vehicles will become increasingly more capable, using self-forming networks and advanced sensors to provide persistent and relevant ISTAR information across the battlespace. (Image courtesy of U.K. MoD.)

Figure 5.12   An armed robotic assault vehicle under trials with the U.S. Army.

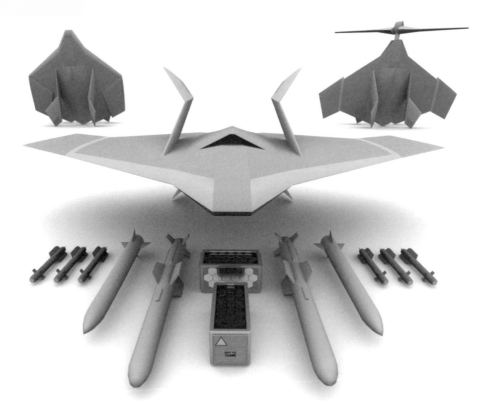

Figure 5.13   Future UAV and UCAV concepts will also permit the delivery of precision effects along with persistent ISTAR in hostile air environments. (Image courtesy of U.K. MoD.)

Figure 5.14 Unmanned, networked devices such as robots will play an increasingly important role in future warfare. Here unmanned air and ground vehicles and remote sensors are shown on display during trials of future U.S. capabilities.

Supported by rapid increases in *artificial intelligence*, unmanned robotic devices will become even more autonomous and will be able to operate either individually or together in a swarm. Their size will reduce significantly and their endurance will improve with advances in *renewable energy sources and fuel cell technologies*.

Larger platforms such as ships and aircraft will become intelligence processing and C2 centers, able to fuse data from multiple networks, interpret the battlespace, and direct the network assets in accordance with unfolding priorities.

Doctrinal aspects

Of course, other than the technological aspects of NEC, much depends on the effectiveness of the NEC capability as a functional system. The effective application of network-enabled principles in the pursuit of military objectives requires aspects such as cultural, training, and doctrinal considerations to be addressed early on in the equipment and system design stages to ensure maximum effectiveness and integrity of the NEC approach.

In particular, the following aspects need to be addressed as part of an overall NEC effectiveness program [4].

Acquisition

Research, development, and acquisition of network-enabled technologies will have an impact on the specification and integration of virtually all equipment programs. While historically platform programs have very much been stand-alone programs, by definition, future platforms will be part of a networked capability and will therefore need to be managed as such. Coordinating the integration of these capabilities represents a considerable challenge for both the military/government procurement organizations and industry.

Experimentation

Development of many new network-enabled technologies relies on operational testing in the field. What may look good in a laboratory may not always provide the intended utility in the field. Coordinated development, experimentation, and demonstration of fielded systems and industry prototypes will be central to accelerating the development of requirements and delivery of capability (Figure 5.15).

Figure 5.15 An experimental stelathy and autonomous robotic spider. (Photo courtesy BAE SYSTEMS.)

Concepts and doctrine

Maximizing the potential of NEC will require further development of operational concepts and doctrine. As technology evolves and the utility provided by NEC-based technology continues to increase, so do the applications to which the technology can be put. It is important that the way the technology is used and exploited maximizes its potential to increase NEC effectiveness (Figure 5.16).

It is particularly important that doctrines exploiting network-enabled capabilities are kept flexible enough to deal with different threat scenarios that may evolve in the future, rather than focus on particular environments which occupy military planners today.

Leadership

Delivering NEC capability requires significant changes from previous generations of warfare in doctrine and approach. The leadership skills and C2 approach required to manage deployed troops through to long-term strategic planning are significantly different in an NEC setting.

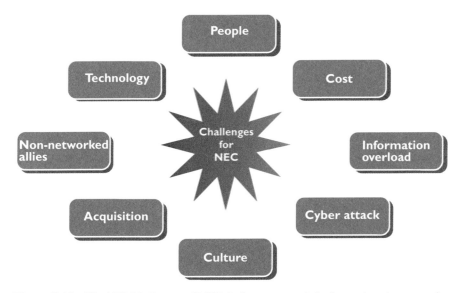

Figure 5.16 The U.K. MoD view of NEC challenges extends far beyond equipment and network considerations [5].

Industry

Industry partners have a key role to play in developing and implementing many aspects of NEC, including experimentation, delivery, and support of systems. As NEC systems become more capable and more complex, civilian contractors will find themselves brought even further towards the front line to support the deployment of such systems.

International partners

Many nations are currently studying and undertaking similar network related/network-centric warfare developments. Coalition partners need to forge relationships with collaborating nations to share research, leverage experimentation, standardize doctrine, and build coalition capability.

Standardization

The proliferation of network-based systems has driven an urgent need to standardize data transmission, storage, and access protocols between collaborating forces. As more advanced forms of networked capability are explored, in particular, around information warfare, the need to be able to classify, catalog, transmit, request, and access information using standard protocols becomes increasingly important.

Training

Training military forces in NEC doctrines is essential to maximize operational effectiveness. With increasingly sophisticated equipment and network-based interactions, training prior to operational deployment is essential if capability is to be maximized and mistakes avoided.

References

[1]   Griggs, Stephen, *Future Wireless Networking Technology and Its Contributions to NCW Capabilities*, Defense Advanced Research Projects Agency, Advanced Technology Office, March 2, 2006.

[2]   Hughes, Todd, (program manager, Information Exploitation Office), "The Mapping Revolution," *DARPATech, DARPA's 25th Systems and Technology Symposium*, Anaheim, CA, August 8, 2007.

[3]   Holland, Charlie, (director of the Information Processing Technology Office), "Decision Dominance," *DARPATech, DARPA's 25th Systems and Technology Symposium*, Anaheim, CA, August 8, 2007.

[4]   Adapted from http://www.iwar.org.uk/rma/resources/uk-mod/nec.htm#d.

[5]   *Network Enabled Capability*, JSP777, U.K. MoD, January 2005.

An F/A-18F Super Hornet assigned to the Blue Diamonds of Strike Fighter Squadron VFA -146, launches from the aircraft carrier USS Ronald Reagan CVN 76 while underway in the Pacific Ocean (June, 2010). (U.S. Navy photo by Mass Communication Specialist Seaman Benjamin C. Jernigan.)

Western Coalitions

The more common "Western" coalitions (Figure A.1) are as follows:

European Union (EU)

The EU's Common Foreign and Security Policy (CFSP) outlined a number of crisis management roles for the organization in the Petersburg Declaration of 1992, including humanitarian and rescue tasks, peacekeeping tasks, and tasks of combat forces in crisis management, including peacemaking.

The EU has many capabilities that complement NATO strengths, with many EU members also being part of the NATO organization. There is an increasing emphasis on the harmonization of European defense standards and policies through the European Defense Agency and other EU-related bodies such as the Western Economic Union (WEU), and a consequent need for collaboration between member states.

Western Economic Union (WEU)

In the Petersburg Declaration of 1992, the EU member states pledged their support for conflict prevention and humanitarian and peacekeeping efforts in cooperation with the United Nations and Organization on Security and

OSCE			
	EAPC		
EU			
Austria	Ireland	Albania	Andorra
Finland	Sweden	Armenia	Bosnia-H
		Azerbaijan	Cyprus
Belgium	Luxembourg	**NATO**	Belarus	Holy See
Czech Republic			Croatia	Liechtenstein
France	Netherlands	United States	Georgia	Malta
Germany	Poland	Canada	Kazakhstan	Monaco
Greece	Portugal		Kyrgyzstan	San Marino
Hungary	Spain		Moldova	Serbia and
Italy	United Kingdom		Russia	Montenegro
			Tajikistan	
Bulgaria	Norway		Turkmenistan	
Estonia	Romania		Ukraine	
Iceland	Slovakia		Uzbekistan	
Latvia	Slovenia			
Lithuania	Turkey **WEU**			

Figure A.1   Formal collaborative structures may also be supplemented by "coalitions of the willing" depending on the political complexities of the planned intervention.

Cooperation in Europe (OSCE) as outlined in the EU's Common Foreign and Security Policy (CFSP). The organization through which this support is delivered is the Western European Union (WEU), which is the defense component of the EU. WEU operations must be endorsed by the United Nations or OSCE and directed by the EU Council of Ministers. As the WEU does not have its own military command structure, it would rely on the NATO infrastructure for military operations and C2 requirements.

Euro-Atlantic Partnership Council

The Euro-Atlantic Partnership Council (EAPC) brings together 49 NATO and partner countries for dialogue and consultation on political and security-related issues including crisis-management and peace-support operations, arms control, international terrorism, defense issues such as planning, policy, and strategy; civil emergency planning and disaster-preparedness, armaments cooperation, nuclear safety, civil-military coordination of air traffic management, and scientific cooperation. The effectiveness of much of this planning will depend on the harmonization of data interchange standards to enable collaboration between collaborating countries.

Organization on Security and Cooperation in Europe (OSCE)

The Organization on Security and Cooperation in Europe (OSCE) is an organization established under the United Nations Charter and comprises 56 states drawn from Europe, Central Asia, and the Americas. The OSCE is the world's largest regional security organization, aiming to achieve cooperative security through political negotiations and decision-making in the fields of early warning, conflict prevention, crisis management, and postconflict rehabilitation. The OSCE's commitments by participating states and mechanisms for conflict prevention and resolution are implemented through the OSCE's multinational field missions (see www.osce.org for more details).

North Atlantic Treaty Organization (NATO)

The North Atlantic Treaty Organization (NATO) is tasked with the protection of its member states through the overt offensive and defensive military operations. It exercises its military responsibilities through the

NATO North Atlantic Council (NAC), in which all member nations are represented.

Although in theory NATO represents a concept of collective defense rather than equipment harmonization, its member states represent a key organization for Western European interoperability standards. Interoperability standards are heavily influenced by the military equipment standards largely defined by the United States.

United Nations (UN)

The goal of saving future generations "from the scourge of war" was enshrined in the UN charter in 1945, and even today the United Nations remains the first choice for authorizing international coalition-based responses.

Other international alliance organizations

In addition to the formal alliances shown earlier, other coalition operations regularly take place in areas such as Afghanistan on disaster relief and antidrug running or antipiracy operations and on military training exercises. In Afghanistan, for example, nearly 30 different nations have volunteered forces under the ISAF banner, and the operational role that those nations can play is very much defined by the compatibility of their networked sensor and effect systems, typically with the standards inherent in NATO systems.

Link-16 Network Message Sets and Network Participation Groups

See Table B.1 for a list of Link-16 network message sets and network participation groups.

Network Management
J0.0 Initial Entry
J0.1 Test
J0.2 Network Time Update
J0.3 Time Slot Assignment
J0.4 Radio Relay Control
J0.5 Repromulgation Relay
J0.6 Communication Control
J0.7 Time Slot Reallocation
J1.0 Connectivity Interrogation
J1.1 Connectivity Status
J1.2 Route Establishment
J1.3 Acknowledgment
J1.4 Communication Status
J1.5 Net Control Initialization
J1.6 Need Line Participation
Group Assignment

Precise Participant Location and Identification
J2.0 Indirect Interface Unit PPLI
J2.2 Air PPLI
J2.3 Surface PPLI
J2.3 Subsurface PPLI
J2.5 Land Point PPLI
J2.6 Land Track PPLI

Surveillance
J3.0 Reference Point
J3.1 Emergency Point
J3.2 Air Track
J3.3 Surface Track
J3.4 Subsurface Track
J3.5 Land Point or Track
J3.6 Space Track
J3.7 Electronic Warfare Product Information

Antisubmarine Warfare
J5.4 Acoustic Bearing and Range

Intelligence
J6.0 Intelligence Information

Information Management
J7.0 Track Management
J7.1 Data Update Request
J7.2 Correlation
J7.3 Pointer
J7.4 Track Identifier
J7.5 IFF/SIF Management
J7.6 Filter Management
J7.7 Association
J8.0 Unit Designator
J8.1 Mission Correlator Change

Weapons Coordination and Management
J9.0 Command
J10.2 Engagement Status
J10.3 Hand Over
J10.5 Controlling Unit Report
J10.6 Pairing

Control
J12.0 Mission Assignment
J12.1 Vector
J12.2 Precision Aircraft Direction
J12.3 Flight Path
J12.4 Controlling Unit Change
J12.5 Target/Track Connection
J12.6 Target Sorting
J12.7 Target Bearing

Platform and System Status
J13.0 Airfield Status Message
J13.2 Air Platform and System Status
J13.3 Surface Platform and System Status
J13.4 Subsurface Platform and System Status
J13.5 Land Platform and System Status

Electronic
J14.0 Parametric Information
J14.2 Electronic Warfare Control/Coordination

Threat Warning
J15.0 Threat Warning

National Use
J28.0 U.S. National 1 (Army)
J28.1 U.S. National 2 (Navy)
J28.2 U.S. National 3 (Air Force)
J28.2 (0) Text Message
J28.3 U.S. National 4 (Marine Corps)
J28.4 French National 1
J28.5 French National 2
J28.6 U.S. National 5 (NSA)
J28.7 U.K.L National
J29 National Use (reserved)
J30 National Use (reserved)

Miscellaneous
J31.0 Over-the-Air Rekeying Management
J31.1 Over-the-Air Rekeying
J31.7 No Statement

Table B.1   J-Series Message Sets

Link-16 network participation groups

NPG 1 Initial entry

NPG 2 Round-trip timing-addressed

NPG3 Round-rip timing-broadcast (RTT-B)

NPG 4 Network management

NPG 5 PPLI A - (C2 units)

NPG 6 PPLI B - (non-C2 units)

NPG 7 Surveillance

NPG 8 Mission management - (mission types) (USN uses for engagement status)

NPG 9 Air control

NPG 10 EW

NPG 11 Unassigned

NPG 12 Voice A [either 2.4 or 16 kilobits per second (kbps)]

NPG 13 Voice B (either 2.4 of 16 kbps)

NPG 14 USN for indirect PPLIs (used for forwarding TADIL A, B units to TADIL J)

NPG 15 Reserved for future joint use

NPG 16 Reserved for future joint use

NPG 17 Unassigned

NPG 18 WC

NPG 19 Fighter-to-fighter net

NPG 20 Non-C2 to non-C2

NPG 21 Engagement coordination

NPG 22 Unassigned [UK Composite A (NPGs 4, 8, and 9)]

NPG 23 Unassigned [UK Composite B (NPGs 7 and 10)]

NPG 24 Unassigned

NPG 25 Reserved for future joint use

NPG 26 Unassigned

NPG 27 Joint PPLI

NPG 28 Unassigned

NPG 29 Free text (residual messages)

NPG 30 Interim JTIDS message standard (IJMS) P message, position

NPG 31 IJMS T message, track report

NPG 400 to 511 user specific lines

Modulation Techniques

Modulation is the process of varying a cyclical waveform in order to use the resulting signal to convey a message. While modulation of RF energy at the speeds required for message transmission is difficult to visualize, modulation is found regularly, albeit at slower speeds in music, where a musician may modulate the volume, timing, and frequency (pitch) of an instrument to produce a musical tune. In communications, an elementary form of modulation is Morse code, where a continuous wave signal is modulated in a form known as on-off keying.

A device that performs modulation is termed a modulator and a device that performs the inverse operation of modulation in order to decode the original signal is known as a demodulator. A device that can perform both modulation and demodulation is termed a modem (short for modulate-demodulate).

Modulation techniques rely on the transmitter and receiver having advance knowledge of the data encoding protocols so that the modulator at the transmitter and the demodulator at the receiver are able to perform inverse operations.

Analog modulation

In analog modulation the analog waveform is constantly modified in response to the input signal. There are two main types of analog modulation:

Amplitude modulation (AM)

Amplitude modulation varies the amplitude of a fixed frequency carrier wave as shown in Figure C.1.

Pulse amplitude modulation is shown in Figure C.2.

Frequency modulation (FM)

Frequency modulation varies the frequency of the carrier waveform in response to the input signal (Figure C.3), which is shown in a digital context in Figures C.4 and C.5.

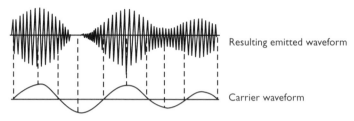

Figure C.1   Amplitude modulation.

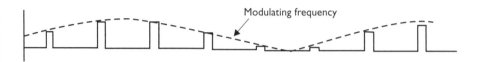

Figure C.2   Pulse-frequency modulation.

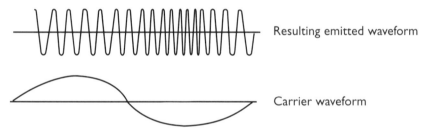

Figure C.3   Frequency modulation.

Phase modulation (PM)

Phase modulation varies the phase shift of the modulated signal.

Digital modulation

In digital modulation techniques, an analog carrier signal is modulated by a digital bit stream. The most common forms of digital modulation include:

- Amplitude shift keying (ASK), where a finite number of amplitudes are used;

Figure C.4 Pulse-frequency modulation.

Freqency Modulation in Practice

This principle is used, for example, in radar altimeters where an analog carrier wave is frequency modulated against a known pattern and transmitted from the aircraft to the ground. As the emiited signal pattern varies in frequency, and due to the time it takes to reach the ground and travel back to the aircraft, the signal will be received at a slightly different point (or frequency) in the pattern. By comparing the time delay between the emitted and received pattern, the distance to teh ground can be calculated.

The principles of Doppler shift and its use in velocity measurement are explained in more detail in the Radar section of this book.

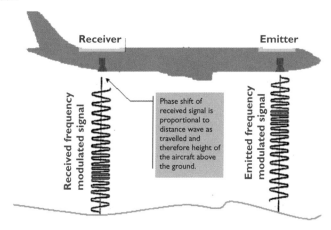

Principle of Operation of Radar Altimeters

Figure C.5 By comparing the time taken for the transmitted signal to return from the target, the distance can be measured, and any change in frequency can also be used to determine target velocity by using the Doppler principle or by measuring the instantaneous change in received carrier frequency compared with the emitted carrier wave frequency.

- Binary phase shift keying (BPSK), the simplest form of phase shift keying using two phases separated by 180°;
- DEQPSK;
- Frequency shift keying (FSK), where a finite number of frequencies are used;
- Minimum shift keying (MSK), a type of continuous-phase frequency-shift keying;

- Offset quadrature phase-shift keying (O-QPSK), a variant of phase-shift keying modulation using four different values of the phase to transmit the signal;
- Phase shift keying (PSK), where a finite number of phases are used;
- QAM, a finite number of at least two phases where at least two amplitudes are used;
- Quadrature phase-shift keying (QPSK), a version of frequency modulation where the phase of the carrier wave is modulated to encode bits of digital information in each phase change;
- Shaped binary phase shift keying (SBPSK), a phase shaping keying system that provides good separation on narrowband channels.

Frequency Classifications

L band	1–2 GHz
S band	2–4 GHz
C band	4–8 GHz
X band	8–12 GHz
Ku band	12–18 GHz
K band	18–26.5 GHz
Ka band	26.5–40 GHz
Q band	30–50 GHz
U band	40–60 GHz
V band	50–75 GHz
E band	60–90 GHz
W band	75–110 GHz
F band	90–140 GHz
D band	110–170 GHz

Table D.1

Radio Society of Great Britain

Designation	Frequency
A	Up to 250 MHz
B	250–500 MHz
C	500–1,000 MHz
D	1–2 GHz
E	2–3 GHz
F	3–4 GHz
G	4–6 GHz
H	6–8 GHz
I	8–10 GHz
J	10–20 GHz
K	20–40 GHz
L	40–60 GHz
M	60–100 GHz

Table D.2   NATO Communication and IEEE Radar Bands

Radar Band	Frequency	Wavelength	Typical Radar Applications
HF	3–50 MHz	6–100m	Very long-range, ground-based, over-the-horizon radars; airborne foliage penetrating radars
VF	50–300 MHz	1–6m	Very long-range, ground-based radars; ground-penetrating radars; counterstealth radar, foliage penetration, and airborne early warning (AEW)
UHF	300 MHz–1 GHz	0.3–1m	Very long range (e.g., ballistic missile early warning), long-range air surveillance, foliage penetrating, counterstealth radar
L	1–2 GHz	15–30 cm	L for long-range; long-range air traffic control and surveillance, secondary surveillance radar
S	2–4 GHz	7.5–15 cm	S for short-range; terminal air traffic control, long-range weather, marine radar
C	4–8 GHz	3.75–7.5 cm	A compromise (hence "C") between X and S bands; weather radar, medium-range surveillance radar
X	8–12 GHz	2.5–3.75 cm	Missile guidance, marine radar, weather, medium-high resolution mapping and ground surveillance
Ku	12–18 GHz	1.67–2.5 cm	Frequency just *under* K band (hence "u"); high-resolution SAR mapping, GMTI, satellite altimetry
K	18–27 GHz	1.11–1.67 cm	From the German *kurz*, meaning short; limited use due to absorption at this frequency by atmospheric water vapor, so Ku and Ka were used instead for surveillance; weather radar
Ka	27–40 GHz	0.75–1.11 cm	Frequency just *above* K band (hence "a"); ground mapping, short-range surveillance, airport terminal and approach radar
mm	40–300 GHz	1–7.5 mm	Millimeter band; terminal guidance imaging seekers, short-range imaging radars, and automotive radar

Table D.3 Radar Bands

The Kill Chain

The kill chain concept (Figure E.1) comprises the following steps:

- **Plan:** Not formally part of the F2TEA process, the step of "planning" is essential if targets are to be identified at the earliest possible stages. Platform-based and networked sensors, along with appropriate effects systems, need to be directed to particular areas of interest, and search and engagement strategies need to be planned in detail prior to any engagement to ensure that the optimum mix of sensors along with the appropriate effects systems is available to maximize the probability of a successful engagement. Planning can either be *deliberate* for addressing fixed targets or *dynamic* for addressing moving or mobile targets.

- **Find:** The task of finding targets typically involves the use of multiple sensors operating at various frequencies across the spectrum and preferably networked to fuse sensor data to enhance the probability of detection. A multispectral sensor detection network is essential for the detection of targets such as moving vehicles, mobile, hardened, or underground facilities, concealed facilities in urban or jungle environments, and, in an asymmetric environment, targets such as leadership and activists at an individual level.

- **Fix:** This involves positive identification of the target and may require the use of several different sensor types or the same type of sensors, sometimes used in different modes to ensure that the target that is about to be engaged is the correct one. Positive identification can be technically challenging if the target is moving, mobile, or concealed above or below ground.

- **Track:** This is a critical step in the kill chain, in particular, for moving, mobile, or obscured targets, and relies on handing over accurate target location coordinates to the Target stage. This is perhaps the most

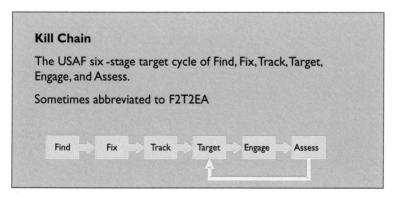

Figure E.1 The kill chain.

technically challenging part of the kill chain for moving targets, as it relies in particular on being able to gather accurate information about the location of the target in time and space and accurately pass time-tagged three-dimensional coordinates to the weapons system for processing in the Target stage.

- **Target:** Targeting is the process for selecting and prioritizing targets and matching appropriate actions to those targets to create specific desired effects that achieve objectives, taking account of operational requirements and capabilities [1].

- **Engage:** Engagement will be carried out by the most appropriate effects system available. Effects will vary from nonlethal to lethal and in a network-enabled environment may be delivered by a separate platform from which the Find, Fix, and Track data is gathered.

- **Assess:** The assessment stage involves the use of a range of appropriate sensors either on-board the platform engaging the target or through other networked sensor systems to assess whether the target has been effectively engaged. Depending on the type of target and its importance to the outcome of the military objectives, reengagement may be required immediately if the engagement opportunity is a transient one, or it may be required hours later if it is a low-value fixed target. Four assessment levels exist depending on the tactical or strategic nature of the target: tactical (similar to the current joint combat assessment), operational (the component-commander level), campaign (the joint-force-commander level), and national (the Secretary of Defense and Presidential level) [1].

The kill probability can be improved by inserting more accurate and timely information about targets into the OODA loop and by better matching the various weapons with the desired effect. An NEC approach to managing the kill chain enables a higher tempo of information provision and enables the right information to be provided to the right place to ensure a higher chance of a successful outcome (Figure E.2).

With the increased availability and need for precision targeting, a variation of the targeting process is called the *effects chain*. The effects chain introduces further refinements to the kill chain cycle by requiring the consideration of the type and precision of the effect required on the target. This enables the commander to select between lethal or nonlethal weapons, to consider the impact of the weapon that will be used against the mission objectives, and to address the need to minimize collateral damage.

While the USAF version of the kill chain (Figure E.3) above is probably the most common version in use, other variations include:

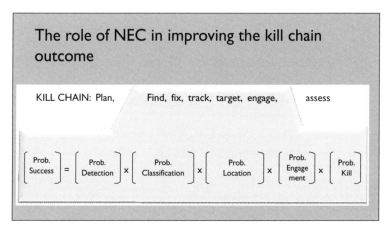

Figure E.2 The role of NEC in improving the kill chain outcome.

- Acquire, Decide, Strike, Assess [2];
- Detect, ID, Track, Decide, Engage, Assess [3].

A functional breakdown of the kill chain is based on the functions provided by nodes described as the Sensor, Decider, and Shooter [2]. Of course, other variations exist to describe specific situations and to highlight the roles played by sensors, effectors, and decision nodes.

Figure E.3 When on-station in a firing position, the AC-130 typically has a kill chain less than 30 seconds against a variety of pop-up ground targets, such as vehicles, mortar positions, and dismounted troops.

References

[1] USAF Air Force Doctrine Document 2-1.9, Targeting.

[2] Haffa, Jr., Robert P., and Jasper Welch, "Command and Control Arrangements for the Attack of Time-Sensitive Targets," *Northrop Grumman Analysis Center Papers*, November 2005.

[3] FORCEnet POM 06 Process, RADM Thomas E. Zelibor, USN, Space, Information Warfare, Command and Control (N61), October 23, 2003.

The Defend Chain

While the kill chain is optimized for offensive engagements, the defend chain (Figure F.1) is optimized to protect the platform from being targeted, and, if targeted, protected from an effective engagement. It is similar in concept to the kill chain, but typically occurs over a shorter time period. It comprises a seven-stage cycle of find, track, prioritize, decoy, target, engage, and maneuver. The stages of target and engage are offensive stages that may be employed depending on the situation and the effects systems available. In more detail, the seven stages comprise:

- **Find:** Finding and sensing threats, whether they are potentially hostile platforms, sensors targeting the platform, or inbound effects, is an essential task of the defensive aids systems fitted to virtually all air, land, and sea platforms. Finding targets at an early stage is essential in order to give the defending platform as much time as possible to confirm identity, track, and take effective action against the threat through the use of sector scans and networked data integration from tactical data link networks such as Link 16 or Link 22. It is worth noting that the stage of fix found in the kill chain process is omitted here because if the object that has been found is hostile, there is generally no need for a further refinement of threat data before moving to the next phase.

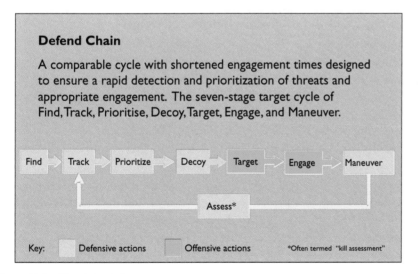

Figure F.1 The defend chain.

- **Track:** Threats need to be tracked to assist with their prioritization and to ensure that the threat posing the threat posing the greatest danger is the one that receives the highest priority response.

- **Prioritize:** Threats need to be tracked to establish their intent and to match their profile to clearly identify the type of threat. The effectiveness of the six-stage cycle is optimized when sensors and effects systems are coordinated across a collaborative network to prioritize the threats to across the network rather than relying on individual platforms to find and engage their own targets.

- **Decoy:** Once it is confirmed that the platform is under attack or in a networked environment in which another target on the network is under attack, active or passive jamming or deception may be employed to prevent the aggressor from tracking the platform or, if track is achieved, from successfully targeting the platform. The use of decoys may be bypassed if the incoming threat is in the terminal phase of the attack and can be engaged directly by hard kill[1] systems.

- **Target:** Depending on the equipment fitted or the effects systems available from a collaborative network, the defending platform may decide to engage the hostile threat to destroy it and prevent it from further engagement. Initially, if a missile has already been launched from the hostile platform, the missile will have engagement priority. Alternatively, if the hostile platform itself is presenting a threat, it will become the priority target. An engagement will ideally employ multiple sensors to ensure that the target is tracked and that its time and space position is accurately known. This is particularly important for defensive engagements, as the engagement of incoming missiles or threat platform requires very precise location and timing information. This stage effectively includes the compressed fix and track stages from the kill chain.

- **Engage:** Offensive action against the threat may involve a range of weapons systems. In a networked environment, collaborative engagement principles may require platforms with appropriate weapons systems to defend another platform that is under attack. The platform under attack will also aim to engage the threat at long, medium, and short ranges, depending on the defensive systems fitted. At long range, the threat platform will be engaged. At medium and short ranges, the missile that has been launched from an offensive platform will be engaged.

- **Maneuver:** Although maneuver is shown at the end of the process, it may, depending on the platform being engaged and the time between the detection and potential impact, be initiated earlier in the sequence. It is likely with modern defensive aids systems that the targeting and engagement sequence will provide significantly more protection than is afforded by maneuvering. At long and medium ranges, the priority will be to engage the launch platform and any missiles that may have been launched. The act of maneuvering will, however, be important in the terminal phase of the engagement as the platform being targeted aims to maneuver to present as small a spectral signature as possible to the sensors on the incoming effects system.

- **Assess:** Assessment is based on whether the actions taken have enabled the platform to avoid the threat. Depending on the outcome of the defensive action, the platform under attack may then wish to reengage the hostile launch platform if time, weapons availability, and operational priorities permit.

Endnote

1. *Hard kill systems* are those that physically destroy incoming threats, while *soft kill systems* are those that destroy threats through deception or electronic warfare means.

List of Acronyms and Abbreviations

σ	radar cross-section, or scattering coefficient of target (m^2)
3D	three-dimensional
A	effective aperture area of receiving antenna
A/A	air-to-air
AAM	air-to-air missile
ABCS	Army Battlespace Command System
ABCTM	Army Brigade Combat Team Modernization programme
ABF	adaptive beam forming
ABL	airborne laser
A/C	aircraft
ACDMA	asynchronous code division multiple access
ACINT	acoustic intelligence
ACTS	advanced communications technology satellite
ADGE	air defense ground environment
AESA	active electronically scanned array
AEW	airborne early warning
AF	Air Force
AFATDS	Advanced Field Artillery Tactical Data System
AFH	adaptive frequency-hopping
A/G	air-to-ground
AGS	air ground surveillance
AM	amplitude modulation
AMF	airborne, maritime, and fixed station applications of the JTRS radio
AOC	air operations center
ASaC	airborne surveillance and control
ASACS	Air Surveillance and Control System
Asat	antisatellite
ASCA	artillery systems cooperation activities
ASK	amplitude shift keying
ASM	air-to-surface missile
ASTOR	airborne stand-off radar
ASuW	antisurface warfare
ASW	antisubmarine warfare

ATM	air traffic management
ATO	air tasking order
ATR	automatic target recognition
B	approximation of receiver noise bandwidth
BC	battle command
BCF	blade chopping frequency
BDA	battle damage assessment
BEAM	bandwidth efficient advanced modulation
BG	battle group
BGISX	Battle Group Information System Exchange
bps	bits per second
BLOS	beyond line of sight
BSOs	battlespace objects
BUG-E	Battlefield Universal Gateway Equipment
BVR	beyond visual range
BVRAAM	Beyond Visual Range Air-to-Air Missile
C2	Command & Control
C2I	Command, Control, and Intelligence
C2V	Command and Control Vehicle (FCS)
C2ISR	Command, Control, Intelligence, Surveillance, and Reconnaissance
C3	Command, Control, and Communications
C4	Command, Control, Communications, Computers
C4ISR	Command, Control, Communications, Computers, Intelligence, Surveillance, and Reconnaissance
C3ISTAR	Command, Control, Communications, Intelligence, Surveillance, Target Acquisition & Reconnaissance
C4ISTAR	Command, Control, Communications, Computers, Intelligence, Surveillance, Target Acquisition, and Reconnaissance
CAOC	combined air operations center
CAP	combat air patrol
CAS	close air support
CBRN	chemical, biological, radiological, nuclear
CC	communications and computers
CC&D	camouflage, concealment, and deception
CDM	code division multiplexing
CDMA	code division multiple access

CEC	collaborative engagement capability
CECM	communications electronic countermeasures
CENTRIXS	Coalition Enterprise Regional Information Exchange System data link
CEP	collaborative engagement processor
CESM	communications electronic support measures
CETPS	Collaborative Engagement Transmission Processing Set
CFSP	Common Foreign and Security Policy
CI	counterintelligence
CIB	common interactive broadcast
CIGSS	Common Imagery Ground/Surface System
CIWS	Close-In Weapons System
CLEW	conventional Link-11 waveform
CLO	counter low observable
CM	countermeasure
CMN	common mission network
CMF	common message format
CNA	computer network attack
CND	computer network defense
CNE	computer network exploitation
CNO	combat network attack or computer network operations
CNR	combat net radio
COB	communications order of battle
COBRA	collection of broadcasts from remote assets
COI	communities of interest
Com/Coms/Comms	communications
COMSEC	communications security
Comint	communications intelligence
Comms	communications
COP	common operating picture
CORBA	Common Object Request Broker Architecture
COTS	commercial off the shelf
CPFSK	continuous-phase frequency-shift keying
CPOF	Command Post of the Future
CRC	control and reporting center
CSMA	carrier-sense multiple access

CUID	country unique identification code
CW	continuous wave
CWAN	coalition wide area network
DAB	digital audio broadcast
DARPA	Defense Advanced Research Projects Agency
DCGS	Distributed Common Ground System
DDS	data distribution system
DIO	defensive information operations
DIRCM	directed infrared countermeasures
DISN	Defense Information Systems Network
DIW	defensive information warfare
DLODs	defense lines of development
DNCS	data net control station
DoD	U.S. Department of Defense
DRFM	digital radio frequency memory
DSCS	Defense Satellite Communications System
DSNs	distributed sensor networks
DSSS	direct sequence spread spectrum
DTMA	dynamic TDMA
DTP	defense technology plan
E	electronic
EA	electronic attack
EBO	effects-based operations
EBP	effects-based planning
ECCM	electronic counter-countermeasures
ECM	electronic countermeasures
EHF	extra high frequency
Elint	electronic intelligence
EM	electromagnetic
EMCON	emission control
EMSEC	emission security
EMI	electromagnetic interference
EMP	electromagnetic pulse
EAM	emergency action message
EO or E/O	electro-optical
EOB	electronic order of battle

EP	electronic protection
EPLRS	Enhanced Position Location and Reporting System
ES	electronic support
ESM	electronic support measures
ET	enhanced throughput
ETA	estimated time of arrival
ETD	estimated time of departure
EU	European Union
EW	electronic warfare
EWOC	electronic warfare operations center
EWSS	electronic warfare support system
f	frequency
F	pattern propagation factor for antenna or radar beams
F	receiver noise figure, generally a figure >1, representing the noise efficiency of the receiver; for an ideal receiver with no noise, F = 1
FAC	forward air controller
FBCB2	Force Battle Command Brigade and Below (C2 System)
FCS	future combat system
FEBA	forward edge of battle area
FEC	forward error correction
FDM	frequency division multiplexing
FDMA	frequency-domain (or frequency division) multiple access
FDOA	frequency difference of arrival
FHSS	frequency hopping spread spectrum
FIACS	fast inshore attack crafts
FIOP	Family of Interoperable Operating Pictures
FISINT	foreign instrumentation signals intelligence
FGA	fine grain analysis
FLOT	forward line of troops
FLET	forward line of enemy troops
FM	frequency modulation
FMCW	frequency modulated continuous wave
FOS	family of systems
FRU	forwarding reporting unit
FSCL	fire support coordination line
ft	feet

G	gain of transmitting antenna
Gbps	gigabytes per second
GBS	Global Broadcast System
GCCS	Global Command and Control System
GEOINT	geophysical intelligence
GIG	Global Information Grid
GMR	Ground Mobile Radio (JTRS main vehicle radio)
GMTI	ground moving target indicator
GPS	Global Positioning System
GSM	Group Special Mobile or Global System for Mobile Communications
HCDR	high capacity data radio
HEL	high energy laser
HERM	helicopter rotor modulation
HF	high frequency
HIDL	high integrity data link
HITB	high interest track broadcast
HMS	handheld, man-packs and small form applications of the JTRS radio
HUD	head-up display
Humint	human intelligence
IA	information assurance
IABM	Integrated Architecture Behavior Model
IBS	Intelligence Broadcast System
ICBM	intercontinental ballistic missile
ICV	infantry carrier vehicle
ID	identification
IDM	Improved Data Modem
IED	improvised explosive device
IEEE	Institute of Electrical and Electronic Engineers
IFDL	intraflight data link
IFF	identification friend or foe
IJMS	Interim Joint Message Standard
IMINT	imagery intelligence
INFOSYS	information and information systems
IO	information operations
IP	Internet Protocol

IPB	intelligence preparation of the battlefield
IR	infrared
IRINT	infrared intelligence
ISAF	International Security Assistance Force
ISAR	inverse synthetic aperture radar
ISM	ISR sensor manager (NCCT)
ISR	identification, surveillance, reconnaissance
ISTAR	identification, surveillance, target acquisition, reconnaissance
IW	information warfare or irregular warfare
J	joint
JBMC2	Joint Battle Management Command and Control
JC2	Joint Command and Control System
JCTN	Joint Composite Tracking Network
JCTN	Joint Cooperative Targeting Network
JDL	Joint Directors of Laboratories
JDN	Joint Data Network
JEM	jet engine modulation or JTRS Enhanced MBITR
JEWC	Joint Electronic Warfare Center
JFAC	Joint Forward Air Controller
JFC	Joint Force Commander
JIM	Joint, Interagency, and Multinational
JOP	Joint Operational Picture
JPADS	Joint Precision Air Drop System
JPN	Joint Planning Network
JSTARS	Joint Surveillance Target Acquisition and Reconnaissance System
JTAC	Joint Tactical Air Controller
JTDLMP	Joint Tactical Data Link Management Plan
JTIDS	Joint Tactical Information Distribution System
JTM	Joint Track Management
JTRS	Joint Tactical Radio System
k	Boltzmann's constant
kbps	kilobits per second
KEC	knowledge-enabled capability
LCS	Littoral Combat Ship
LADAR	laser radar

LAN	local area network
LandWarNet	land warfare network
LASINT	laser intelligence
LD	laser designator
LED	light emitting diode
LF	low frequency
LRF	laser range finder
LIDAR	Light Detection And Ranging
LOS	line of sight
LPD	low probability of detection
LPD	landing platform dock
LPI	low probability of intercept
LPRF	low pulse repetition frequency
LR	long range
LRR	laser retro-reflection
LRSP	long-range surveillance patrols
LWR	laser warning receiver
MADL	multifunction advanced datalink
MADSN	Mobile Agent Distributed Sensor Network
MALD	Miniature Air Launched Decoy
MANET	mobile ad hoc network
MANPADS	MAN Portable Air Defence Missile Systems
MASINT	measurement and signature intelligence
MASN	mission area subnetwork
MAW	missile approach warner
MBDA	Matra BAE Dynamics Alenia
MBITR	JTRS-Enhanced Multiband Intrateam Radio
Mbps	megabytes per second
MC4	Medical Communications for Combat Casualty Care
MCCP	mobile command and control post
MCNs	model-based communication networks
MCS	mounted combat system
MDS	minimum detectable signal
MFR	multifunction radar
MIDS	multifunctional information distribution system
MILDEC	military deception

MILORD	Moyen d'Identification Lointaine d'Objectifs Radar Désignés
MIMO	multiple input, multiple output
MIRVs	multiple independent reentry vehicles
MLS	Microwave Landing System
MFD	multifunction display
MFR	multifunction radar
MMA	multimission aircraft
mmW	millimeter wave
MoP	modulation on pulse
MOUT	military operations in urban terrain
MPA	maritime patrol aircraft
MP-RTIP	Multiplatform Radar Technology Insertion program
MPA	maritime patrol aircraft
MPRF	medium pulse repetition frequency
MR	maritime reconnaissance
MRA	maritime reconnaissance attack
mSec	millisecond (1/1,000th of a second)
MSK	minimum shift keying
MTI	moving target indication
MUM	manned and unmanned
MUOS	Mobile User Objective System (JTRS radio system)
NADGE	NATO Air Defense Ground Environment
NATO	North Atlantic Treaty Organization
Nav	navigation
NBC	nuclear biological chemical
NBCR	nuclear biological chemical radiological
NCCT	network-centric collaborative targeting
NCDL	networked common data link
NCES	network-centric enterprise services
NCO	network-centric operations
NCOIC	Network Centric Operations Industry Consortium
NCS	net control station
NCTR	noncooperative target recognition
NCW	network-centric warfare
NEC	network-enabled capability
NETINT	network intelligence

NEW	network-enabled warfare
NESI	Net-Centric Enterprise Solutions for Interoperability
NFA	no-fly areas
NGO	nongovernmental organizations
NILE	NATO Improved Link-11 data link (also known as Link 22)
NIPRNET	Nonsecure Internet Protocol Router Network
NLOS	nonline of sight
NLOS-C	nonline of sight—canon
NLOS-LS	nonline of sight—launch system
NLOS-M	nonline of sight—mortar
nm	nautical miles
NOC	network operations center (United States)
NPGs	network participation groups
NTR	net time reference
NUCINT	nuclear intelligence
OATM	Open Architecture track manager
OIO	offensive information operations
OIW	offensive information warfare
OODA	observe, orient, decide, act
OPTINT	optical intelligence
ORBAT	order of battle
OSI	Open Systems Interconnection model
OSCE	Organization on Security and Co-operation in Europe
OSINT	open source intelligence
Pd	probability of detection or power density (radar equation)
Pn	power of noise receiver at the antenna from a reflected signal
Pt	antenna transmitter power
PESA	passive electronically scanned array
PHOTINT	photographic intelligence
PIM	platform interface module (NCCT)
PIP	primary injection point (GBS)
POI	probability of intercept
PPI	plan position indicator
PPLI	precise participant location and identification
PRF	pulse repetition frequency
PRI	pulse repetition interval

Prob	probability
PROM	propeller rotor modulation
PSK	phase shift keying
PSYOPS	psychological operations
PUs	participating units
R	distance to target from receiver or distance from receiver to target where transmitter and receiver are colocated
Rr	distance from target to receiver
Rt	distance from transmitter to target
RAAF	Royal Australian Air Force
RADINT	radar intelligence
RAIDRS	Rapid Attack Identification Detection Reporting System
RAM	radar absorbent material
RCS	radar cross-section
RESM	radar electronic support measures
RF	radio frequency
RGPO	range gate pull off
RINT	unintentional radiation intelligence
RMV	Recovery and Maintenance Vehicle (FCS)
ROA	restricted operations areas
ROI	region of interest
RPC	remote procedure calling
RSTA	reconnaissance, surveillance, and target acquisition
RSV	reconnaissance and surveillance vehicle
RTOF	remote towable optical fiber
RU	reporting unit
RWR	radar warning receiver
Rx	receive(r)
SA	situational awareness
SAASM	selective availability/antispoofing module (for GPS systems)
SADL	situational awareness data link
SAR	synthetic aperture radar
SCDL	surveillance and control data link
SDR	software defined radio
SEAD	Suppression of Enemy Air Defenses
SEC	sector operation center
SEI	specific emitter identification

SGLS	Space-Ground Link System (satellite)
SIAP	Single Integrated Air Picture
SIFF	successor IFF
SFF	small form factor (JTRS Radio)
SHF	super high frequency
Sigint	signals intelligence
SINCGARS	Single-Channel Ground to Air Radio System
SIPRNET	secret IP router network
SLEW	Single Tone Link-11 Waveform
S/N	signal-to-noise ratio (also SNR)
SNC	system network controller
SOAP	Simple Object Access Protocol
SOCs	sector operation centers
SOF	Special Operations Forces
SOS	system of systems
SOSCOE	System-of-Systems Common Operating Environment (FCS)
SRF	shaft rotation frequency
SRW	soldier radio waveform
SSSB	ship-shore-ship buffer
SSM	surface-to-surface missile
STAP	Space-Time Adaptive Processing
STDL	satellite (or submarine) tactical data link
SUGV	small unmanned ground vehicle (FCS)
Surf	surface
SWAN	(NATO) secret wide area network
T	receiver temperature (in degrees Kelvin)
TACAN	tactical airborne navigation
TACC	tactical air command center
TACS	tactical air command system
TADIL	Tactical Digital Information Link (U.S. DoD abbreviation)
TBM	theater ballistic missile
TCA	track continuity areas
TCP	Transmission Control Protocol
TCT	time-critical target or time-critical targeting
TCS	transformational communications system
TDL	tactical data link

TDM	time-division multiplexing
TDMA	time-division multiple access or time-domain multiple access
TDOA	time difference of arrival
TEMPEST	Transient Electromagnetic Pulse Emanation Standard and Exploitation
TG	task group
Tgt	target
TIBS	Theater Information Broadcast System
TMIP	Theater Medical Information Program
TRANSEC	transmission security
TST	time-sensitive target or time-sensitive targeting
TTNT	tactical targeting networking technology
T/U-UGS	tactical/urban UGS
Tx	transmit(er)
UAS	Unmanned Air System
UAV	unmanned air vehicle
UBDL	underwater broadband data link
UCAV	unmanned combat aerial vehicle
UDDI	universal description, discovery, and integration
UFO	UHF follow-on satellite system
UGS	unattended ground sensor
UGV	unmanned ground vehicle
UHF	ultra high frequency
UN	United Nations
URAV	unmanned reconnaissance aerial vehicle
USJFCOM	United States Joint Forces Command
USMC	United States Marine Corps
U-UGS	urban UGS
UUV	unmanned underwater vehicle
VGPI	velocity gate pull-in
VGPO	velocity gate pull-off
VLF	very low frequency
VMF	variable message format
VoIP	Voice over Internet Protocol
VHF	very high frequency
WAN	wide area network

WDM	wavelength division multiplexing
WCDMA	wideband code division multiple access
WEU	Western Economic Union
WIN-T	warfighter Information Network–Tactical
WMD	weapons of mass destruction
WNR	wideband network radio
WNW	wideband network waveform
XML	Extensible Markup Language

About the Author

Richard S. Deakin is the chief executive officer of the United Kingdom's National Air Traffic Services (NATS). He has over 25 years of experience in the aerospace and defense industry with extensive experience in networked systems, radar, and electronic warfare. His wide-ranging industrial experience in France, the United Kingdom, and on deployed operations in Afghanistan includes responsibility for the United Kingdom's Watchkeeper UAV program, NATO Air Defence, and Air C2 programs, as well as ground-based, naval and airborne radar, and signals intelligence programs. He is currently responsible for one of the world's largest networked real-time system of systems, as head of the United Kingdom's air traffic management system.

A chartered engineer, Richard holds a first class honors degree in aeronautical engineering from Kingston University and an M.B.A. from Cranfield School of Management. He is also a fellow of the Royal Aeronautical Society, and was awarded an honorary doctorate in engineering from Kingston University in January 2007.

Index

Target Acquisition in Communication Electronic Warfare Systems, Richard A. Poisel

For further information on these and other Artech House titles,
including previously considered out-of-print books now available through our
In-Print-Forever® (IPF®) program, contact:

Artech House
685 Canton Street
Norwood, MA 02062
Phone: 781-769-9750
Fax: 781-769-6334
e-mail: artech@artechhouse.com

Artech House
16 Sussex Street
London SW1V 4RW UK
Phone: +44 (0)20-7596-8750
Fax: +44 (0)20-7630-0166
e-mail: artech-uk@artechhouse.com

Find us on the World Wide Web at: www.artechhouse.com